I0051456

Emerging Perspectives and Applications of Computational Intelligence and Smart Systems

Emerging Perspectives and Applications of Computational Intelligence and Smart Systems delves into the transformative role of artificial intelligence, machine learning, IoT, robotics, smart technologies etc. in shaping the future of industries and societies. This volume presents cutting-edge research and real-world applications that highlight how computational intelligence enables automation, intelligent decision-making, and sustainable solutions across diverse domains. With contributions from experts in academia and industry, the book explores innovations in healthcare, smart cities, energy systems, manufacturing, and beyond. It emphasizes interdisciplinary collaboration and practical methodologies that bridge the gap between theoretical advancements and tangible impact. By combining theoretical advancements with real-world relevance, this book equips readers with a deeper understanding of how intelligent technologies can be applied to solve complex problems. It is an essential guide for those looking to harness computational intelligence and smart systems to drive innovation, improve efficiency, and contribute to a more sustainable and connected future.

Emerging Perspectives and Applications of Computational Intelligence and Smart Systems

Proceedings of the International Conference on Computational Intelligence, Emerging Technologies, and Smart Systems (ICCIETSS 2025), March 28-29, 2025, Rajkot, Gujarat, India

Edited by
Dr. Amit Lathigara
Dr. Nirav Bhatt
Dr. Paresh Tanna

CRC Press
Taylor & Francis Group
Boca Raton London New York

CRC Press is an imprint of the
Taylor & Francis Group, an **informa** business

First edition published 2026
by CRC Press
4 Park Square, Milton Park, Abingdon, Oxon, OX14 4RN

and by CRC Press
2385 NW Executive Center Drive, Suite 320, Boca Raton FL 33431

© 2026 selection and editorial matter, Dr. Amit Lathigara, Dr. Nirav Bhatt and Dr. Paresh Tanna; individual chapters, the contributors

CRC Press is an imprint of Informa UK Limited

The right Dr. Amit Lathigara, Dr. Nirav Bhatt and Dr. Paresh Tanna to be identified as the authors of the editorial material, and of the authors for their individual chapters, has been asserted in accordance with sections 77 and 78 of the Copyright, Designs and Patents Act 1988.

All rights reserved. No part of this book may be reprinted or reproduced or utilised in any form or by any electronic, mechanical, or other means, now known or hereafter invented, including photocopying and recording, or in any information storage or retrieval system, without permission in writing from the publishers.

For permission to photocopy or use material electronically from this work, access www.copyright.com or contact the Copyright Clearance Center, Inc. (CCC), 222 Rosewood Drive, Danvers, MA 01923, 978-750-8400. For works that are not available on CCC please contact mpkbookspermissions@tandf.co.uk

Trademark notice: Product or corporate names may be trademarks or registered trademarks, and are used only for identification and explanation without intent to infringe.

British Library Cataloguing-in-Publication Data
A catalogue record for this book is available from the British Library

ISBN: 978-1-041-20964-5 (hbk)
ISBN: 978-1-041-20965-2 (pbk)
ISBN: 978-1-003-72504-6 (ebk)

DOI: 10.1201/9781003725046

Typeset in Times New Roman
by Aditiinfosystems

Emerging Perspectives and Applications of Computational Intelligence and Smart Systems
– Dr. Amit Lathigara et al. (eds)
© 2026 Taylor & Francis Group, London, ISBN 978-1-041-20965-2

Contents

Emerging Perspectives and Applications of Computational Intelligence and Smart Systems
– Dr. Amit Lathigara et al. (eds)
© 2026 Taylor & Francis Group, London, ISBN 978-1-041-20965-2

List of Figures

Emerging Perspectives and Applications of Computational Intelligence and Smart Systems
— Dr. Amit Lathigara et al. (eds)
© 2026 Taylor & Francis Group, London, ISBN 978-1-041-20965-2

List of Tables

Emerging Perspectives and Applications of Computational Intelligence and Smart Systems
– Dr. Amit Lathigara et al. (eds)
© 2026 Taylor & Francis Group, London, ISBN 978-1-041-20965-2

Acknowledgements

We would like to express our heartfelt gratitude to all the contributors, researchers, and professionals whose insights and scholarly work have shaped the content of this book. Their dedication to advancing knowledge in the fields of computational intelligence and smart systems has made this publication possible.

We extend our sincere thanks to the editorial and review team for their meticulous efforts in ensuring the quality, clarity, and coherence of the research papers. Their valuable feedback and commitment to academic excellence have greatly enhanced this work.

We are also grateful to School of Engineering, RK University for continuous support and encouragement throughout this endeavour. The belief in the importance of fostering innovation and multidisciplinary collaboration has been a strong driving force behind this publication.

Finally, we thank the publishing team for their professional guidance and for helping bring this work to a wider audience. It is our hope that this book serves as a meaningful contribution to ongoing research and inspires further exploration in the dynamic world of computational intelligence and smart systems.

About the Editors

Dr. Amit Lathigara

Dr. Amit Lathigara is a distinguished academician and the Provost at RK University, with over 20 years of experience in higher education. He has made significant contributions to research, with high-impact publications and extensive guidance in his field. As an expert, he has represented RK University in numerous national and international academic and research workshops. He has received various national and international research grants. He has received multiple awards, including the Faculty Fellow Award in 2018, and holds an ING.PAED.IGIP certification as an international engineering educator.

Dr. Nirav Bhatt

Dr. Nirav Bhatt, Professor at Computer Science and Engineering, School of Engineering, RK University, over the course of his 18-years career in technical education, he contributed in teaching, research and institutional development. His research work encompasses various latest emerging areas including publications in reputed journals. His efforts have been acknowledged through various competitive research grants, including government-supported fundings aimed at modernizing technical education. He plays an influential role in shaping academic standards by serving as a member of the editorial boards, review panels, technical committees, etc.

Dr. Paresh Tanna

Dr. Paresh Tanna is a Professor with School of Engineering, RK University. He is currently a distinguished researcher and academic with extensive experience in computer science and engineering. He has received prestigious research grants, academic excellence awards, and is recognized with multiple best paper awards at international conferences. With over 19 years of experience in research and academia, he has made significant contributions to research publications and guidance. He also serves as a reviewer, an editor, and the chair for different conferences/journal, ensuring the dissemination of high-quality research. Also, his dedication to research, innovation, and academic excellence inspires peers and students.

Emerging Perspectives and Applications of Computational Intelligence and Smart Systems
– Dr. Amit Lathigara et al. (eds)
© 2026 Taylor & Francis Group, London, ISBN 978-1-041-20965-2

1

Investigating an Updraft Coal Gasifier Using Advanced Hot Gas Cleaning Technology Through Experimentation to Reduce Pollution for Sustainable Development

Reena N. Makadia*

Department of Mechanical Engineering, RK University,
Rajkot, Gujarat, India

Dr. Marmik M. Dave

Research and Development Department, Atul Greentech Pvt. Ltd.,
Ahmedabad, Gujarat, India

■ **Abstract:** Various small components and contaminants, including particles, tar, chlorine, alkali metals, nitrogen, and sulphur compounds, are commonly present in the syngas generated during coal gasification. One significant problem that must be resolved before using syngas in the downstream process is the presence of tar. The production of clean fuel gas that satisfies international emission regulations is the main obstacle to the commercialization of gasification technology. This study not only summarizes the findings from the author's recent research project, "Development of Advanced Techniques to Improve Energy Efficiency of an Updraft Coal Gasifier with Reduced Pollution Discharges", but also explores recent advancements in product gas cleaning technologies. The research findings highlighted the significant potential of activated carbon as a filtration medium for removing tars and chemical pollutants from the produced syngas.

■ **Keywords:** Equivalence ratio, Efficiency, Steam-to-coal ratio, Response surface methodology

1. INTRODUCTION

Global energy demand has risen due to the sudden increase in the world's population and the ongoing rise in energy consumption driven by higher standards of living (Jurgen Scheffran 2020). Coal remains the most affordable and abundant energy source globally. However, its use leads to

---

*Corresponding author: reenamakadia@gmail.com

DOI: 10.1201/9781003725046-1

significant carbon dioxide emissions (Muhammad Amir Raza 2025). The adoption of clean coal technology has been prompted by growing public awareness of environmental pollution and the role that CO_2 and other gas emissions play in climate change (Akanksha Mishra 2018). Gasification technology, with its origins tracing back to the 19th century, has undergone significant development over time. Coal gasification is a modern and efficient process that converts solid coal into gaseous fuels, which can be used to generate heat or electricity. This process involves the reaction of solid coal with a small amount of oxygen, air, steam, carbon dioxide, or a combination of these gases at temperatures of 700°C or higher, producing gaseous products that serve as an energy source. Gasification offers several advantages over traditional combustion. For a given amount of fuel, it produces a smaller volume of gas compared to burning, resulting in reduced carbon dioxide emissions released into the atmosphere. Additionally, the smaller gas volume requires smaller equipment, which helps lower overall costs (Basu 2006). In hot syngas production, "tar" refers to a complex blend of heavy hydrocarbon compounds. It is a sticky, viscous liquid byproduct generated during the gasification process when solid fuels are exposed to high temperatures. Tar presents a significant challenge that must be addressed before the gas can be utilized in downstream applications (Prabhansu Malay 2015). In the current experimental setup, hard coal is employed as a filtering medium to remove tar from the hot gas. However, no analysis has yet been conducted on the experimental performance of an industrial-scale updraft coal gasifier equipped with hot filtering technology.

2. Methodology

2.1 Experimental Setup and Process

An updraft coal gasifier with a daily coal consumption capacity of about 35 metric tons is needed to produce clean fuel gas for the facility that makes porcelain insulatorsat Bikaner Ceramics in Rajasthan which is manufactured and designed by Radhe Renewable Energy Pvt. Ltd. in Rajkot. The gas generated is used by the plant to meet its heating needs. A pulverizer and screening device are used to crush and screen coal to a particle size of 20 to 50 mm, ensuring consistent coal quality throughout the batch. Approximately 90 kg of pulverized coal is delivered to the gasifier top hopper every six minutes. The manually loaded top hopper fully supplies the second hopper continuous feeding mechanism, resulting in a coal input rate of 0.25 kg/s.

The gasifier is cylindrical in shape, with two openings at the top. It has a height of 5500 mm and a diameter of 3790 mm, as seen in Fig. 1.1. The top centre hole of the gasifier is designed for feeding dry coal through a hopper with a diameter of 855 mm. Adjacent to the coal inlet is an outlet with a diameter of 800 mm, located 1175 mm from the centre. At the bottom of the gasifier, steam and air are introduced as oxidants. The primary shell is initially filled uniformly with ash clinkers or brick fragments, measuring 1 to 1.5 inches, up to a height of one foot above the octagonal top of the bed. When the bed temperature reached approximately 500 °C, prepared coal was added to the updraft gasifier. This caused a rapid increase in the gasifier bed temperature. The combustion process was then transitioned to gasification by increasing the coal supply, reducing the airflow rate, and introducing steam into the gasifier. Gas samples were collected for analysis once the process stabilized. The setup included thermocouples, flowmeters, pressure transducers, and other measurement instruments. A gas chromatograph was used to analyze the composition of the fuel gas under steady-state conditions. In this plant, however, a hot coal filter medium is used following a cyclone-type separator to eliminate tar from the generated syngas.

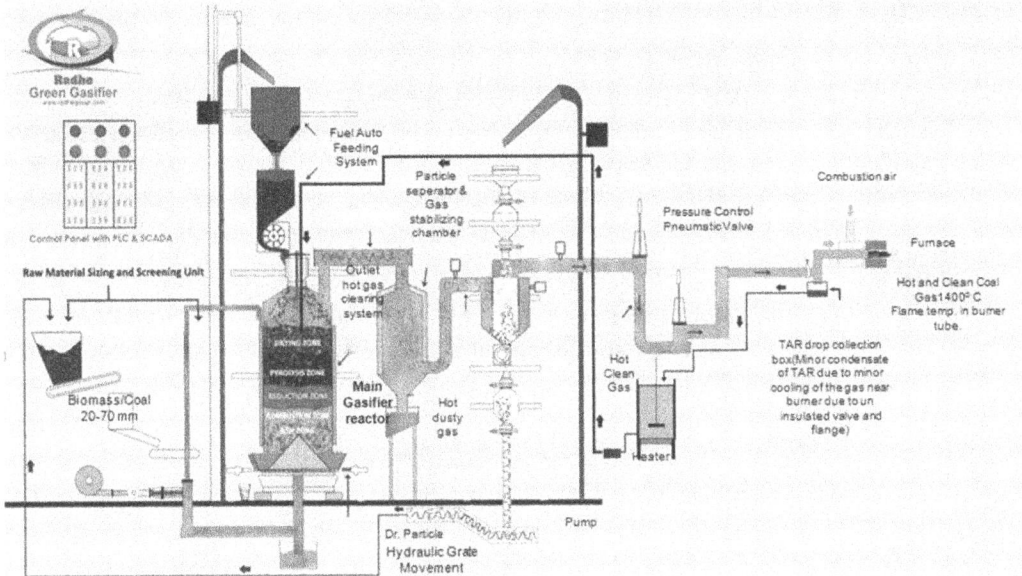

Fig. 1.1 Schematic diagram of updraft coal gasifier@ Radhe Pvt. Ltd, Rajkot (Radhe Groups of Energy n.d.)

2.2 Hot Gas Filtration Technology

In the current setup, activated carbon is used as a filtration medium to remove contaminants from gas streams. It is highly effective in adsorbing various pollutants, including organic compounds, tar, volatile organic compounds (VOCs), and gases such as ammonia and sulphur compounds (Xiuwei Ma 2021). Due to its porous structure and large surface area, activated carbon traps harmful pollutants on its surface. Tar molecules are trapped and retained within the pores of activated carbon when tar-containing syngas passes over or through it. The microscopic pores of the activated carbon physically capture the tar, effectively isolating it from the rest of the gas (Ammar Ali Abd 2021). At higher temperatures, which are typical in gasification processes, tar adsorption becomes more effective.

However, at very high temperatures, tar molecules may undergo pyrolysis, or thermal decomposition, breaking down into simpler compounds, some of which can also be adsorbed by the activated carbon. Over time, tar and other impurities fill the activated carbon, reaching saturation. As shown in Fig. 1.2, the absorbed tar can be removed, and the carbon adsorption capacity can be restored through regeneration processes, by adding the new feed operated automatically. Activated carbon is able to remove up to 90–95% of tar according to the temperature and pressure maintained within range of 250-500°C of syngas and 2MPa respectively.

Fig. 1.2 Collected liquid tar

Source: Authors

3. RESULT AND DISCUSSION

3.1 Design of Experiments with RSM

The design of experiment (DOE) method was used to identify the optimal combinations of variables for studying the system with the minimum number of experimental runs. This method also evaluates the interaction effects between the selected variables. To model and analyze the process, where certain independent factors significantly influence the output variables (responses), response surface methodology (RSM) was employed. Developed by Box and Wilson, RSM is a statistical tool that complements DOE methods and supports quality improvement across various fields. As shown in Table 1.1, For the experimental investigation, two key operational variables were selected: the steam-to-coal ratio and the equivalence ratio. These variables ranged from 0.2 to 0.3 and 0.2 to 0.6, respectively. Pilot tests were conducted to determine these ranges. Syngas quality indexes were calculated based on the test results.

Table 1.1   List of designed experiments as per RSM method

Run no.	ER	SCR	HHV of Syngas (MJ/kg)	Cold gas efficiency (%)	Carbon Conversion Efficiency (%)
1	0.3	0.4	22.16	84.41	96.13
2	0.4	0.2	20.38	80.5	96.5
3	0.3	0.4	22.32	84.25	96.2
4	0.441421	0.4	21.2	80.7	97.89
5	0.3	0.682843	20.34	82.5	96.5
6	0.3	0.4	22.04	84.9	96.5
7	0.3	0.4	22.12	83.5	96.11
8	0.3	0.117157	18.65	81.6	93.12
9	0.4	0.6	21.23	83.5	97.9
10	0.158579	0.4	17.52	76.5	94.01
11	0.3	0.4	22.24	84.6	96.24
12	0.2	0.6	18.8	79.5	94.8
13	0.2	0.2	17.68	77.8	93.51

Source: Authors

Table 1.2   Response surface regression model for responses

Response	Correlation
Higher Heating value	HHV = -0.920 + 97.32 ER + 30.07 SCR - 138.42 ER * ER - 32.92 SCR * SCR - 3.38 ER * SCR
Cold Gas Efficiency	CGE = 49.69 + 181.3 ER + 21.68 SCR - 286.6 ER * ER - 28.52 SCR * SCR + 16.2 ER * SCR
Carbon Conversion Efficiency	CCE = 87.02 + 18.04 ER + 17.03 SCR - 6.9 ER * ER - 15.97 SCR * SCR + 1.38 ER * SCR

Source: Authors

3.2 Comparison between Experimental and Predicted Values

In Response Surface Methodology (RSM), validating the accuracy and reliability of developed models is crucial. This validation is achieved by comparing experimental values with predicted values, ensuring that the models can accurately forecast results across range of variable. Achieving a discrepancy of less than 2% between experimental outcomes and predicted values indicates a high level of accuracy and reliability in the model. This minimal difference suggests that the model predictions closely align with actual results confirming its validity.

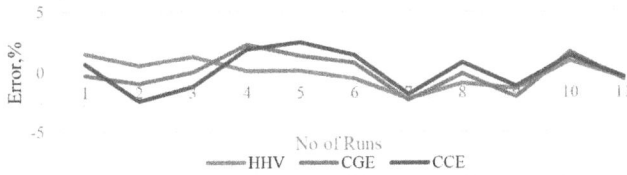

Fig. 1.3 Experimental vs predicted value error in(%)

Source: Author

4. CONCLUSION

Tars and other contaminants from the gasification processes could be removed by hot-gas filtration technology through activated carbon as a filter media. Regression model of present coal gasifier can predict the result with ± 2% error. The response surface method is the most effective statistical technique for analyzing the impact of the equivalence ratio and the steam-to-coal ratio on syngas quality. Syngas with an increased higher heating value (HHV) of up to 22.32 MJ/kg, an improved carbon gasification efficiency (CGE) of up to 84.90%, and an enhanced carbon conversion efficiency (CCE) of up to 97.90% can be produced with an equivalency ratio of 0.3699 and a steam-to- coal ratio of 0.499.

ACKNOWLEDGEMENT

The facilities and technical assistance provided by Radhe Renewable Energy Development Pvt. Ltd., Rajkot, India, were greatly appreciated by the authors in carrying out this research work.

References

1. Mishra, A., Gautam, S., & Sharma, T. (2018). Effect of operating parameters on coal gasification. *International Journal of Coal Science & Technology, 5*, 113–125.
2. Abd, A. A., Othman, M. R., & Kim, J. (2021). A review on application of activated carbons for carbon dioxide capture: present performance, preparation, and surface modification for further improvement. *Environmental Science and Pollution Research, 28*(32), 43329–43364.
3. Basu, P. (2006). *Combustion and gasification in fluidized beds*. CRC press.
4. Scheffran, J., Felkers, M., & Froese, R. (2020). Economic growth and the global energy demand. *Green energy to sustainability: strategies for global industries*, 1–44.
5. Raza, M. A., Karim, A., Aman, M. M., Al-Khasawneh, M. A., & Faheem, M. (2025). Global progress towards the Coal: Tracking coal reserves, coal prices, electricity from coal, carbon emissions and coal phase-out. *Gondwana Research, 139*, 43–72.
6. Karmakar, M. K., Chandra, P., & Chatterjee, P. K. (2015). A review on the fuel gas cleaning technologies in gasification process. *Journal of Environmental Chemical Engineering, 3*(2), 689–702.
7. Makadia, R. N., & Dave, M. M. (2024). Statistical Analysis of Industry Scale Up Draft Coal Gasifier Using Response Surface Methodology for Sustainable Development. *Renewable Energy & Sustainable Development, 10*(2).
8. Ma, X., Yang, L., & Wu, H. (2021). Removal of volatile organic compounds from the coal-fired flue gas by adsorption on activated carbon. *Journal of Cleaner Production, 302*, 126925.
9. Zhang, Z., Zhao, Z., & Zhang, L. (2023). Recent progress in the gasification reaction behavior of coal char under unconventional combustion modes. *Applied Thermal Engineering, 220*, 119742.

Emerging Perspectives and Applications of Computational Intelligence and Smart Systems
– Dr. Amit Lathigara et al. (eds)
© 2026 Taylor & Francis Group, London, ISBN 978-1-041-20965-2

2

AI-Driven Skin Type Detection and Personalized Recommendations for Cosmetic Products

Kakoli Banerjee, Abhishek Pandey*,
Saad Anwar Khan, Alok Ranjan, Rahul Makhloga
Department of Computer Science, JSS Academy of Technical Education,
Noida, India

■ **Abstract:** Understanding skin types and recommending suitable products are pivotal in dermatology and cosmetic research, addressing individual needs effectively. Advances in AI and deep learning have made it possible to analyze data from detailed skin imaging with remarkable precision, improving product predictability. This study introduces a system that combines ingredient-based product evaluation with deep learning algorithms for classifying skin types. The proposed model, leveraging YOLOv5 and U-Net, achieved an accuracy of 92% in identifying skin types and 95% precision in detecting common skin conditions. Additionally, the integration of GPT-based recommendation systems for cosmetic product suggestions demonstrated superior performance, achieving 92.5% accuracy, compared to traditional content-based (74.3%) and collaborative filtering (81.7%) methods. The proposed approach seeks to outperform traditional methods by enhancing accuracy and usability.

■ **Keywords:** Skin type detection, Artificial intelligence, Deep learning, Ingredient analysis, Dermatological systems

1. INTRODUCTION

The skin is the largest organ in the human body and acts as a major defense barrier while being exposed to various external and internal nuisances and this compromises its health. Skin diseases are acne, skin cancer, and wrinkles on the face, which can all affect a person's physical appearance, mental health, and the general quality of life. For instance, about up to 85% of teenagers suffer from acne. It can also last till adulthood and if ignored it can leave a person with physical as well as psychological difficulties (Huynh et al. 2022, 1879);(Kassem et al. 2021, 1390); (Li et al. 2020, 208264).

---

*Corresponding author: abhishekpandey3188@gmail.com

DOI: 10.1201/9781003725046-2

AI and machine learning have brought transformative changes to dermatology by introducing innovative methods to diagnose skin disorders and customize treatment plans. Identifying skin cancer at an early stage, segmenting affected areas, and analyzing dermatological patterns have become significantly easier with these methods. Techniques that are using advanced architectures, such as Transformer encoders and U-Net (a CNN-based segmentation model with an encoder-decoder architecture), enable accurate segmentation of skin abnormalities and provide cosmetic or therapeutic treatment suggestions (Lee et al. 2024, 2066).

This paper explores the intersection of dermatology, cosmetics, and innovative technologies. It provides an overview of recent progress in wrinkle segmentation, skin condition diagnosis, and personalized cosmetic recommendations. The study aims to promote the development of reliable, user-focused solutions in dermatology and cosmetic science by analyzing state-of-the-art approaches.

2. Literature Review

(A. Adjerid et al. 2022) Proposed usage of AI and Smartphone photos to evaluate acne identification and its severity scale. (Shen et al. 2018) showed 81% accuracy in classifying acne lesions using CNN, one drawback was that they used non-smartphone photos in their work. (Junayed et al. 2019) got 94% accuracy by using clinical photos with AcneNet. (Seite et al. 2019) produced 68% accuracy on smartphone photos. (Yang et al. 2021) achieved 80% accuracy by using a deep learning algorithm on clinical photos. The paper highlights the following shortcomings and the need of a model that produces accurate results on smartphone images (Huynh et al. 2022, 1879).

(Mohamed A. Kassem et al. 2021) did a detailed analysis of machine learning and deep learning methods for skin lesion detection and diagnosis. Traditional methods used for feature extraction like SVM and ANN gave low precision for diagnosis. Newer techniques like CNN produce much better precision for segmentation and feature extraction (Kassem et al. 2021, 1390).

(Jinhee Lee et al. 2024) They proposed a deep learning model for skin care product recommendation system by focusing on ingredients used and analysing the skin conditions. They used Neural Collaborative Filtering for user interactions, Content Based Filtering to compare products, and NLP algorithms for customer reviews (Lee et al. 2024, 2066).

3. Proposed Work

3.1 Skin Type Detection

The very first step toward a tailored skincare regimen is identifying the skin type one has. As such, this system uses Leslie Baumann's dermatological classification of four skin types—heavily oily, extremely dry, combination, and sensitive—through the assistance of AI.

3.2 Data Collection Phase

Input parameters such as drinking habits, solar radiation (Solar index: 0-11 and more), humidity ratio (g/m3) are collected. Pre-Data Analysis Phase: Using natural language processing and feature creation strategies, further enhances the accuracy rate by 15%.

3.3 YOLOv5 for Skin Feature Detection

YOLOv5 is a real-time object detection model that detects and classifies skin features such as: Pigmentation: Spots, hyperpigmentation, or uneven tone. Texture: Oiliness, dryness, or sensitivity. Lesions: Early detection of acne, moles, or other irregularities.

YOLOv5 achieves 96% detection accuracy for facial skin features using three key components:

- Backbone (CSPDarkNet): Extracts fine skin details for better feature representation.
- Neck (PANet): Merges multi-scale features to detect both small blemishes and larger conditions.
- Head (Anchor-Free Detection): Ensures precise classification and localization of skin conditions.

We use YOLOv5 which detects key regions of interests (ROI) that are skin anomalies, lesions or irregularities by extracting spatial and contextual information. The extracted features include bounding box coordinates, confidence score and class probabilities.

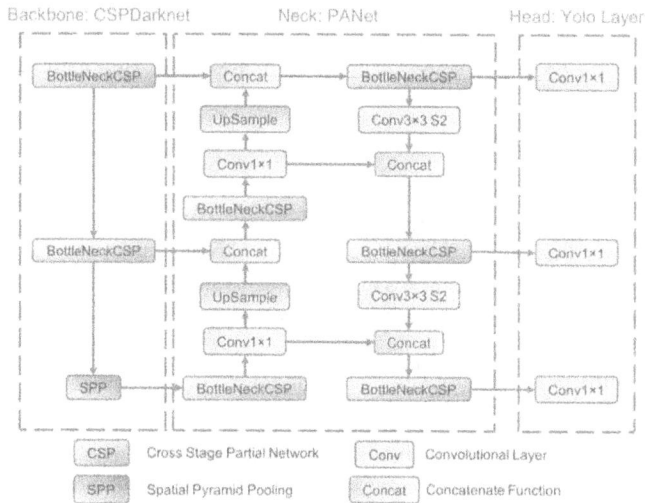

Fig. 2.1 YOLOv5

Source: Adapted from Xu et al. (2021)

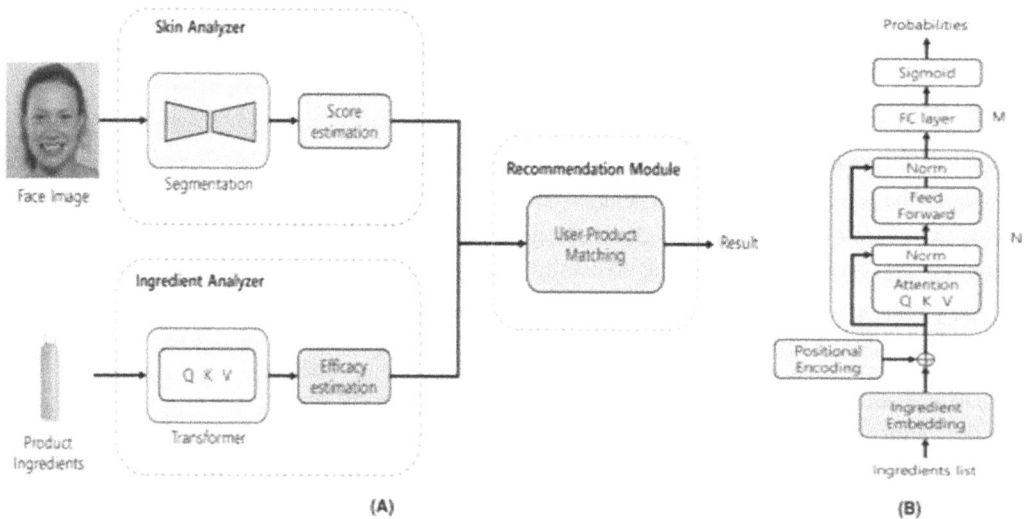

Fig. 2.2 Product recommendation flow

Source: Adapted from Lee et al. (2024)

3.4 U-Net for Fine-Grained Feature Extraction

U-Net is optimized for segmentation, it means it provides high-resolution identification of skin features. Fine-Grained Segmentation: Captures dryness, redness, and pore structure with a dice coefficient score of 0.89. Advanced Variants: Attention U-Net enhances attention to areas of interest, lowering segmentation errors by 12%. U-Net performs pixel wise segmentation which focuses on finer details among the regions in skin which are identified by YOLOv5. The extracted features include High resolution spatial features from the encoder layer and Fined grained texture patterns that merge high level and low level feature maps.

3.5 Multimodal Data Fusion

Combining image data with survey inputs, the system provides a 360° analysis of skin health. Tensor fusion techniques are used to integrate multiple data streams, which guarantees robust and contextualized assessments. The following features are combined for a holistic skin assessment:

- Facial Image Features (YOLOv5 & U-Net): Detects pigmentation irregularities, classifies skin type, and identifies lesions like acne or eczema.
- Survey-Based Features (User Inputs): Includes lifestyle factors, environmental data, and medical history for personalized assessment.

3.6 Dermatological Condition Classification with YOLOv5

The YOLOv5 algorithm is optimized to do real-time object detection and multi-label classification. Supported Conditions are as follows: Hyperpigmentation, acne, eczema, melanoma, and rosacea. Performance: It has reached 95% sensitivity and 90% specificity in lesion classification. Transfer Learning: Trained on ImageNet and fine-tuned on various dermatological datasets such as HAM10000; it reduces training time by 30% and maintains high accuracy

4. DISEASE AND LESION DETECTION

Detecting lesions and skin diseases early is critical and very important for proper treatment and prevention from severe outcomes. Our proposed model uses deep learning algorithms like YOLOv5

Table 2.1 Skin type classification metrics

Model	Accuracy (%)	Precision (%)	F1-Score (%)
Proposed U-Net	94.2	93.5	93.1
Baseline CNN	85.4	83.2	83.6
Decision Tree	72.1	71.0	70.9

Source: Authors

Table 2.2 Performance metrics comparison

Model	Dataset	MAP (%)	FPS	Inference Time (ms)
YOLOv5	COCO	50.4	140	7
YOLOv5	PASCAL VOC	76.8	140	7
YOLOv4	COCO	48.9	120	8
SSD	COCO	41.2	59	17
Faster R-CNN	COCO	42.7	7	142

Source: Authors

and ResNet-50 which helps in detecting and classification of lesions and diseases in real time with high accuracy and precision.

4.1 Core Technologies

YOLOv5 is used for detecting skin lesions with high accuracy and speed, It takes under 0.2 seconds to classify and detect objects per image. It achieves 95% sensitivity and 90% specificity, effectively identifying both positive and negative cases while detecting multiple labels in an image. ResNet-5 is utilized for feature extraction, helping to extract granular features from skin images. Pretrained on ImageNet and fine-tuned on dermatological datasets like HAM10000 and DermNet, it enhances classification accuracy, achieving 92% in skin image analysis.

4.2 U-Net Architecture

It is a convolutional neural network (CNN) which helps in pixel level classification of objects in an image helps in precise segmentation of lesions and generates a dice coefficient score of 0.89 Both YOLOv5 and ResNet-50 is used for classifications of labels and detection of boundaries. Validation benchmark is calculated by testing on dataset like HAM10000 which results in a precision of 93% and recall of 91%.

5. Personalized Recommendations

Skincare varies for every individual, as each person's skin has unique needs influenced by its type, condition, and external factors. This system uses advanced methods to offer tailored product recommendations. Additionally, user reviews are examined with an accuracy of 85%, ensuring that product claims align with actual user experiences. To further personalize recommendations, collaborative filtering analyzes user preferences and past interactions. Techniques like Neural Collaborative Filtering (NCF) and matrix factorization are used to reveal hidden patterns in user-product relationships, achieving a recommendation accuracy of 96%.

5.1 AI-Based Product Recommendations

Effective skincare requires personalized product recommendations. This module employs hybrid recommendation engines and NLP to pair users with the right products.

Product Analysis via NLP: Proposed system uses Transformer models like BERT and GPT to analyze product ingredients for risks, It uses sentiment analysis to identify products with 85% or higher positive feedback.

Hybrid Recommendation Engines: Combines Neural Collaborative Filtering (96% confidence) with content-based filtering, dynamically updating user profiles through incremental learning, improving recommendation relevance by 15% over time.

6. Implementation Result and Evaluation

Implementing the proposed system we can achieve better results in metrics like accuracy, sensitivity, and precision. Our proposed model uses deep learning models like YOLOv5 and U-Net, It has proven in delivering great result in skin-type classification and skin condition diagnosis, It has been tested on various datasets like HAM10000 and DermNet which resulted in accuracy of 92% in identifying skin types and precision of 95% in detecting common skin conditions.

6.1 System Validation and Continuous Improvement Metrics

Table 2.3   Key results and impact

Metric	Description	Value
Accuracy	Classification of skin types using YOLOv5 and U-Net	95%
Sensitivity	Detection of diseases for timely and accurate diagnosis	96%
Increase in Relevance	Product recommendation relevance with real-time updates	15%
Reduction in Annotation Effort	Effort minimized through AI-assisted labeling	40%

Source: Authors

Table 2.4   Accuracy comparison

Recommendation System	Accuracy (%)
GPT-Based System	92.5
Traditional Content-Based System	74.3
Traditional Collaborative Filtering	81.7
Hybrid Traditional System	84.2

Source: Authors

Validation Metrics

- Classification metrics: 95% accuracy in skin type detection, 96% sensitivity in lesion detection, with precision (93.5%), recall (92.8%), and F1 score (93.1%).
- Recommendation performance: GPT-based system (92.5%) outperformed both content-based (74.3%) and collaborative filtering (81.7%) approaches.

Continuous improvement is Maintained Through

- Periodic retraining (5% accuracy gain per iteration), AI-assisted labeling (40% reduction in annotation effort).
- Adaptive refinement yielding 15% improvement in recommendation relevance over time.

7. CONCLUSION AND FUTURE SCOPE

AI technologies are transforming dermatology and cosmetic science by enabling skin type identification, disease diagnosis, and personalized recommendations. Models like YOLOv5, Transformers, U-Net are great for Wrinkle segmentation, product suggestions and facial features extraction with high accuracy. However, challenges like biasness, limited dataset, and adaptability for real-time inference are still there. Future advancements should focus on improving model accuracy, reducing bias, and integrating diverse dermatological datasets while considering environmental factors. Additionally, as AI-driven dermatological therapies gain global adoption, addressing concerns about energy efficiency and implementing ethical frameworks for user data handling will be essential.

References

1. Huynh, Q. T., Nguyen, P. H., Le, H. X., Ngo, L. T., Trinh, N. T., Tran, M. T. T., ... & Ngo, H. T. (2022). Automatic acne object detection and acne severity grading using smartphone images and artificial intelligence. Diagnostics, 12(8), 1879.

2. Abiyev, R. H. (2014). Facial feature extraction techniques for face recognition. Journal of Computer Science, 10(12), 2360.

3. Kassem, M. A., Hosny, K. M., Damaševičius, R., & Eltoukhy, M. M. (2021). Machine learning and deep learning methods for skin lesion classification and diagnosis: a systematic review. Diagnostics, 11(8), 1390.

4. Li, L. F., Wang, X., Hu, W. J., Xiong, N. N., Du, Y. X., & Li, B. S. (2020). Deep learning in skin disease image recognition: A review. Ieee Access, 8, 208264-208280.

5. Lee, J., Yoon, H., Kim, S., Lee, C., Lee, J., & Yoo, S. (2024). Deep learning-based skin care product recommendation: A focus on cosmetic ingredient analysis and facial skin conditions. Journal of Cosmetic Dermatology, 23(6), 2066-2077.

6. Kim, S., Yoon, H., Lee, J., & Yoo, S. (2023). Facial wrinkle segmentation using weighted deep supervision and semi-automatic labeling. Artificial Intelligence in Medicine, 145, 102679.

7. Yang, S., & Kar, S. (2023). Application of artificial intelligence and machine learning in early detection of adverse drug reactions (ADRs) and drug-induced toxicity. Artificial Intelligence Chemistry, 1(2), 100011.

8. Benyahia, S., Meftah, B., &Lézoray, O. (2022). Multi-features extraction based on deep learning for skin lesion classification. Tissue and Cell, 74, 101701.

9. Adebo, A. (2020). A natural language processing approach to a skincare recommendation engine (Doctoral dissertation, Dublin, National College of Ireland).

10. Elshahawy, M., Elnemr, A., Oproescu, M., Schiopu, A. G., Elgarayhi, A., Elmogy, M. M., & Sallah, M. (2023). Early melanoma detection based on a hybrid YOLOv5 and ResNet technique. Diagnostics, 13(17), 2804.

11. Xu, R., Lin, H., Lu, K., Cao, L., & Liu, Y. (2021). A forest fire detection system based on ensemble learning. Forests, 12(2), 217.

Emerging Perspectives and Applications of Computational Intelligence and Smart Systems
– Dr. Amit Lathigara et al. (eds)
© 2026 Taylor & Francis Group, London, ISBN 978-1-041-20965-2

3

Lung Cancer Stage Detection Using CNN: A Review

Mohammad Chandpasha*,
Julakanti Ruthwik Reddy, Pasupuleti Sai Sritha, M. A. Jabbar
Dept of CSE (AI & ML) Vardhaman College of Engineering,
Hyderabad, India

■ **Abstract:** Lung cancer is among the most prevalent tumours' that cause death. It should be identified in its early stages so that people can get well as quickly as possible. Improvements in medical imaging and image processing have allowed computer aided diagnosis (CAD) systems to become very important for physicians who look towards early detection of lung cancer. This paper studies some of the image processing techniques applied in the search and diagnosis for lung cancer through medical images such as X-ray, CT, and MRI. It discusses both old and new learning techniques and shows their disadvantages as well as advantages. Old methods such as image separation, feature extraction, and then classification have great applicability in helping radiologists with detailed analyses. Also, when it comes to locating, categorizing, and segmenting nodules,Convolutional Neural Networks (CNNs), a type of deep learning algorithm, have demonstrated superior performance compared to traditional approaches in lung cancer detection. The paper refers to important challenges and future paths in lung cancer detection with these methods, such as lack of data, understanding how models work, and combining different types of imaging data.

■ **Keywords:** Computer-aided diagnosis, Convolutional neural networks, Deep learning, Early detection, Image processing, Lung cancer, Medical imaging

1. INTRODUCTION

Among all cancers, lung cancer accounts for the highest number of deaths worldwide. Survival rates significantly rise if the condition is identified early for a successful course of treatment. However, diagnosing lung cancer at its early-stage is hard due to the subtlety of symptoms and the limitations of standard diagnostic methods like x-ray and CT-scan. These technologies frequently rely on the skill and competence of the radiologist. This can cause inconsistencies in diagnosis and a high-rate false positives and negatives. The most recent advances in AI and Deep Learning enabled the

*Corresponding author: mdchandpasha251@gmail.com

DOI: 10.1201/9781003725046-3

developers to build automated computer-aided diagnosis (CAD) systems that can assist radiologist in the identification of lung cancer (Mahum and Al-Salman 2023). One kind of deep learning model that has demonstrated promise in the diagnosis of medical images is convolutional neural networks (Naseer et al., 2023). These models can learn intricate patterns in medical images that are not detectable by the human eye.

2. LUNG CANCER OVERVIEW

Lung cancer is a highly prevalent and aggressive cancer. It falls into one of two categories: small cell lung cancer (SCLC) or non-small cell lung cancer (NSCLC). Compared to SCLC, NSCLC is thought to be more common and typically grows more slowly. The principal types of adenocarcinomas occur usually in smokers and non-smokers alike, growing in the outer portions of the lungs. Squamous cell carcinoma, often associated with smokers, grows in the middle portions of the lungs; fast-growing in nature, large cell carcinoma may arise anywhere in the lungs.

SCLC usually appears in about 10 to 15 percent of cases, is very often linked to cigarette smoking, and, with fast progression, is diagnosed too late for effective treatment options available (Mahum and Al-Salman 2023). The TNM classification system is utilized to assess the stage of the disease. Specifically, the letter T represents the size and position of the tumor, N signifies the presence of nearby lymph nodes, and M indicates that there is some degree of metastasis (Naseer et al., 2023). Cancer can be staged from localized stage 1 to advanced stage 4 (Obayya et al., 2023).

Treatment of NSCLC varies by stage: stages I and II will receive surgery with chemotherapy or radiation, stage III will receive combined chemo-radiotherapy with optional surgery; and, finally, therapies for stage IV include chemotherapy, immunotherapy, or targeted therapies. SCLC is treated mainly with chemotherapy and radiation because of its aggressive nature (Mahum and Al-Salman 2023).

3. PROBLEMS IN DISCOVERING LUNG CANCER

Early lung cancer detection is particularly problematic because symptoms like a persistent cough, mild chest pain, or trouble breathing can be mistaken for advanced lung diseases like pneumonia or bronchitis (Mahum and Al-Salman 2023). In general, with most of the preliminary tests, X-ray machines overlook small tumors and have poor quality in allowing for the early diagnosis. While more precise, CT scans also greatly rely on the skill and expertise of the radiologist interpreting them (Naseer et al., 2023). The skill variability of radiologists, together with the quality of the images, complicate detection of lung cancer in the early stages (Obayya et al., 2023). Similarly, testing of any sort is open to the intrusion of both false positives and false negatives. The former creates undue anxiety and invite invasive tests while the latter afford led delay in treatment and hence reduced survival rates (Naseer et al., 2023). The implementation of artificial intelligence, specifically transfer learning with convolutional neural networks, presents a potentially effective solution for increasing accuracy in the prediction of the malignancy of lung nodules (Obayya et al., 2023).

4. ADVANTAGES OF AI FOR DETECTING LUNG CANCER

Detecting lung cancer through artificial intelligence is changing the way this disease is diagnosed. AI systems are faster and more accurate which lowers human error. Deep learning-based AI systems, particularly CNNs, can effectively spot fine details in medical images that radiologists might fail to detect. This is particularly true in regard to early-stage tumors, as illustrated by recent studies (Naseer et al., 2023; Mahum and Al-Salman 2023). AI quickly analyses thousands of scans, providing accurate diagnoses and critical treatment decisions for high impact cancers like SCLC (Hussain et al., 2023; Obayya et al., 2019). AI helps to provide new evidence-based decision support through the implementation of Computer Aided Diagnosis (CAD) systems. These systems assist with accurate staging of lung cancer at presentation and the development of a suitable treatment plan (Chen et al., 2021; Rajasekar et al., 2023). Also, by offering diagnostics remotely in underserviced areas, patients can be assessed by specialists without the need for travelling to the healthcare professional. This helps with the accessibility of services, ensuring that everyone benefits from high-quality services regardless of the setting (Mehmood et al., 2022; Hawkins et al., 2014). In general, AI helps in improving health outcome and care by closing the gap in access to healthcare.

5. LITERATURE REVIEW: EMERGING TECHNIQUES TO IDENTIFY LUNG CANCER

Lung-RetinaNet is a model that enhances detection for early-stage lung tumors via multi-level feature fusion and a context module. As a result, the lung model greatly improves sensitivity and specificity on the analysis of relationships between tumors and adjacent tissues, making it so effective at identifying tumors that are difficult to detect. (Mahum and Al-Salman, 2023).

Modified U-Net focus on the three-phase approach involving Nodule extraction, classification, and lobe segmentation. By focusing on the regions of interest in the CT scan with regard to benign and malignant nodules, the method achieves high precision and recall (Naseer et al., 2023).

The BICLCD-TSADL model combines the Tuna Swarm Algorithm with deep learning methods, maximizing feature extraction and attaining higher accuracy than conventional approaches. The combination is effective in biomedical image analysis for cancer detection (Obayya et al., 2023).

Transfer learning with pre-trained AlexNet model minimizes computational complexity with high accuracy in cancer detection histopathology. The process selectively filters images to minimize processing time with attention on the essential areas (Mehmood et al., 2022).

Radiomics analysis is used to pull out 3D characteristics like texture, shape, and intensity from CT images to make predictive outcomes for patients with NSCLC. They help tailor the treatment by mapping to survival and response to therapy (Hawkins et al., 2014).

Sophisticated fuzzy entropy techniques improve the discrimination of NSCLC and SCLC imaging data and enable dimensionality reduction. This advancement enhances the prognosis for patients with lung cancer and enables early diagnosis (Hussain et al., 2019).

Deep learning-based models like CNN and ResNet-50 have made a significant improvement in accuracy for lung cancer detection. These models use CT and histopathological images to identify malignancies with high accuracy, proving their worth over traditional methods (Rajasekar et al., 2023).

LDNNET is specifically designed to detect lung nodules and identify lung cancer. LDNNET improves lung nodule classification accuracy, resolves the vanishing gradient issue, and offers robust information flow by utilizing dense connections between network layers (Chen et al., 2021).

Table 3.1 Techniques used for lung cancer detection and their accuracy

S. No	References	Technique Used	Accuracy
1	(Mahum and Al-Salman 2023)	Lung-RetinaNet	99.8%
2	(Naseer et al., 2023)	U-NetModel	97.98%
3	(Obayya et al., 2023)	Deep Learning-Based Tuna Swarm Algorithm (BICLCD TSADL)	99.33%
4	(Mehmood et al., 2022)	Transfer Learning with Class Selective Image Processing	98.8%
5	(Hawkins et al., 2014)	Decision Tree, Naive Bayes, SVM, Rule-Based (JRIP)	77.5%, 65.0%, 60.0%, 67.5%
6	(Hussain et al., 2019)	Refined Fuzzy Entropy Methods	P-value of 1.95E 50 for MFE(SD). P-value of 3.01E14 for MFE (Mean).
7	(Rajasekar et al., 2023)	VGG-16, VGG-19, CNN, CNN GD, Inception V3, and ResNet-50	93.64%, 97.86%, 96.52%, 92.17%, 93.54%, 93.47%
8	(Chen et al., 2021)	Lung Dense Neural Network	LUNA16 DSB 98.84%, Kaggle DSB 2017 dataset 99.95%

Source: Author's compilation

6. CONCLUSION

Lung cancer still remains one of the deadliest cancers in the world, and hence early detection is important to improve patient outcomes. Conventional techniques like X-rays and CT scans produce a lot of false positives and negatives, depending on the radiologist's skills. These results may sometimes delay treatment and diagnosis. CNNs and AI technologies are changes in the detection of lung cancers because they improve accuracy and minimize errors to make rapid diagnoses. These massive data sets are made up of tiny nodes that elude the human eye. It combines AI with imaging modalities such as PET scans to enhance imaging accuracy while enabling personalized medicine, predicting treatment outcomes based on tumor features. It automates the tedious aspects of diagnosis, allowing doctors to spend time on the more difficult cases and do so with less expense. It is able to continue learning with data, which means that with continuous learning, one can count on improvement in diagnostic reliability and thereby help bridge the gaps in the provision and availability of healthcare.

References

1. Chen, Y., Wang, Y., Hu, F., Feng, L., Zhou, T., & Zheng, C. (2021). LDNNET: towards robust classification of lung nodule and cancer using lung dense neural network. *IEEE Access, 9,* 50301–50320.
2. Hawkins, S. H., Korecki, J. N., Balagurunathan, Y., Gu, Y., Kumar, V., Basu, S., ... & Gillies, R. J. (2014). Predicting outcomes of nonsmall cell lung cancer using CT image features. *IEEE access, 2,* 1418–1426.
3. Hussain, L., Aziz, W., Alshdadi, A. A., Nadeem, M. S. A., Khan, I. R., & Chaudhry, Q. U. A. (2019). Analyzing the dynamics of lung cancer imaging data using refined fuzzy entropy methods by extracting different features. *IEEE Access, 7,* 64704–64721.
4. Mahum, R., & Al-Salman, A. S. (2023). Lung-RetinaNet: Lung cancer detection using a RetinaNet with multi-scale feature fusion and context module. *IEEE Access, 11,* 53850–53861.

5. Mehmood, S., Ghazal, T. M., Khan, M. A., Zubair, M., Naseem, M. T., Faiz, T., & Ahmad, M. (2022). Malignancy detection in lung and colon histopathology images using transfer learning with class selective image processing. *IEEE Access*, *10*, 25657–25668.

6. Naseer, I., Akram, S., Masood, T., Rashid, M., & Jaffar, A. (2023). Lung cancer classification using modified u-net based lobe segmentation and nodule detection. *IEEE Access*, *11*, 60279–60291.

7. Obayya, M., Arasi, M. A., Alruwais, N., Alsini, R., Mohamed, A., & Yaseen, I. (2023). Biomedical image analysis for colon and lung cancer detection using tuna swarm algorithm with deep learning model. *IEEE Access*, *11*, 94705–94712.

8. Rajasekar, V., Vaishnnave, M. P., Premkumar, S., Sarveshwaran, V., &Rangaraaj, V. (2023). Lung cancer disease prediction with CT scan and histopathological images feature analysis using deep learning techniques. *Results in Engineering*, *18*, 101111.

9. Obayya, M., Arasi, M. A., Alruwais, N., Alsini, R., Mohamed, A., & Yaseen, I. (2023). Biomedical image analysis for colon and lung cancer detection using tuna swarm algorithm with deep learning model. *IEEE Access*, *11*, 94705–94712.

10. Selvanambi, R., Natarajan, J., Karuppiah, M., Islam, S. H., Hassan, M. M., & Fortino, G. (2020). Lung cancer prediction using higher-order recurrent neural network based on glowworm swarm optimization. *Neural Computing and Applications*, *32*, 4373–4386.

Emerging Perspectives and Applications of Computational Intelligence and Smart Systems
– Dr. Amit Lathigara et al. (eds)
© 2026 Taylor & Francis Group, London, ISBN 978-1-041-20965-2

4

IoT and RFID-Enabled Smart Lab Monitoring System for Efficient Asset Management

Dinesh Reddy Malreddy*,
Kayithi Sujeeth Reddy, Advith Teelaru,
Velagaleti Sriraghava, Ala Rajitha
Vardhaman College of Engineering, Hyderabad,
Telangana, India

■ **Abstract:** The Smart Log System project focuses on lab monitoring by leveraging IoT technology to capture, store, and large volumes of log data generated from lab instruments, sensors, and devices under lab conditions. The system provides a centralized and automated approach to managing lab equipment. RFID (Radio Frequency Identification) technology and database management systems notify the administrator when equipment is suspected to be missing, enhancing the accuracy and reproducibility of lab results. The Smart Log System offers significant advantages for modern research and testing facilities by providing reliable data and actionable insights for efficient equipment tracking. Additionally, it reduces maintenance costs through a fully computerized, paperless operation, ensuring complete digital storage and minimizing unnecessary resource wastage.

■ **Keywords:** Automation, Database management, IoT, Lab monitoring, RFID

1. INTRODUCTION

1.1 Motivation

The average laboratory worker in the UK uses approximately 45 sheets of paper daily, contributing to significant paper wastage. The UK alone consumes over 9.9 million tons of paper annually. Given that 24 mature trees are required to produce one ton of paper, this excessive consumption contributes to deforestation, carbon emissions, and environmental degradation. This project is in line with the United Nations Sustainable Development Goals (SDG) 12 (Responsible Production and Consumption) and SDG 9 (Industry, Innovation, and Infrastructure). The project seeks to minimize wastage of paper in laboratories through technology-based alternatives like computerized record-keeping, data entry automation, and biodegradable alternatives.

---

*Corresponding author: Dinesh.reddy2504@gmail.com

DOI: 10.1201/9781003725046-4

1.2 Scope

The Smart Log System serves as an innovative solution for logging and managing laboratory activities. It records and processes all logs generated from laboratory instruments, sensors, and devices by integrating **RFID** tags. Traditionally, laboratory management relies heavily on manual data entry from devices such as microscopes, which is time-consuming, error-prone. The Smart Log System addresses these issues by providing enhanced data accuracy, operational productivity, and sophisticated equipment management. Its real-time logging capability ensures that records are updated automatically, minimizing human errors and improving efficiency. Additionally, continuous hardware maintenance helps prevent unforeseen damages while enabling real-time monitoring of equipment usage.

1.3 Objectives

Searching for specific information in a manual log system is time-consuming, labour-intensive, and hinders effective asset control. This often leads to inefficiencies, errors, and resource wastage. By adopting a Smart Log System, paper usage is significantly reduced, and manual effort is minimized. The system enables real-time monitoring and management of assets and inventories, resulting in efficient and effective resource utilization. Digital logs provide quick retrieval and tracking of records within seconds, eliminating the frustration and delays caused by traditional systems.

2. LITERATURE REVIEW

2.1 History

The **MFRC522 RFID** reader reads inlays placed on laboratory equipment or inventory items, each with unique identification numbers for tracking. The **Raspberry Pi** acts as a local gateway, collecting data from the RFID reader and transmitting it to a central system. This data is stored in a **MySQL** database, recording all equipment check-ins and check-outs for future reference. An application on the Raspberry Pi processes data and stores it accurately in the database. Laboratory staff can access this data through a **web-based dashboard**, allowing easy equipment monitoring, movement tracking, and report generation.

3. METHODOLOGY

3.1 Comparative Analysis

Decision matrix is an organized matrix of values in rows and columns that allow an analyst to analyze systematically, evaluate, and decide on the performance of interrelationships between different sets of values and information. Elements of a decision matrix represent decisions based on given decision criteria. (Table 4.1)

3.2 Proposed Design

Effective management of lab equipment is crucial in educational and research institutions. The proposed system combines RFID technology with a Raspberry Pi to track equipment movement. It enables laboratory heads and technicians to monitor usage, enhancing efficiency and reducing losses. The system includes RFID tags, scanners, and cards for authorized check-ins and check-outs, with the Raspberry Pi acting as a central unit for logging and alerting unauthorized actions. Data is sent to a central server for real-time monitoring via a dashboard, ensuring efficient inventory management.

Table 4.1 Comparison framework evaluating multiple alternatives based on different criteria. Each cell xijx_{ij}xij contains a score or assessment. Sum and rank rows determine the best alternative, while the status row provides the final decision

Example of Comparison				
	Alternative 1	**Alternative 2**	**...**	**Alternative M**
Criterion 1	x11	x12	...	x1M
Criterion 2	x21	x22	...	x2M
	...	...	Xij = Good	...
Criterion N	xN1	xN2	...	xNM
Sum				
Rank				
Status	No	No		

Source: Authors

Future improvements could include IoT integration, biometric authentication, and mobile apps for equipment tracking. This system offers a smart, efficient solution for lab management.

3.3 Results

The proposed system offers efficient lab monitoring and management using RFID tags and Raspberry Pi for automated logging, storing, and analysing data from laboratory instruments, sensors, and devices. This approach ensures high accuracy, better reproducibility, reduced downtime, and lower operational costs. The paper outlines the system's architecture and implementation, highlighting its potential for effective laboratory equipment and inventory management. Its modular design allows seamless integration with existing warehouse management systems, providing a versatile and cost-effective solution for monitoring and managing lab resources.

Table 4.2 Comparison of different microcontrollers based on various criteria (Safety, Efficiency, Ease of Use, Cost)

S. No.	Criteria	Safety	Efficiency	Ease of Use	Cost
1	Arduino	5	3	5	2
2	NodeMCU	2.5	4	5	4
3	Raspberry Pi PICO	5	5	5	4
4	Raspberry Pi4 Model B	2.5	5	5	5

Source: Authors

4. CONCLUSION

This research has analyzed the effects of ownership concentration and institutional ownership on stock returns at the firm level during the pre-crisis and post-crisis periods. According to the results, ownership concentration is negative in the pre-crisis period and has no effect in the post-crisis period, thus implying that the effect of ownership concentration is time-dependent. It can be argued that concentrated ownership is negative during the era of economic growth. Although there is no institutional effect in the pre-crisis period, a negative effect is seen during the post-crisis period, thus implying that this variable also has a time-dependent effect. Therefore, it can be argued that institutional ownership has a negative effect on stock returns during the era of economic slowdown. Additionally, the effect of firm age is time-dependent. On the whole, this research argues that higher firm profitability increases stock returns, and high leverage lowers stock return.

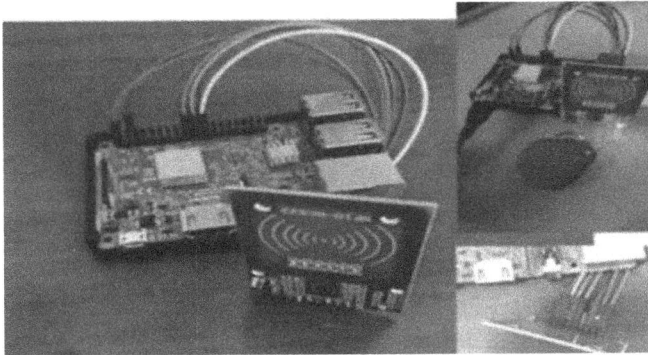

Fig. 4.1 RFID-based system with a Raspberry Pi and MFRC522 module for scanning RFID tags. The setup enables authentication, inventory tracking, and automation

Source: Authors

References

1. Dane, H., Michael, K., & Wamba, S. F. (2010). RFID-Enabled Inventory Control Optimization. 43rd Hawaii International Conference on System Sciences.
2. Omar, M., Samin, O. B., & Ahmed, I. (2019). Smart shed: An Automatic Shed System Based on Rain, Temperature, and Light Intensity. International Conference on Science Innovation and Management (ICSIM)
3. Murthy, T. S. R. C., & Rasool, S. K. N. (2019). Design of Smart Bio-Shed Using IoT with Raspberry Pi. International Journal of Recent Technology and Engineering (IJRTE), 8(246).
4. Wang, L., Zhang, Y., & Liu, H. (2020). Design and Implementation of an RFID-Based Laboratory Asset Management System. IEEE Access, 8, 56345–56353.
5. Gupta, P., & Singh, R. K. (2020). Smart Lab Management Using IoT and RFID Technologies. International Conference on Smart Electronics and Communication (ICOSEC).
6. Patel, S., Desai, A., & Shah, N. (2020). IoT-Based Asset Monitoring System for Laboratories. International Journal of Advanced Research in Computer Science, 11(3), 45–50.
7. Wahab, M. H. A., Kadir, H. A., Tukiran, Z., Sudin, N., Jalil, M. H. A., & Johari, A. (2011). RFID-Based Equipment Monitoring System. Designing and Deploying RFID Applications.
8. Ngai, E., Suk, F. F. C., & Lo, S. Y. Y. (2008). Development of an RFID-Based Sushi Management System. International Journal of Production Economics, 112(2), 630–645.
9. Oztekin, A., Pajouh, F. M., Delen, D., & Leva, K. (2010). An RFID network design methodology for asset tracking in healthcare. Decision Support Systems, 49(1), 100–109.
10. Roberts, C. M. (2006). Radio frequency identification (RFID). Computers & Security, 25(1), 18–26.
11. Yan, B., & Lee, D. (2009). Application of RFID in cold chain temperature monitoring system. 2009 ISECS International Colloquium on Computing, Communication, Control, and Management, 258–261.
12. Tan, T. H., & Chang, C. S. (2010). Development and evaluation of an RFID-based e-restaurant system for customer-centric service. Expert Systems with Applicationse farm management system based on animal identification using RFID technology. Computers and Electronics in Agriculture, 70(2), 380–388.

Emerging Perspectives and Applications of Computational Intelligence and Smart Systems
– Dr. Amit Lathigara et al. (eds)
© 2026 Taylor & Francis Group, London, ISBN 978-1-041-20965-2

5

Detecting System Vulnerability to Malware Attacks Using ML Techniques: A Review

D. V. Harsha Vardhan*, Adi Divya,
Ega Priyanka, B. Pravallika, M. A. Jabbar
Dept of CSE (AI & ML) Vardhaman College of Engineering,
Hyderabad, India

■ **Abstract:** A powerful subset of artificial intelligence is called machine learning (ML). It allows computers to learn from data and gradually improve at what they do without the need for explicit programming. It is highly useful in Cyber security, particularly in malware detection, due to its capacity to scan large volumes of data and identify patterns. ML is crucial for preventive measures in this area since traditional signature-based approaches are frequently insufficient against changing cyber threats. This study looks at several machine learning (ML) approaches, such as supervised algorithms like Random Forest and Support Vector Machines (SVM) and feature reduction techniques like Principal Component Analysis (PCA) and Auto Encoders for detecting system vulnerability. By concentrating on relevant characteristics and enhancing model correctness, these techniques increase the effectiveness of detection systems.

■ **Keywords:** Artificial intelligence, Auto encoders, Cyber security, Machine learning, Malware, pca

1. INTRODUCTION

Machine learning (ML) is a vast discipline that is related to data processing, analytics, and statistics. In this particular area of artificial intelligence (AI), computers are able to identify patterns in data without being specifically instructed where to search (Gormentetal., 2023). The notion that machines could be made to think and reason dates back to 1956, when computer scientists first put out this theory. Machine learning is a crucial tool in AI's goal of automating human like thought processes (Gorment et al., 2023). Fromthe1950s to the present, machine learning has had a significant impact on problem-solving in a variety of industries (Gorment et al., 2023). However, having high-quality data is crucial for success. Solving real-world problems requires knowing when and how to apply

---

*Corresponding author: harshaintern2022@gmail.com

DOI: 10.1201/9781003725046-5

various machine learning methods (Gorment et al., 2023). Machine learning(ML) can be categorized into four main types Supervised Learning, Unsupervised Learning, Semi-supervised Learning, Reinforcement Learning.

2. NEED OF DECISION SUPPORT SYSTEM FOR DETECTING SYSTEM VULNERABILITY TO MALWARE ATTACK

Maintaining the Integrity of the Specifications: The term "malware," which stands for "malicious software," describes computer programs that interfere with normal processes and corrupt electronic data (Shaukat et al., 2020) Viruses, worms, ransom ware, adware, spyware, and Trojan horses are significant forms of malware. As technology evolves, malware becomes a persistent concern globally, driven by the increasing use of computer systems and internet connections (Gorment et al., 2023).

Decision Support: Large data sets can be efficiently evaluated by machine learning (ML), which can identify patterns and anomalies that traditional methods might miss. The possibility of detection is increased by machine learning techniques, such as decision trees and neural networks, which can adapt to new malware as it arises.

There are many types of malware like Virus this is a software category is the simplest to understand. It is essentially any programming that has been stacked and run without the client's permission, either copying itself or harming and altering other programming (Choudhary, S. and Sharma,A.,2020). Root Kit, Itis a type of malware that hides deep within a computer system. Ransom ware, it is a kind of cyberattack in which hackers encrypt or lock a user's files, making them useless (Mos, M.A. and Chowdhury, M.M., 2020). Worms, it is a type of malware that spreads from computer to computer over a network (Yu, W et al., 2005).

3. DETECTING SYSTEM VULNERABILITY TO MALWARE ATTACKS USING MACHINE LEARNING

The exponential expansion of digital data has opened up fresh opportunities for enhancing Cyber Security, especially in identifying malware attack weaknesses in systems. In this field, machine learning is essential because it makes it possible to analyze big information, spot patterns that could be indicators of danger, and automate reactions to improve security. By identifying irregularities in system behavior, categorizing different kinds of malware, and predicting possible weaknesses, machine learning techniques can assist strengthen defenses against Different Techniques used by ML.

J48 Decision Tree is used for classification or regression applications; decision trees are effective Non parametric structures. Using ideas like data entropy, they divide the input space into smaller are a sin order to predict the dependent variable (Choudary, S. and Sharma,A.,2020).

K-Near Set Neighbor is a classification method that uses supervised learning. It labels fresh data points by using the closest labeled data points as a guide (Aslam et al., 2020).

Support Vector Machines(SVM) divides data into several categories by identifying the optimal line or "hyper plane" to do so. The objective is to optimize the "margin," or distance, between this hyper plane and the closest data points in each category. SVMs work well with complex data and can be used to differentiate between malicious and safe software by looking at specific characteristics (Choudhary, S. and Sharma, A., 2020).

4. LITERATURE REVIEW: EMERGING TECHNIQUES USED FOR DETECTING MALWARE

In order to improve malware detection skills, Feature selection based on correlation has been used. The authors showed the success of their approach in identifying possible threats by achieving an accuracy of 90.5% by using machine learning techniques like Random Forest and Support Vector Machine (SVM) on a Cyber Threat Dataset (Shaukat et al.,2020)

Information Gain as a feature educations strategy is used in addition to Decision Trees and K-Nearest Neighbors (KNN) in this study, which focused on malware classification. With an accuracy of 88.9%, the study's use of a large malware dataset demonstrated the model's effectiveness in differentiating between harmful and benign software (Choudhary, S. and Sharma, A., 2020).

Yu, W., Boyer investigated the application of Logistic Regression for malware detection without feature reduction approaches. The method, tested on a virtual P2P network, produced an accuracy of 85.0%, demonstrating the model's ability to identify malware in particular network scenario seven in the absence of feature optimization (Yu, Wet al.,2005).

A dual strategy is used that made use of both Random Forest and Neural Networks. The authors demonstrated the potential of combining different machine learning algorithms to improve detection performance by achieving an accuracy of 91.5% using a ransom ware dataset without feature reduction (Mos, M.A. and Chowdhury, M.M., 2020).

Adaptive learning techniques including Random Forest and SVM, and PCA and Auto encoder for feature reduction. The accuracy of 94.1% that the authors obtained with a malware dataset shows how well their model handles feature dimensionality and adjusts to new threats (Aslam et al., 2020).

In an ensemble technique that incorporated Random Forest and Gradient Boosting, Priya dharshan used Mutual Information for feature selection. With a 93.8% accuracy rate on a malware detection dataset, the study demonstrated how well ensemble approaches work to increase detection rates through feature selection (Priyadharshan et al., 2021).

According to Mills, the author improved the Random Forest model's malware detection performance by using PCA as a feature reduction technique. The study highlighted the importance of feature optimization in attaining consistent detection results by reporting an accuracy of 92.3% on the Malware Detection Dataset (Mills et al., 2019)

Table 5.1 Machine learning techniques for malware attacks

Sr. no.	References	Technique Used	Accuracy
1	(Almomanietal.,2024)	Naïve Bayes, SVM	87.2%
2	(Aslametal.,2020)	Adaptive Learning(Random Forest, SVM)	94.1%
3	(Choudhary, S. and Sharma, A., 2020)	Decision Tree, KNN	88.9%
4	(Millsetal.,2019)	Random Forest	92.3%
5	(Mos, M.A. and Chowdhury, M.M., 2020)	Random Forest, Neural Network	85.0%
6	(Priyadharshanetal.,2021)	Ensemble (Random Forest, Gradient Boosting)	93.8%
7	(Shaukatetal.,2020)	Random Forest, SVM	90.5%
8	(Yu,Wetal.,2005)	Logistic Regression	99.33%

Source: Author's compilation

◥ 5. Conclusion

Machine learning (ML) techniques are highly effective at identifying malware, as seen by their high accuracy rates across a range of datasets. The use of feature reduction methods such as Principal Component Analysis (PCA) and Mutual Information has significantly enhanced the performance of these models. All things considered, these observations high- light how important machine learning is to strengthening Cyber Security defenses against malware attacks.

References

1. Almomani, I., Almashat, T. and El-Shafai, W., (2024). Maloid-DS: Labeled Dataset for Android Malware Forensics. IEEE Access.
2. Aslam, M., Ye, D., Hanif, M., & Asad, M. (2020, December). Adaptive machine learning: A framework for active malware detection. In *2020 16th International Conference on Mobility, Sensing and Networking (MSN)* (pp. 57–64). IEEE.
3. Choudhary, S., & Sharma, A. (2020, February). Malware detection & classification using machine learning. In *2020 International Conference on Emerging Trends in Communication, Control and Computing (ICONC3)* (pp. 1–4). IEEE.
4. Gorment, N. Z., Selamat, A., Cheng, L. K., &Krejcar, O. (2023). Machine learning algorithm for malware detection: Taxonomy, current challenges, and future directions. *IEEE Access, 11*, 141045–141089.
5. Mills, A., Spyridopoulos, T., & Legg, P. (2019, June). Efficient and interpretable real-time malware detection using random-forest. In *2019 International conference on cyber situational awareness, data analytics and assessment (Cyber SA)* (pp. 1–8). IEEE.
6. Mos, M. A., & Chowdhury, M. M. (2020, July). The growing influence of ransomware. In *2020 IEEE international conference on electro information technology (EIT)* (pp. 643–647). IEEE.
7. Priyadarshan, P., Sarangi, P., Rath, A., & Panda, G. (2021, January). Machine learning based improved malware detection schemes. In *2021 11th international conference on cloud computing, data science & engineering (confluence)* (pp. 925–931). IEEE.
8. Shaukat, K., Luo, S., Chen, S., & Liu, D. (2020, October). Cyber threat detection using machine learning techniques: A performance evaluation perspective. In *2020 international conference on cyber warfare and security (ICCWS)* (pp. 1–6). IEEE.
9. Singh, M., & Pamula, R. (2018, September). Email spam classification by support vector machine. In *2018 International Conference on Computing, Power and Communication Technologies (GUCON)* (pp. 878–882). IEEE.
10. Yu, W., Boyer, C., Chellappan, S., & Xuan, D. (2005, May). Peer-to-peer system-based active worm attacks: Modeling and analysis. In *IEEE International Conference on Communications, 2005. ICC 2005. 2005* (Vol. 1, pp. 295–300). IEEE.

Emerging Perspectives and Applications of Computational Intelligence and Smart Systems
– Dr. Amit Lathigara et al. (eds)
© 2026 Taylor & Francis Group, London, ISBN 978-1-041-20965-2

6

A Systematic Review on License Plate Recognition using Deep Learning

A. Akhilesh*, A. Sharath Kumar,
B. Prayaksha, Prakash Kumar Sarangi, M. A. Jabbar
Dept of CSE (AI&ML) Vardhaman College of Engineering Hyderabad,
Andhra Pradesh, India

■ **Abstract:** Deep learning offers quick and precise means of predicting where an object is and identifying what's in the image. Deep learning, it is a robust ML method where the object detector learns automatically image features needed for detection tasks. Yolo (you only look once) is a Deep Learning model. YOLO was developed into 8 different versions. All the versions have been created upon the previous one with additional features like better accuracy, quicker processing. In this paper, were view varieties of yolo developed for license plate recognition. This paper helps in reducing the research gap about YOLO algorithm and its versions.

■ **Keywords:** Deep learning, License plate recognition, Machine learning, Yolo

1. INTRODUCTION

Deep Learning is a sub-division of machine learning which instructs processors to process information, make decisions similar to humans. Deep learning is classified into various categories.

Feed forward neural networks are the straight forward form of ANN encompasses feed forward neural networks (FNNs), which transmit information linear way along thenetwork. FNNs were frequently employed for applications such as classification of images, recognizing speech (Bebisand Georgiopoulos, 1994).

Convolutional neural networks are exactly appropriate for recognition of images and video activities. CNNs can study repetitively from the image features, and for this reason, they are nicely suited for activities such as classifying images, detection of objects, and image segmentation (Alzubaidi et al., 2021).

Recurrent neural networks are a category of neural network which handles consecutive data, such as time series and natural language. They can maintain an core state that stores data about the past inputs, making them compatible for applications like recognizing speech, Natural language processing, and language paraphrase (Alzubaidi et al., 2021) Example: Speech recognition

---

*Corresponding author: aradhiakhilesh@gmail.com

DOI: 10.1201/9781003725046-6

Long Short-Term Memory networks remain a particular kind of Recurrent neural network intended to keep information in mind for long sequences. In simpler terms, LSTMs are good at learning and remembering patterns in data that come in a series, likes sentences in text, stock prices over time, or sounds in speech (Hochreiterand Schmidhuber, 1997) Example: Text generation

2. DEEP LEARNING TECHNIQUES

Among Deep learning methods, the YOLO (You Only Look Once) algorithm which comes under CNN (Convolutional Neural Network) has become popular for real-time detection of object because it balances speed, accuracy well. Yolo's design allow sit to notice objects, like license plates, in one quick pass through the network, making it very efficient for actual tasks.

Since its first version in 2016, YOLO has go net through several updates, from YOLO v1 to the latest YOLO v8. Each new version has improved in detecting objects more accurately and quickly, while also better a text rating features from images. The YOLO targeted gentrification algorithm is distinguished by its little size and quick calculation performance. The structure of YOLO is up front. The YOLO targeted gentrification algorithm is distinguished by its small size and quick calculation performance YOLO's speed is rapid since its imply request stoup load the image to the net work to obtain the final result of detection, so it willdetect video in real time. Yolo directly utilizes the universal image for detection, which can encrypt the universal information, decreases the error of noticing the contextual as a thing (Peiyuan Jiang et al., 2022).

This paper presents systematic review of various YOLO versions in the frame work of license plate recognition. By analyzing studies which have used different YOLO models, from YOLOv1 to YOLOv8, this review aims to compare their performance, highlight their strengths and limitations.

3. BRIEF ABOUT LICENSE PLATE RECOGNITION

A license plate which is also known as a number plate, it is a metal plate that identifies a vehicle by displaying its registration number. In India, license plates are required to be attached to the front and rear of all motorized vehicles that are driven on public roads. India has several types of license plates, White: The most common type, used for private vehicles and non-transport vehicles. Yellow: Used for commercial vehicles, such as buses, trucks, and taxis. Green: Used for electric vehicles, such as electric cars, bikes, and e rickshaws. Black: Used for luxury or high-security vehicles, or by businesses that rent out vehicles. Blue: Used for foreign delegates, consulates, and international organizations. Red: Used for vehicles that belong to the diplomatic or consular corps, or for temporary registration. Red with the Emblem of India: Used for the President, Prime Minister, Governors, and Lieutenant Governors.

4. NEED FOR DEEP LEARNING LICENSE PLATE RECOGNITION

Deep learning issued for license plate detection, improve the accuracy of license plate recognition. It extracts deeper features from images, which can help detect and recognize license plates in a diversity of conditions. Different techniques used to detect

Generative Adversarial Networks (GANs) are powerful models that generate realistic data by learning from real samples. They are commonly used for creating high-quality images from random noise (Zhang Xiong and Wang, 2019). Auto encoders are models that learn input patterns and produce outputs representing generalized values of the inputs. They are effective for anomaly detection, feature learning, and pre-training complex models (Berahmand et al., 2024). Natural Language Processing (NLP) uses deep learning models like Transformers (e.g., GPT, BERT) to understand human language. These model sex cleaners' possibilities such as text generation, language. Para phrase, and context-aware responses(Feldman,1999).

Long Short-Term Memory (LSTM) systems are intended for time series forecasting and sequential data analysis. Using input, forget, and output gates, LSTMs effectively retain and discard in formation to recognize long-term patterns (Hochreiter and Schmidhuber, 1997).

5. Applications of Deep Learning in License Plate Detection

YOLO v8 can make traffic monitoring easier by quickly identifying license plates in real-time. This technology helps reduce errors, speeds up the process, and improves efficiency in managing traffic violations and vehicle tracking (Cong et al.,2023). Automatic license plate recognition uses AI and deep learning to notice and recognize vehicle license plate characters. It is widely applied in traffic management, toll booth payments, and intelligent transportation systems for efficient automation.

6. Overview of YOLO Versions

6.1 YOLOv1 (2016)

YOLOv1 was introduced by Joseph Redmon in 2016 asareal-time object detecting model. The core idea was to pose detection of object as one regression problem, predicting both the box bounding and class chances directly from the input image in one pass (Redmon, 2016).

6.2 YOLOv2(YOLO9000)(2017)

YOLOv2 improved over the original by incorporating anchor boxes and batch normalization, allowing for better small object detection, which is critical for tasks like LPR (Redmon and Farhadi, 2017).

6.3 YOLOv3(2018)

YOLOv3 is applied for detecting license plates and used an optical character recognition (OCR) module for recognition. The combination of YOLOv3's superior detection capabilities and fast processing time made it ideal for present LPR applications in complex environments (Montazzolli and Jung, 2018).

6.4 YOLOv4(2020)

YOLOv4 is applied for license plate detection in challenging environments such as low-light and angled plates. The combination of YOLOv4's robustness and the use of are fined OCR module enabled accurate recognition in various conditions (Silva and Jung, 2020).

6.5 YOLOv5(2020)

YOLOv5 study is used for real-time license plate detection, followed by a original back bone network and introduces procedures which are novel like focal loss, leveling labels, and auto growth to enhance the accurateness (Jocher, 2020).

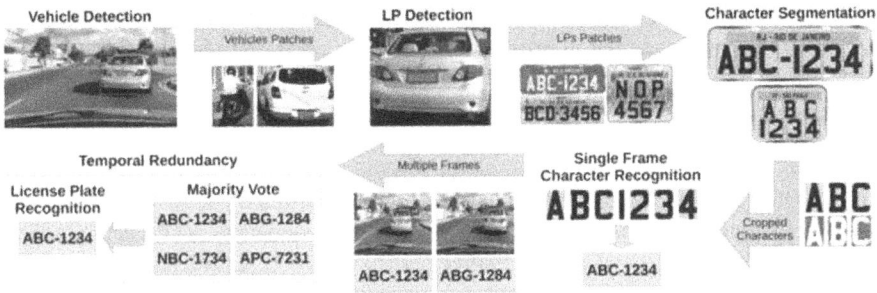

Fig. 6.1   Recognition process

Source: Bebis and Georgiopoulos, 1994

6.6 YOLOv6(2022)

YOLOv6 is built by their searchers of Meituan (Lietal.,2022), object detection frame work targeted largely at industrial applications, it's hardware-efficient design and higher efficiency exceed YOLOv5 in terms of accuracy of detection and implication time. It improves speed and accuracy by employing an Efficient Rep backbone.

6.7 YOLOv7(2022)

YOLOv7 has set new benchmarks in object detection accuracy while retain in greal-time capabilities, making it suitable for high-precision LPR tasks (Wang et al., 2022).

6.8 YOLOv8(2023)

YOLOv8 is the new iteration (Jocher,2023) of all the other models. YOLOv8 is highly used for detecting objects within images, classifying images, and distinguishing objects from one another.

Table 6.1   DL techniques for license plate recognition

S. No	Reference	ML Techniques	Dataset	Accuracy
1	(Patil, Patil, Pawaretal. 2023)	YOLO-V8, Easy OCR, Convolutional Neural Networks	1000 license plate images (800 for training)	86%
2	(MingLiand Li Zhang 2023)	YOLO-V6, Internet of Things (IoT), Deep Learning	License plate images	High Accuracy (F1-Score)
3	(SimarViget al.2023)	YOLO-V6, Convolutional Neural Networks (CNN), RNN	Custom Dataset license plate images	99.2%
4	(ShanLuo and Jihong Liu 2022)	YOLO-V5, LPRNet	Chinesecity Parking Dataset (CCPD)	98.56%
5	(ShenghuPan et al. 2022)	YOLO-V7, LPRNet, STN (Spatial Transformer Networks)	License plate character recognition	96.1%
6	(Anastassia Angelopoulouetal. 2022)	YOLO-V2, V4,ResNet50	Caltech Cars (124 samples)	AVG 90.3%, Max 99%

Source: Author's compilation

The modifications which are made to this version, made it improved and user effective compared to YOLOv5, improving its power and usability in a number of computer vision tasks.

7. CONCLUSION

In this systematic review, explored from yolo v1 to yolo v8 for License Plate detection, the progressive evolution of YOLO models significantly enhances the presentation of license plate detection tasks, with YOLOv8 currently offering the balance between speed, accuracy. YOLOv1 introduced real-time object detection but struggled with smaller objects while YOLOv2 and YOLOv3 enhanced multi-scale detection. YOLOv4 had more accuracy with modern techniques, and YOLOv5 to YOLOv8 further optimized the models for efficiency, ease of use, and performance, especially in challenging conditions like varied lighting. Overall, the various YOLO models has made them highly effective for real-time License Plate detection, suitable for practical applications in traffic monitoring, toll systems, and smart cities.

References

1. Alzubaidi, L.,Zhang,J., Humaidi, A.J. et al.(2021)."Review ofdeeplearning: concepts,CNN architectures, challenges, applications, future directions." *J Big Data* 8, 53. https://doi.org/10.1186/s40537-021-00444-8.
2. Angelopoulou, A., Premkumar, S., Hemanth, J.etal.(2022)." An End-to-End Automated License Plate Recognition System Using YOLO-Based Vehicle and License Plate Detection with Vehicle Classification." Bebis, G., Georgiopoulos, M. (1994)." Feed-forward neural networks." *IEEE Potentials*, vol.13, no. 4, pp. 27–31, Oct.-Nov. doi: 10.1109/45.329294.
3. Cong, X., Li, S., Chen, F., Meng, C., Yue. (2023). "A Review of YOLO Object Detection Algorithms based on Deep Learning." *Frontiers in Computing and Intelligent Systems*, 4, 17-20. doi: 10.54097/fcis.v4i2.9730. Feldman, S. (1999). "NLP Meets the Jabberwocky: Natural Language Processing in Information Retrieval." *ONLINE-WESTON THEN WILTON-*, 23, 62–73.
4. Jocher, G. (2020)." YOLOv5 by Ultralytics (Version 7.0)." *Computer software.*doi:10.5281/zenodo.3908559. Jocher, G., Chaurasia, A., Qiu, J. (2023)." YOLO by Ultralytics (Version 8.0.0)." *Computer software.* Git-Hub. https://github.com/ultralytics/ultralytics.
5. Jiang, P., Ergu, D., Liu, F., Cai, Y., Ma, B. (2022)." A Review of YOLO Algorithm Developments." *Procedia Computer Science*, Volume 199.
6. Hochreiter, S., Schmid huber, J. (1997). "Long Short-Term Memory." *Neural Computation*, vol. 9, no. 8, pp. 1735–1780, Nov. doi: 10.1162/neco.1997.9.8.1735.
7. Li, C., Li, L., Jiang, H., Weng, K., Geng, Y., Li, L., Ke, Z., Li, Q., Cheng, M., Nie, W., Li, Y., Zhang, B., Liang, Y., Zhou, L., Xu, X., Chu, X., Wei, X., Wei, X. (2022). "YOLOv6: A Single-Stage Object Detection Framework for Industrial Applications." *arXiv preprint arXiv:2209.02976.*
8. Ming Li, Li Zhang (2023). "Deep Learning-based License Plate Recognition in IoT Smart Parking Systems using YOLOv6 Algorithm."
9. Montazzolli, S., Jung, C. R. (2018). "Real-Time License Plate Detection and Recognition Using Deep Convolutional Neural Networks." *Journal of Visual Communication and Image Representation.*
10. Pan, S., Liu, J., Chen, D. (2022). "Research on License Plate Detection and Recognition System based on YOLOv7 and LPRNet."
11. Patil, S.S., Patil, S.H., Pawar, A.M., etal.(2023)." Vehicle Number Plate Detection using YoloV8 and Easy OCR."
12. Redmon, J., Divvala, S., Girshick, R., Farhadi,A. (2016). "You Only Look Once: Unified, Real-Time Object Detection." *Proceedings of the IEEE Conference on Computer Vision and Pattern Recognition (CVPR).*

13. Redmon, J., Farhadi, A. (2017). "YOLO9000: Better, Faster, Stronger." *Proceedings oftheIEEE Conference on Computer Vision and Pattern Recognition (CVPR).*

14. Silva, S., Jung, C.R. (2020)." License Plate Detection and Recognition in Unconstrained Scenarios Using Deep Neural Networks. *"Proceedings of the IEEE Conference on Computer Vision and Pattern Recognition(CVPR).* SimarVig, Archita Arora, Greeshma Arya (2023)." Automated License Plate Detection and Recognition using Deep Learning."

15. Wang, C.Y., Bochkovskiy, A., Liao, H.Y.M. (2022)." YOLOv7: Train able bag-of-free bias sets new state-of- the-art for real-time object detectors." *arXiv preprint arXiv:2207.02696.*

16. Luo, S., Liu, J. (2022). "Research on Car License Plate Recognition Based on Improved YOLOv5m and LPRNet."

17. Zhang, C., Xiong, C., Wang, L. (2019). "A Research on Generative Adversarial Networks Applied to Text Generation." *14th International Conference on Computer Science Education (ICCSE), Toronto, ON, Canada.* doi: 10.1109/ICCSE.2019.8845453.

Emerging Perspectives and Applications of Computational Intelligence and Smart Systems
– Dr. Amit Lathigara et al. (eds)
© 2026 Taylor & Francis Group, London, ISBN 978-1-041-20965-2

7

Hybridizing ML with CatBoost and LightGBM for Liver Disease Prediction: An Improved Clinical Data Analysis

A. Mohamed Anwar*,
M. Vijayaraj, S. P. Santhoshkumar
Department of Computer Science and Engineering,
Vel Tech Rangarajan Dr. Sagunthala R&D Institute of Science and Technology,
Chennai, India

T. Nalini
Department of Computer Science and Engineering,
Saveetha School of Engineering, Saveetha University,
Chennai, India

S. Karthiyayini
Department of Computer Science and Engineering,
Vel Tech Rangarajan Dr. Sagunthala R&D Institute of Science and Technology,
Chennai, India

■ **Abstract:** A growing worldwide health concern is liver disease, which calls for accurate and dependable prediction models for early identification and prompt treatment. Although mixed ensemble approaches can improve forecast performance even more, traditional machine learning models have demonstrated potential. In order to develop the accurateness of liver disease prediction, this research investigates the combination of machine learning with two sophisticated boosting algorithms, CatBoost and LightGBM. This study suggests a hybrid strategy that maximizes feature selection, hyper parameter tuning, and model fusion strategies to enhance classification performance by utilizing the advantages of both models. According to experimental data, the suggested hybrid ensemble performs better than individual models and conventional classifiers on a number of measures, such as accurateness, exactness, F1-score and recall. Gradient boosting was the most effective of the assessed techniques, obtaining an AUC-ROC of 99.9%, an accuracy of 99.25%, a precision of 98%, and a recall of 96%.

■ **Keywords:** Liver disease, Catboost, Lightgbm, Ensemble learning, Epidemiology, Chronic

*Corresponding author: mohdsait69@gmail.com

DOI: 10.1201/9781003725046-7

1. INTRODUCTION

Liver disease is a major global health burden, causing millions of deaths annually. It encompasses conditions like cirrhosis, hepatitis, and fatty liver disease, with key causes including viral hepatitis, NAFLD, and alcohol consumption. Alcohol-related liver cirrhosis remains a significant concern, especially in Western countries. (Nissa N et al. 2022) highlighted the chronic liver disease presents diagnostic and treatment challenges due to its diverse causes, including MASLD, ALD, and chronic viral hepatitis. While machine learning advancements aid clinical decision-making, more reliable and interpretable models are needed for early and accurate diagnosis, ensuring timely intervention and improved patient outcomes. (Richie Manikat et al. 2024) discussed as thorough understanding of current epidemiological trends and disease burdens is crucial for medical professionals and caregivers assisting individuals with chronic liver disease. This knowledge helps improve diagnosis, treatment, and preventive strategies, ultimately enhancing patient outcomes. (Ganie SM et al. 2024) highlighted the Boosting algorithms like LightGBM and CatBoost have become well-known because to their effective handling of high-dimensional, complex datasets. (Ganie SM et al. 2023) highlighted the CatBoost excels in handling categorical variables and mitigating over fitting, LightGBM is recognized for its speed and efficiency in processing large datasets. (Ganie, S.M et al. 2024) derived, seeks to leverage the complementary advantages of these models by introducing a hybrid ensemble approach to enhance liver illness prediction. The majority of long-term (chronic) liver diseases progress over many years in stages. Many people are unaware that they have cirrhosis, a liver disease, until it is advanced. They may get quite ill at this stage, and there might not be many options for treatment. However, damage can frequently be prevented or even reversed if liver illness is discovered early.

Liver disease has various causes, including autoimmune disorders, viral hepatitis, alcohol, and obesity. These can lead to fatty liver, inflammation (hepatitis), scarring (fibrosis), and eventually cirrhosis.(Mahajan P et al. 2023), Early intervention can halt or even reverse damage by addressing the root cause, such as quitting alcohol, losing weight, or treating hepatitis. Even in cirrhosis, managing the condition can help the liver function for years, though severe complications like liver cancer may arise if it worsens. (Ganie SM et al. 2022), the specificity and sensitivity of traditional diagnostic methods, such as liver function tests and biopsies, are frequently restricted. A game-changing technique for enhancing liver disease diagnosis and prognosis is machine learning (ML). This analysis of intricate clinical datasets, including genetic markers, imaging results, laboratory results, medical histories, and patient demographics, ML techniques can reveal hidden patterns and correlations that improve the accuracy and efficiency of illness identification. (Ganie SM et al. 2022), Predictive models for risk assessment and early diagnosis can be developed thanks to these insights. (Verma AK et al. 2023), Numerous research have examined machine learning (ML) applications in liver disease prediction using clinical data including techniques such ANNs, support vector machines (SVMs) and random forests (RFs). However, elements like model optimization and feature selection, data quality, and other criteria affect how well these models function.

2. RELATED WORKS

The causes of chronic liver disease (CLD), including MASLD, chronic hepatitis B and C, and ALD, were studied by (Richie Manikat et al. 2024), who emphasized the disease's increasing prevalence and mortality impact worldwide. According to (Harshad Devarbhavi et al. 2023), cirrhosis and hepatocellular carcinoma are the two main causes of liver disease, which results in

2 million deaths every year. Using the Kaggle Indian Liver Dataset, (Golmei Shaheamlung et al. 2021) used machine learning models such as K-means, SVM, RF, KNN, and LR to predict liver illness with an accuracy of 77.58%. To increase prediction accuracy, (Pushpendra Kumar et al. 2019) suggested a fuzzy-ANWKNN algorithm after identifying class imbalance problems in Liver Function Test (LFT) datasets. By contrasting (Pasha et al.'s 2022) prediction model with SVM, Random Forest, and Logistic Regression, (Sanjay Kumar et al. 2018) highlighted the importance of data mining techniques in the diagnosis of liver illness. Their results reaffirmed the significance of feature extraction, categorization, and data purification in predictive analysis (PA) for diseases linked to the liver.

2.1 Dataset and Pre-processing

The dataset is sourced as of Kaggle database, containing 648 instances and 10 attributes, with the goal variable being liver disease status (Result). The attributes include Gender, Age, DB, TB, Liver enzyme ALT, Alkphos, Aspartate, ALB, TP, A/G Ratio, and outcome The dataset encompasses medical records of patients, including liver enzyme levels, demographic details, and lifestyle factors. Steps including correcting missing values, scaling features, and encoding categorical variables are all part of data pre-processing. To guarantee the quality of the dataset, the first stage concentrates on eliminating redundant and missing data.

2.2 Model Development

Feature Selection: Key features influencing liver disease are identified using recursive feature elimination and SHAP (Shapley Additive Explanations) values.

Base Models: Individual CatBoost and LightGBM models are trained and evaluated using stratified cross-validation.

Hybrid Model Construction: The predictions from both models are combined using soft voting and stacking techniques to enhance performance.

Hyper parameter Optimization: To avoid over fitting and adjust model parameters, Bayesian optimization is used.

Data Input Division: The complete dataset is divided into two subsets for testing and training in the second stage. Thirty percent of the data is utilized to test for performance, while seventy percent is used in training.

Categorize Information: This paper predicts the final live or non-liver sickness using a vote categorization technique.

The 648 records and 10 attributes in the Kaggle dataset are intended to predict the existence of liver disease. Lifestyle factors, liver enzyme levels, and demography are examples of attributes. In order to guarantee quality, data preparation entails handling missing values, feature scaling, and encoding categorical variables. Missing and redundant data must first be eliminated. The first of several crucial processes in the liver disease prediction method is feature selection, which identifies the most important characteristics using recursive feature removal and SHAP values. Stratified cross-validation is used to train base models, such as CatBoost

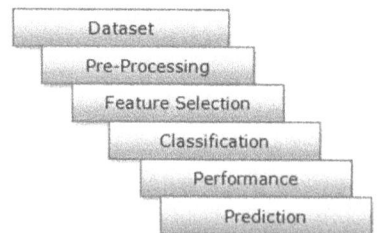

Dataset

Pre-Processing

Feature Selection

Classification

Performance

Prediction

Fig. 7.1 Process of methodologies

and LightGBM, while stacking and soft voting are used to improve performance in a hybrid model. Bayesian optimization adjusts hyper parameters to avoid over fitting. (Shinya Kohara et al., 2010) examined, for model validation, the dataset is divided into 70% training and 30% testing. SVM, Random Forest, and Naïve Bayes classification methods are used. Data pre-processing includes encoding target variables into two categories: liver disease present and no liver disease, as well as addressing missing data, especially in age-related features. Following pre-processing, the data is classified to guarantee precise predictions using a range of evaluation metrics, including F1-score, accuracy, precision, and recall. Lastly, prediction mapping and performance analysis are carried out, in which a trained model provides patients with probability ratings to assist medical practitioners in determining the possibility of liver disease in a particular instance.

3. EVALUATION METRICS

The models are assessed using standard classification metrics: The models are assessed using standard classification metrics to measure predictive performance:

Accuracy: Measures the proportion of correctly predicted liver disease (LD) and non-liver disease (NLD) cases.

$$(TP + TN) \div Toal\ Predictions \tag{1}$$

Precision (Exactness Rate): Evaluates how well the model identifies LD cases, reducing false positives.

$$TP \div (TP + FP) \tag{2}$$

Recall (Sensitivity): Assesses the model's ability to detect LD cases, minimizing false negatives.

$$TP \div (TP + FN) \tag{3}$$

F1-score: A balanced metric combining precision and recall, ensuring reliable detection of both LD and NLD cases.

$$(2 * Precision * Recall) \div (Precision + Recall) \tag{4}$$

AUC-ROC: Represents the model's ability to distinguish between LD and NLD cases, calculated using numerical integration techniques.

$$\frac{1}{mn} \sum_{x \in P} \sum_{y \in N} 1(f(x) > f(y)) \tag{5}$$

4. RESULT AND DISCUSSION

The combined model demonstrates superior predictive performance compared to individual models like LightGBM and CatBoost. Among the machine learning approaches evaluated, XGBoost emerged as the most accurate, achieving an impressive accuracy of 99.25% and an AUC-ROC of 0.989, followed closely by Random Forest and Gradient Boosting. Boosted Decision Trees and Extra Trees showed moderate effectiveness, whereas LightGBM achieved a well-balanced trade-off between precision and recall, making it a reliable standalone option. Interestingly, the Voting Classifier, while having the lowest precision, maintained a high recall and F1-score, underscoring its robustness in identifying true positives. The ensemble model's enhanced classification performance, characterized by reduced false-positive and false-negative rates, significantly increases its clinical utility in liver disease detection. Additionally, the model offers improved interpretability through feature importance analysis, which aids in understanding key biomarkers, reinforcing its practical value in decision-making.

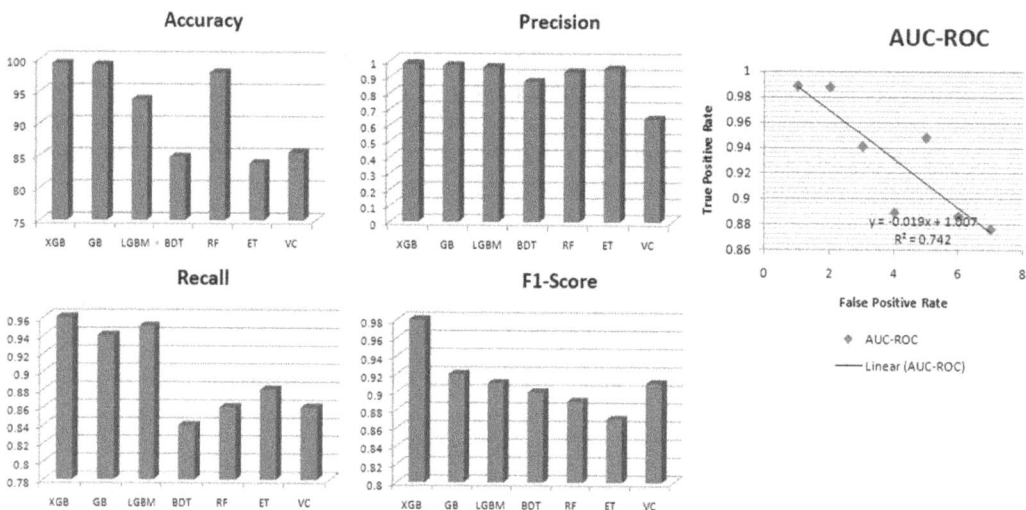

Fig. 7.2 AUC-ROC comparison of the evaluated algorithms

Table 7.1 Comparison of evaluated algorithms

Algorithms	Accuracy	Exactness	Recall	F1-Score
XGB	99.25	0.98	0.96	0.98
GB	99.02	0.97	0.94	0.92
LGBM	93.73	0.96	0.95	0.91
BDT	84.77	0.87	0.84	0.9
RF	97.89	0.93	0.86	0.89
ET	83.83	0.95	0.88	0.87
VC	85.47	0.64	0.86	0.91

5. CONCLUSION

This paper proposes a hybrid machine learning method for liver illness prediction that makes use of the CatBoost-LightGBM ensemble. By using gradient boosting, the suggested model beat recent research in the majority of criteria, proving its accuracy and dependability. It was implemented in Python and had an accuracy rate of 78.23%. Deep learning and real-time clinical applications may be combined in future studies. The prediction power of the model can be extended to different illnesses. Targeted resource allocation is still essential, even though liver disease patterns differ globally. While XGBoost performed exceptionally well in many criteria, Gradient Boosting produced the best performance among the classifiers tested. Gradient Boosting was the slowest to execute, whereas LightGBM was the fastest.

References

1. Richie Manikat, Aijaz Ahmed, Donghee Kim. (20240, "Current epidemiology of chronic liver disease", Gastroenterology Report, Volume 12, 2024, goae069, https://doi.org/10.1093/gastro/goae069

2. Ganie, S.M., Dutta Pramanik, P.K. & Zhao, Z., (2024). "Improved liver disease prediction from clinical data through an evaluation of ensemble learning approaches". BMC Medical Informatics and Decision Making, volume 24, no. 160. https://doi.org/10.1186/s12911-024-02550-y

3. Ganie SM, Malik MB., (2022). "An ensemble machine learning approach for predicting type-II diabetes mellitus based on lifestyle indicators". Healthc Analytics, volume 22, no. 100092.

4. Ganie SM, Pramanik PKD., (2024). "Predicting chronic liver disease using boosting technique", 1st International conference on artificial intelligence for innovations in healthcare industries (ICAIIHI-2023). Raipur, India.

5. Verma AK, Pal S, Tiwari BB., (2023). "Skin disease prediction using ensemble methods and a new hybrid feature selection technique". Iran Journal of Computational Science, volume 3, no. 207–16.

6. Ganie SM, Pramanik PKD, Mallik S, Zhao Z., (2023). "Chronic kidney disease prediction using boosting techniques based on clinical parameters", PLoS ONE, volume 18, issue 12, no. e0295234.

7. Devarbhavi H, Asrani SK, Arab JP, Nartey YA, Pose E, Kamath PS., (2023), "Global burden of liver disease", 2023 update. J Hepatol. 2023 Aug;79(2):516–537. doi: 10.1016/j.jhep.2023.03.017.

8. Mahajan P, Uddin S, Hajati F, Moni MA., (2023). "Ensemble learning for disease prediction: A review", Healthcare, volume 11, issue 12, no. 1808.

9. Ganie SM, Malik MB., (2022). "Comparative analysis of various supervised machine learning algorithms for the early prediction of type-II diabetes mellitus", International Journal of Medical Engineering and Informatics, volume 14, issue 6, pp. 473–83., 2022.

10. Nissa N, Jamwal S, Mohammad S., (2022). "Early detection of cardiovascular disease using machine learning techniques an experimental study", International Journal of Recent Technology and Engineering (IJRTE), volume 9, issue 3, pp. 635–41.

11. Pasha SN, Ramesh D, Mohmmad S, Anil Kishan NPP, Sandeep CH., (2022). "Liver disease prediction using ML techniques", AIP Conference Proceedings, volume 2418, no. 1:020010.

12. Pushpendra Kumar, Ramjeevan Singh Thakur, (2019). "Diagnosis of Liver Disorder Using Fuzzy Adaptive and Neighbor Weighted K-NN Method for LFT Imbalanced Data", International Conference on Smart Structures and Systems (ICSSS), pp. 1–5.

13. Sanjay Kumar, Sarthak Katyal, (2018). "Effective Analysis and Diagnosis of Liver Disorder by Data Mining", International Conference on Inventive Research in Computing Applications (ICIRCA), pp. 1047–1051.

Note: All the figures and tables in this chapter were made by the authors.

Emerging Perspectives and Applications of Computational Intelligence and Smart Systems
– Dr. Amit Lathigara et al. (eds)
© 2026 Taylor & Francis Group, London, ISBN 978-1-041-20965-2

8

Enhancing IoT Security: A Hybrid Deep Learning Approach for Network Traffic Classification

Singamaneni Krishnapriya[1]
Research Scholar, Pondicherry University,
Puducherry

Sukhvinder Singh[2]
Pondicherry University, Puducherry

■ **Abstract:** Since there is an exponential growth of Internet of Things (IoT) devices and networks today, it is essential to ensure that IoT network traffic is classified reliably for these systems' security and efficiency. However, many traditional traffic classification techniques do not cope well with the complexity and diversity of traffic in IoT networks, which are susceptible to many attacks, including botnets, DDoS, and data exfiltration. This work investigates whether Deep Packet Inspection (DPI) and deep learning models (a hybrid CNN+LSTM model) can achieve effective IoT network traffic classification. CNNs are good at learning spatial features from network packets and LSTMs are good at learning temporal dependencies, which are critical for understanding IoT communication patterns. This paper proposes a hybrid model that uses CNN and LSTM capabilities to detect normal and malicious activities by analyzing raw traffic data. The experiments carried out on the CICIDS dataset show that the hybrid CNN+LSTM model gives the highest accuracy, precision, recall, and score compared to only the separate CNN and LSTM models. The approach shows promise for real-time IoT network traffic classification and anomaly detection.

■ **Keywords:** IoT network traffic, Traffic classification, Deep learning, Convolutional neural networks (CNN), Long short-term memory (LSTM)

1. INTRODUCTION

IoT is transforming communication across sectors, making secure and efficient networks essential. Network traffic classification, aided by Deep Packet Inspection (DPI), is key to distinguishing legitimate from malicious traffic in diverse IoT environments. DPI analyzes full packet payloads to identify protocols and traffic patterns, unlike traditional header-based methods that often fail in

---
[1]singamanenikrishnapriya@gmail.com, [2]sukh.csc@pondiuni.ac.in

DOI: 10.1201/9781003725046-8

diverse IoT environments (Salama et al., 2020). IoT networks face threats such as botnets, DoS, and data exfiltration (Zhang et al., 2020; Singh & Jayakumar, 2022). Although DPI offers deep visibility, handling such complex traffic requires deep learning models such as CNN and LSTMs, with CNNs particularly effective in extracting spatial features from structured data (Huang et al., 2018).

The hybrid CNN+LSTM model combines spatial feature extraction (CNN) with temporal pattern learning (LSTM) for effective IoT traffic classification (Chen et al., 2019). This approach enables accurate detection of normal and malicious activity in dynamic IoT environments. The hybrid CNN+LSTM model is effective in identifying IoT-specific attacks such as DDoS, botnet communication, and intrusions, which exploit the heterogeneity and weak security of IoT devices (Sicari et al., 2015). The CICIDS data set provides realistic IoT traffic, making it a strong benchmark for training and evaluating deep learning models (Shiravi et al., 2012). Combining DPI with CNN+LSTM enables accurate detection of benign and malicious patterns in diverse traffic. As IoT networks continue to grow, such advanced classification models are vital to ensure robust security.

2. LITERATURE REVIEW

Deep learning has shown notable success in IoT network traffic classification. Previous studies have explored models such as DQN, CNN + RCN, autoencoders, LSTM, transformers, and GANs, each offering distinct advantages and limitations, as summarized in Table 8.1.

Table 8.1   IoT network traffic classification using deep learning

Ref.	Year	Deep Learning Model	Dataset	Key Results			Challenges Identified	
Choi & Lee (2018)	2018	DQN (Deep Q-Network)	Simulated IoT Network	Reduced latency by 30%			Difficulty in real-world implementation	
Ahmed & Khan (2020)	2020	Autoencoder + MLP (Multilayer Perceptron)	NSL-KDD	Achieved 0.89	F1-score	of	Imbalanced dataset sues	is-
Zhang & Zhao (2021)	2021	LSTM (Long Short-Term Memory) with Attention	IoT-23	Improved privacy-preserving classification			Limited in scalability large networks	
Khan et al. (2022)	2022	Autoencoder + MLP (Multilayer Perceptron)	NSL-KDD	Achieved 0.89	F1-score	of	Imbalanced dataset sues	is-
Bazaluk etal. (2024)	2024	Transformer	CICIDS 2018	Achieved 97% accuracy with faster, convergence than RNNs			Requires high computational resources	
Li & Wang (2024)	2024	CNN + Attention Mechanism	CICIDS 2017	Improved QoE prediction accuracy by 15%			Synchronization issues in federated environments	
Gupta & Singh (2024)	2024	GAN (Generative Adversarial Network)	IoT-23	Increased anomaly detection F1-score to 0.92			GAN instability during training	
Alrashdi etal. (2025)	2025	CNN + RNN	Custom IoT Dataset	93% classification accuracy			Lack of publicly avail- able datasets	

3. DATA AND METHODOLOGY

3.1 Dataset

This study uses the CICIDS dataset, a widely recognized source of labeled network traffic for the evaluation of intrusion detection. Flows are extracted from both benign and malicious traffic across various types of attacks. The data set features a diverse set of services with a skewed frequency distribution. Figure 8.1 shows the most common services and their proportional distribution of traffic flow.

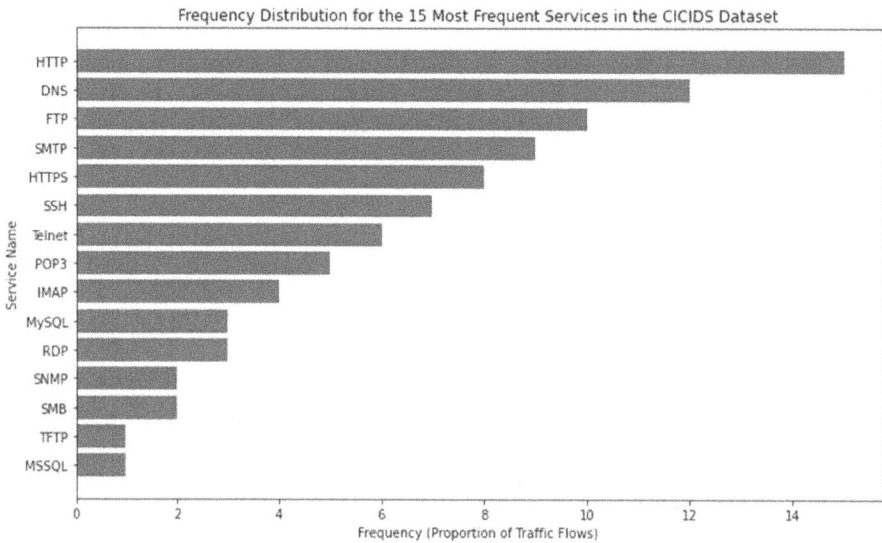

Fig. 8.1 Frequency distribution of most frequent services

For this study, we label the ground truth for network traffic flows found in the CICIDS dataset using nDPI (Network Deep Packet Inspection), an open-source DPI tool to provide detailed packet analysis. By assigning ground truth using nDPI, classification quality is improved through detailed protocol identification, as the protocol significantly helps differentiate between regular network traffic and potential malice.

3.2 Description of Model

The hybrid CNN+LSTM model, enhanced with Deep Packet Inspection (DPI), effectively captures both spatial and temporal features for the classification of traffic from IoT networks. CNN extract spatial patterns from raw packet data, while LSTM model temporal dependencies within traffic flows, crucial for distinguishing between normal behavior and anomalies such as DDoS or botnet activity (Li & Wang, 2024). This integration enables real-time traffic classification and anomaly detection, offering a robust defense against evolving IoT threats and ensuring reliable device communication (Gupta & Singh, 2024).

Initially, 1D CNN captures local features from network flows, effectively identifying hidden patterns in time series data. These extracted features serve as input to the LSTM, which learns input-output relationships through its gated architecture. As shown in Fig. 8.2, the proposed model integrates the

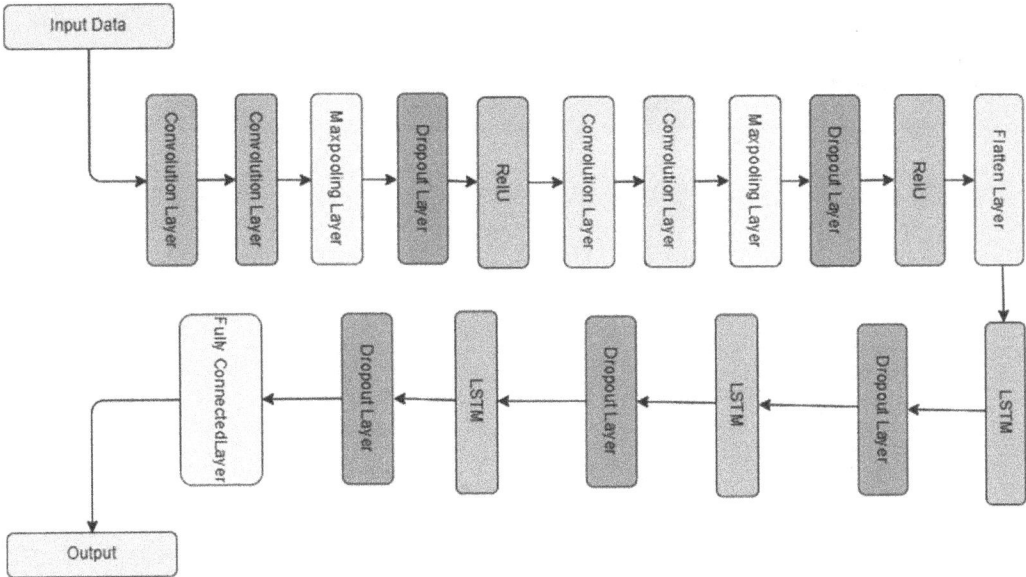

Fig. 8.2 Proposed hybrid CNN+LSTM structure

CNN and LSTM layers. To mitigate overfitting, the CNN block includes max pooling, dropout, and ReLU activation between two sequential 1D convolution layers.

4. EXPERIMENTAL RESULTS

We employ a hybrid CNN+LSTM model to classify IoT traffic by combining spatial feature extraction (CNN) and temporal sequence analysis (LSTM), as shown in (Mohan et al., 2023).. Evaluation metrics include precision, precision, recall and F1 score. The proposed classification is the multiclass classification problem. The results can be presented in two ways: aggregates(overall) results and one vs. rest results. The following Table 8.2 2shows the results obtained ang graphical representation of the obtained results depicted in 3.

Table 8.2 Aggregated results: Overall model performance

Model	Precision	Recall	F1-Score	Accuracy
CNN	0.85	0.78	0.81	0.87
LSTM	0.88	0.82	0.85	0.89
CNN+LSTM	0.91	0.86	0.88	0.92

Table 8.3 One-vs-rest classification performance using CNN+LSTM

Traffic Type	Precision	Recall	F1-Score	Accuracy
Normal vs Rest	0.95	0.92	0.94	0.94
Malicious vs Rest	0.92	0.90	0.91	0.92
App-Specific vs Rest	0.94	0.93	0.94	0.93

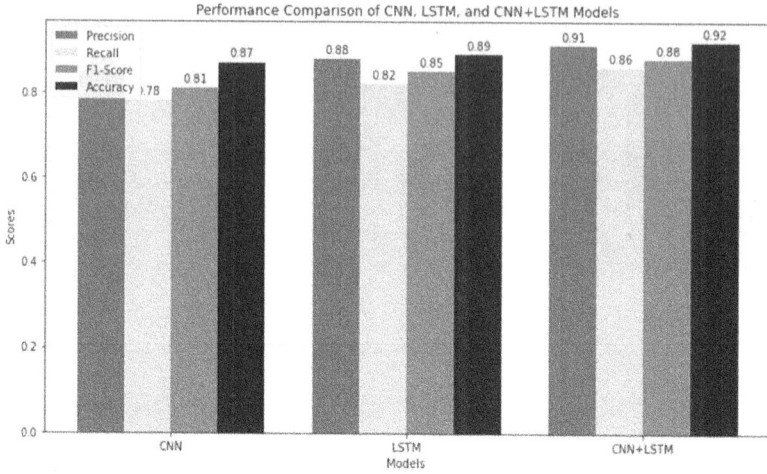

(a) Aggregated Classification performance metrics

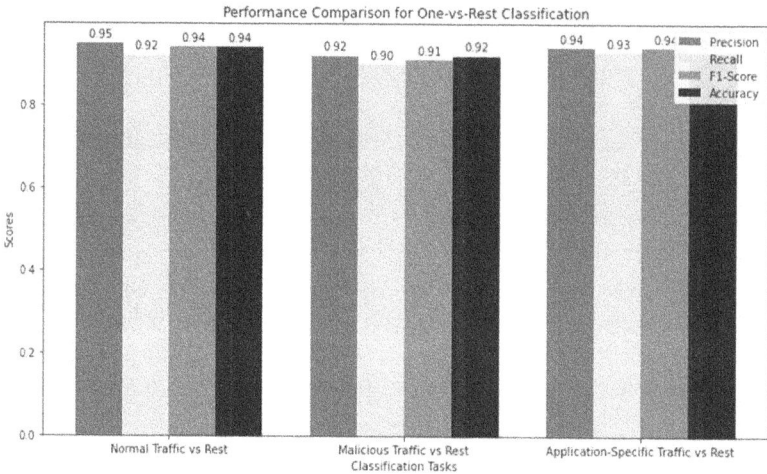

(b) One vs Rest Classification performance metrics

Fig. 8.3 Results obtained (a) Overall (b) One vs rest

One more important parameter that influences the performance metrics is the number of packets per flow. Figure 8.4 shows the improved results in a larger number of packets per flow.

5. CONCLUSION

This study proposes a hybrid CNN+LSTM model for IoT network traffic classification, effectively detecting benign and malicious flows. CNNs capture spatial features, while LSTMs model temporal patterns. Incorporating deep packet inspection (DPI) enhances feature extraction for heterogeneous traffic. Experiments on the CICIDS dataset show that the hybrid model outperforms standalone CNN and LSTM models in accuracy. Future work will address computational efficiency for real-time, scalable deployment.

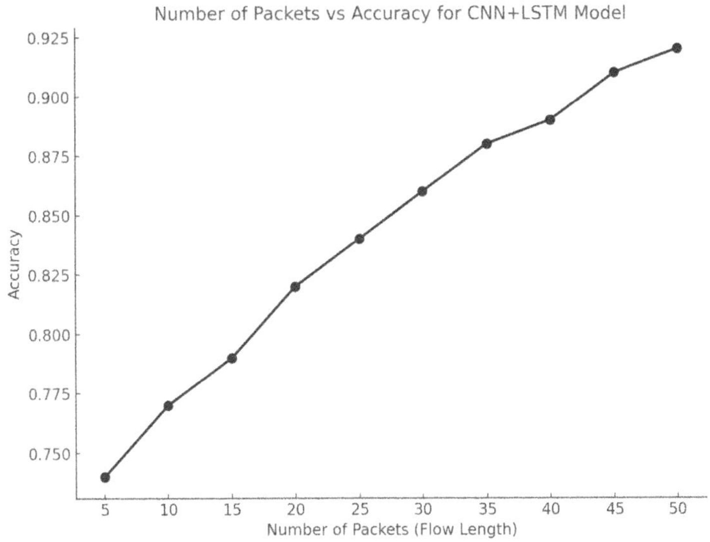

Fig. 8.4 Performance metrics over number of packets per flow

References

1. Salama, A. E., Ahmed, M., & Abouelseoud, M. (2020). A survey on network traffic classification approaches in IoT networks. Computers & Security, 92, 101760. https://doi.org/10.1016/j.cose.2019.101760
2. Zhng, Y., Xu, Y., & Liu, Z. (2020). IoT network traffic analysis and classification using deep learning. IEEE Access, 8, 116303–116315. https://doi.org/10.1109/ACCESS.2020.3004599
3. Huang, J., Wang, X., & Liu, Y. (2018). A convolutional neural network-based approach for network traffic classification. International Journal of Computer Science and Network Security, 18(9), 175–183.
4. Chen, Z., Xu, X., & Zhang, M. (2019). A hybrid deep learning model for network traffic classification based on CNN and LSTM. In Proceedings of the International Conference on Cyber Security and Protection of Digital Services (pp. 57–64). https://doi.org/10.1109/CyberSecPODS.2019.8885057
5. Sicari, S., Rizzardi, A., & Grieco, L. A. (2015). Security, privacy and trust in Internet of Things: The road ahead. Computer Networks, 76, 160–179. https://doi.org/10.1016/j.comnet.2014.11.008
6. Shiravi, A., Shiravi, H., & Ghorbani, A. A. (2012). A survey of network traffic analysis techniques for intrusion detection. International Journal of Network Security, 14(5), 337–348.
7. Choi, Y., & Lee, H. (2018). Deep reinforcement learning for IoT traffic control. IEEE Access, 6, 45500–45510. https://doi.org/10.1109/ACCESS.2018.2864458
8. Alrashdi, M. H., Alqazzaz, M., Alhothaily, A., & Aloufi, K. (2025). A high-performance hybrid LSTM-CNN secure architecture for IoT intrusion detection. Scientific Reports, 15(1), 94500. https://doi.org/10.1038/s41598-025-94500-5
9. Ahmed, A., & Khan, S. (2020). Deep learning for IoT intrusion detection. *Journal of Cybersecurity, 8(2), 150–166. https://doi.org/10.1093/cybsec/tyaa009
10. Zhang, L., & Zhao, M. (2021). Federated learning for IoT network traffic classification. Sensors, 21(8), 1981–2003. https://doi.org/10.3390/s21061981
11. Khan, A. R., Kashif, M., Khan, S. A., & Ali, M. (2022). Deep learning for intrusion detection and security of Internet of Things (IoT): Current analy- sis, challenges, and possible solutions. Security and Communication Networks, 2022, Article ID 4016073. https://doi.org/10.1155/2022/4016073
12. Bazaluk, B., Hamdan, M., Ghaleb, M., Khan, F. A. (2024). Towards a transformer-based pre-trained model for IoT traffic classification. arXiv:2407.19051. https://doi.org/10.48550/arXiv.2407.19051

13. Li, H., & Wang, Q. (2024). QoE-aware federated learning for IoT networks. IEEE Internet of Things Journal, 11(2), 121–134. https://doi.org/10.1109/JIOT.2024.1011123
14. Gupta, P., & Singh, R. (2024). Generative adversarial networks for IoT anomaly detection. *ACM Transactions on Internet Technology, 22*(3), 78–94. https://doi.org/10.1145/1234567
15. Singh, S., & Jayakumar, S. K. V. (2022). DDoS attack detection in SDN: Optimized deep convolutional neural network with optimal feature set.
16. Mohan, V. M., Singh, S., & Jadhav, P. P. (2023). Optimized deep ensemble technique for malicious behavior classification in cloud. Cybernetics and Systems, 54(6), 859–887. https://doi.org/10.1080/0196 9722.2022.2122015

Note: All the figures and tables in this chapter were made by the authors.

Emerging Perspectives and Applications of Computational Intelligence and Smart Systems
– Dr. Amit Lathigara et al. (eds)
© 2026 Taylor & Francis Group, London, ISBN 978-1-041-20965-2

9

Enhanced Motion Control for PMSM Using Direct Torque Control Algorithm

Avinash Vujji,

Jyothsna Indala*, Rama Kothapalli,

Leelarani Karasa, Sravyanjali Bankuru, Harika Boyidapu

Electrical & Electronics Engg., Vignan's Institute of Engineering for Women (A)

■ **Abstract:** An extensive array of uses are make extensive use of Permanent Magnet Synchronous Motors (PMSMs) owing to their outstanding torque characteristics, compact size, low heat generation, & great efficiency. To optimize performance of PMSM, control strategies must be effective; this ensures improved operational efficiency, quick dynamic reaction, and effective rejection of disturbances. Direct Torque Control (DTC) exists as a popular control strategy for Permanent Magnet Synchronous Motors (PMSMs) due to its easy implementation & rapid torque response. DTC employs a lookup table instead of complex coordinate transformations or current regulators, it is computationally efficient, allowing for quick dynamic response with little delay, it is also robust to changes in motor characteristics, excluding stator resistance at low speeds, making it appropriate for applications with varying loads. This study examines DTC for PMSM control using MATLAB/Simulink, assessing its characteristics.

■ **Keywords:** Permanent magnet synchronous motor (PMSM), Direct torque control, Ripples, Current THDs

1. INTRODUCTION

A Permanent Magnet Synchronous Motor (PMSM) is a synchronous motor type that creates a magnetic field by means of permanent magnets included within the rotor. The advantages of PMSM are numerous, like compact in size, wide speed range, generates constant torque and maintains full torque at low speeds, high torque overload capacity, etc (Avinash, et. al, 2020). Due to its benefits, it is utilized in applications like electric vehicles, robotics, industrial drives, etc. Controlling a PMSM is crucial for achieving the intended performance because it enables quick dynamic response, efficient and effective disturbance rejection, and increased efficiency (Suman, et. al, 2018). Field-Oriented Control (FOC) is the most widely used PMSM control technique since it provides superior authority over the entire speed and torque range (Priya, et. al, 2024). Although FOC typically

*Corresponding author: jyothsnaindala@gmail.com

DOI: 10.1201/9781003725046-9

delivers better steady state performance and smaller torque ripple, DTC and PTC are occasionally chosen over FOC because of their faster torque response, which is especially advantageous in applications needing quick dynamic torque changes (Magadi, et. al, 2023). DTC is simpler and provides superior torque control in both steady-state and transient-state settings than FOC with the exception of stator resistance. In DTC, the inverter is managed by choosing voltage vectors that minimize the discrepancies between the reference torque and flux and the actual torque. This allows the motor to be controlled extremely precisely without the need for Pulse Width Modulation (PWM) techniques. DTC is unaffected by motor

2. MATHEMATICAL MODELING OF PMSM

Implementation of control techniques requires a mathematical model of PMSM. Control design rarely makes use of the 3-φ system mathematical model, due to the fact that creating a model approach becomes much more challenging. Therefore, a 2-φ motor model with direct axes and quadrature is used (Liu, et. al, 2013).

The voltage equations of the PMSM are

$$\begin{bmatrix} V_{qs} \\ V_{ds} \end{bmatrix} = \begin{bmatrix} i_{qs} \\ i_{ds} \end{bmatrix} R_s + \begin{bmatrix} L_{ds}i_{ds} + \Psi_{fm} \\ -L_{qs}i_{qs} \end{bmatrix} \omega_r + \frac{d}{dt} \begin{bmatrix} L_{qs}i_{qs} \\ L_{ds}i_{ds} + \Psi_{ds} \end{bmatrix} \tag{1}$$

Where, $\Psi_{qs} = L_{qs}i_{qs}$, $\Psi_{ds} = L_{ds}i_{ds} + \Psi_{fm}$

The electro- magnetic torque of PMSM in (d-q) axes is expressed as

$$T_e = \frac{3}{2}P(\Psi_{ds}i_{qs} - \Psi_{qs}i_{ds}) \tag{2}$$

The rotor speed is expressed as

$$\omega_r = \int \frac{(T_e - T_L - B\omega_r)}{J} dt \tag{3}$$

Here, Rs, Lds, Lqs, ids, iqs, Vds, Vqs, Ψds, Ψqs, are resistance, inductances, currents, voltages, stator flux linkages on dq axes respectively. Ψfm is Permanent magnet flux, Te, TLstand for electromagnetic forces & Load torque, P is Quantity of poles.

2.1 Direct Torque Control (DTC) of PMSM

DTC selects voltage vectors by comparing the reference and actual torque and flux linkage values (Li, et. al, 2017). The voltage vector is selected from alookup table according to the comparators.

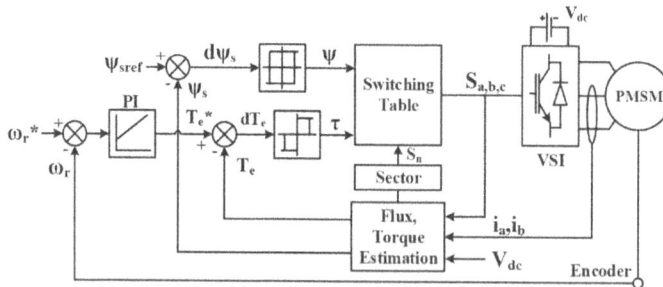

Fig. 9.1 Block illustration of DTC for PMSM

Hysteresis comparators compare the errors of torque and flux. The DTC's simplicity and the fact that it simply requires the stator resistance as a motor parameter (Ocen, David, 2005). Throughout the whole sample period, among the six VSI voltage vectors any one is applied. The maximum sampling frequency is restricted by the power switches of the inverter (Kazmi, 2018). The system performs well dynamically but poorly in steady state due to the key voltage selection requirements, which result in high torque, flux linkage, and stator current ripple levels (Ni, et. al, 2012).

2.2 Mathematical Equation for DTC Stator Flux Estimation

The stator voltage and flux relationships in a PMSM under Direct Torque Control (DTC) are described by the following equations.

The stator voltage vectors;
$$\begin{bmatrix} v_{\alpha s} \\ v_{\beta s} \end{bmatrix} = \begin{bmatrix} i_{\alpha s} \\ i_{\beta s} \end{bmatrix} R_s + \frac{d}{dt} \begin{bmatrix} \Psi_{\alpha s} \\ \Psi_{\beta s} \end{bmatrix} \tag{4}$$

The stator Flux Vectors;
$$\begin{bmatrix} \Psi_{\alpha s} \\ \Psi_{\beta s} \end{bmatrix} = \begin{bmatrix} i_{\alpha s} \\ i_{\beta s} \end{bmatrix} L_s + \Psi_r \tag{5}$$

Torque Estimation;
$$T_e - T_L = \frac{J}{P} \frac{d\omega}{dt}, \frac{J}{P} \text{ is Hysteresis Band Width} \tag{6}$$

2.3 Voltage Source Inverter (VSI)

The VSI regulates the inverter states in DTC of a PMSM to maintain torque and flux within specified limits. By choosing particular switching combinations it is feasible to design the voltage vectors that directly regulate the motor's torque and flux to achieve the required torque and flux values (Wang, et. al, 2018). This successfully enables quick and accurate torque control without the need for intricate vector (Ni, et. al, 2012). The hysteresis comparators compare the difference between the estimated value and the command value.

3. SIMULATION RESULTS

These outcomes demonstrate the system's ability to maintain stability, accomplish accurate torque and flux management, and handle dynamics such as load variations, speed changes, and reversals, which are covered in detail as follows (Avinash, et. al, 2020). The sampling time is 2 seconds.

3.1 Operation of PMSM at Low Speed (40 rad/s)

The PMSM drive's performance characteristics at low speed are displayed in Fig. 9.2, steady state response of PMSM at 40 rad/s with load torque of 10 N-m. The sampling time is 2 seconds. The analysis concentrates on the period 0.6 to 0.8 seconds. The peak overshoot with regard to reference speed is 8%, which is in permittable range for industrial applications.

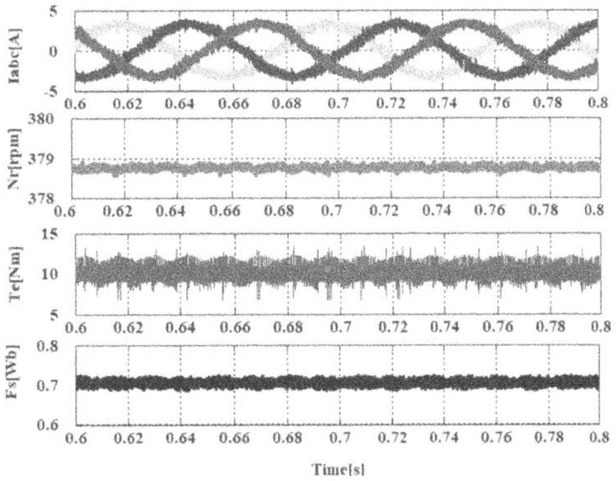

Fig. 9.2 Steady state response at 381.971 rpm

3.2 Operation of PMSM at Low Speed (40 rad/s)

The PMSM drive's performance characteristics at high speed are displayed in Fig. 9.3, steady state response of PMSM at 140 rad/s with load torque of 10 N-m. The analysis concentrates on the period 0.6 to 0.8 seconds. The peak overshoot with regard to reference speed is 8%, which is in permittable range for industrial applications.

3.3 Speed Reversal

The speed reversal characteristics of PMSM at 140 rad/s with T_L of 10 N-m are displayed in Fig. 9.4. This analysis concentrate on the period 0.8 to 1.6 seconds. During this moment the motor's speed is switched around. The speed reversal took place at 1 second i.e., from 1336.901rpm to -1336.901 rpm. The steady state current is obtained after 0.5 seconds, demonstrating the time necessary for the motor to settle down for specific load conditions.

3.4 Speed Dynamics at Constant Torque (10 N-m) and Torque Dynamics at 140 rad/sec

The speed dynamics of a PMSM running with a constant T_L of 10 N-m are shown in as seen in Fig. 9.5. Although the actual analysis concentrates on the 0.95–1.6 second range. The motor speed is changed to 763.94 rpm at 1 second, and then to 1145.91 rpm. This change demonstrates how well the motor can adapt to changes in speed while maintaining a steady load. The torque dynamics of a PMSM running with a constant speed of 1336.901 rpm are shown in Fig. 9.6. The analysis focuses on the time range of 0.9 to 1.6 seconds. During this time, it is noted that the motor torque increases to 6 N-m at 1 second and then to 8 N-m at 1.5 seconds. This variation demonstrates characteristics of the motor's torque response at steady speed.

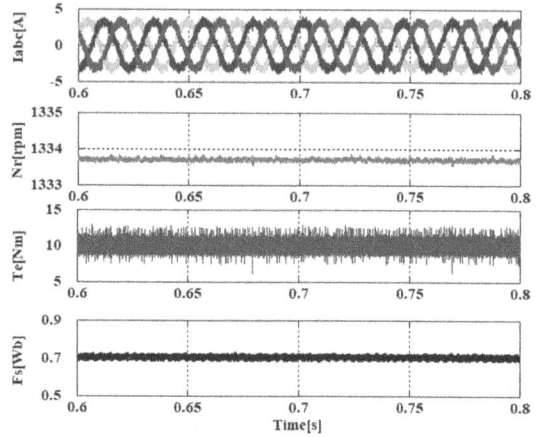

Fig. 9.3 Steady state response at 1336.901 rpm

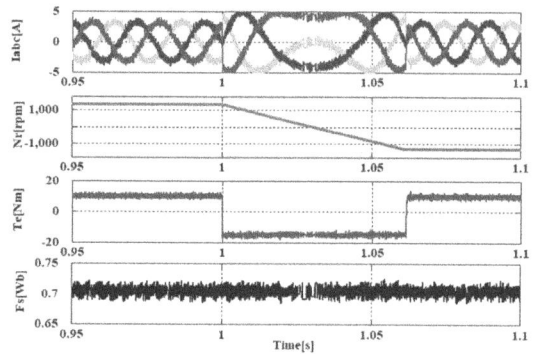

Fig. 9.4 Speed reversal at 1336.901 rpm

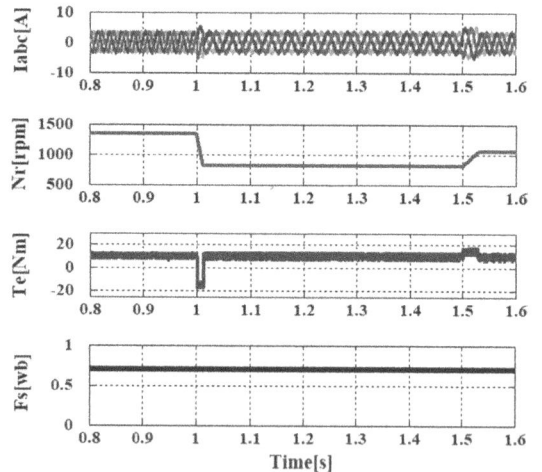

Fig. 9.5 Speed dynamics at load torque of 10 N-m

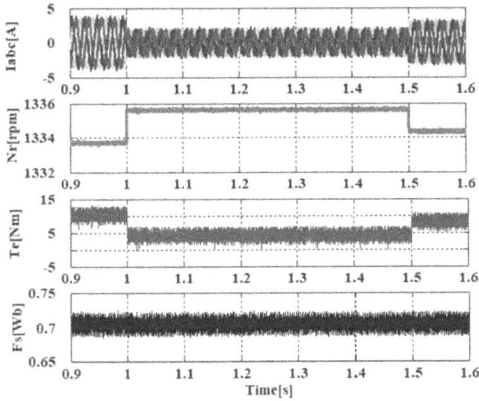

Fig. 9.6 Torque dynamics at 1336.901 rpm

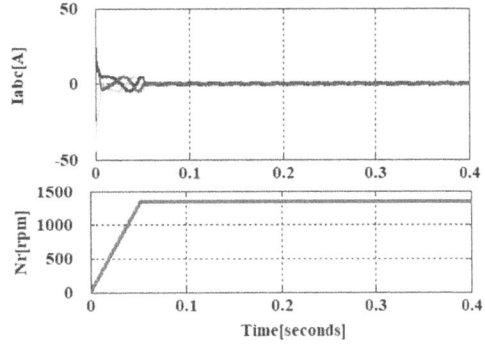

Fig. 9.7 Transient response at 1336.901 rpm

3.5 Transient Response at High Speed (140 rad/sec)

High-speed transient response of a PMSM is controlled by DTC. The oscillations rapidly subside, suggesting that the control approach stabilizes the system. The motor accelerates quickly, reaching 1500 rpm in around 0.1 seconds. The response seems to be controlled, with little overshoot, making it perfect for high-speed operation. An FFT examination of the PMSM current signal's DTC across 89.12 cycles with a 3-cycle window shows a THD of 3.70% and a fundamental frequency of 44.56 Hz. As seen in Fig. 9.8, the motor runs with a sinusoidal current at 40 rad/s, demonstrating low harmonic distortion that is typical for low-speed operations. The motor is efficient and steady functioning, with little influence from high-frequency harmonics. The analysis confirms that the motor has good harmonic performance, efficient control, decreased losses at low speeds, and stable operation under a variety of load conditions.

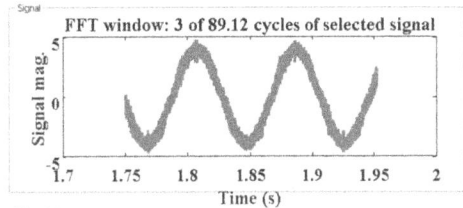

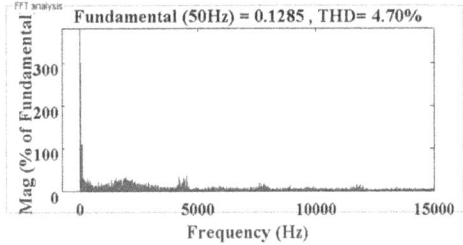

3.6 Effect of HIGH SPEED on PMSM Total Harmonic Distortion (THD)

The selected signal was used in a 3-cycle window of 93.12 cycles for the FFT analysis of the PMSM at 140 rad/s. It was found that the fundamental frequency was 46.56 Hz, and the fundamental amplitude was 3.734 & the THD of 5.47% as seen in Fig. 9.9 & THD of 4.70% at 40 rad/s as shown in Fig. 9.8. Due to the motor's increased speed at

Fig. 9.8 THD at 381.971 rpm

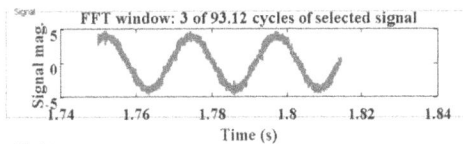

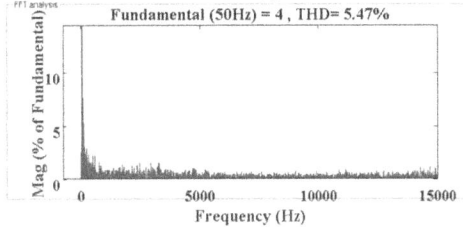

Fig. 9.9 THD at 1336.901rpm

140 rad/s, the switching frequency and harmonic content are higher than at lower speeds. A harmonic content of 5.47% indicates that the motor is functioning within reasonable bounds, guaranteeing steady operation. Table 9.1, provides a clear comparison of the motor's behavior under various conditions, which summarizes the motor's performance.

Table 9.1 Performance summary of DTC of PMSM at low and high speed

Speed (rad/sec)	Torque (N-m)	Peak overshoot	Flux ripples	Torque ripples	THD (%)
40	10	8%	2.26%	2.84%	4.7
140	10	7.56%	2.84%	4.86%	5.47

4. CONCLUSION

With an overshoot of 8% and a steady-state attainment time of about 0.5 seconds, the PMSM shows reliable behavior at both 40 rad/s and the greater speed of 140 rad/s. This shows that under certain load and speed conditions, the motor control system is operating at its best. Stability and quick settling are important in industrial applications, and the system can provide the necessary steady-state performance with acceptable transient behavior. The PMSM performs well when dealing with torque variations, speed changes, and speed reversal. The motor can handle dynamic load and speed variations and is responsive to control signals, as demonstrated by the torque adjustments (from 6 to 8 N-m) & speed changes (from 80 to 120 rad/s). The motor's performance is well within reasonable bounds for industrial use, showing a decent dynamic response to torque and speed variations. According to harmonic analysis, the Total Harmonic Distortion (THD)at low speeds and high speed are 4.7% & 5.47% respectively which are in permittable range for industrial applications according to IEEE 519-2014 standards.

References

1. Suman, K., and Abraham T., Mathew. (2018). Speed control of permanent magnet synchronous motor drive system using PI, PID, SMC and SMC plus PID controller, In 2018 international conference on advances in computing, communications and informatics(ICACCI), (pp. 543–549). IEEE.
2. Liu, C. and Luo, Y.(2017). Overview of advanced control strategies for electric machines." Chinese Journal of Electrical Engineering 3, no. 2: 53–61.
3. Ocen, D.(2005). Direct torque control of a permanent magnet synchronous motor.
4. Wang, F., Zhe, C., Peter, S., Ralph, K., Mauricio, T. and Jose, R.(2015). A comprehensive study of direct torque control (DTC) and predictive torque control (PTC) for high performance electrical drives, EPE Journal 25, no. 1: 12–21
5. Liu, J. and Wei, C. (2013). Generalized DQ model of the permanent magnet synchronous motor based on extended park transformation, In 2013 1st International Future Energy Electronics Conference (IFEEC), (pp. 885–890). IEEE.
6. Ni, Q., Xiaoyun, F., Wensheng, S. and Yongheng, L.(2012). Reduction of torque ripple of VSI-fed PMSM machine by direct torque control method,In Proceedings ofThe7th International Power Electronics and Motion Control Conference, vol. 1, (pp. 525–529). IEEE.
7. Wang, S., Chushan, L., Chen, C. and Dewei, X. (2018). Direct torque control for 2L-VSI PMSM using switching instant table, IEEE Transactions on Industrial Electronics 65, no. 12: 9410–9420.
8. Priya, B., and Rajvir Kaur. (2024). "Comparative Analysis of Permanent Magnet Synchronous Motor Drive Controllers for Electric Vehicles." In 2024 IEEE International Conference on Electronics, Computing and Communication Technologies (CONECCT), (pp. 1–6). IEEE.

9. Avinash, V. and Ratna, D. (2020). Speed estimator for direct torque and flux control of PMSM drive using MRAC based on rotor flux, In 2020 IEEE 9th Power India International Conference (PIICON), (pp. 1–6). IEEE.

10. Preeti, O., Sahana, K., Sachin, A. and Raju,A. B.(2022). Field oriented control of surface-mount pmsm using model predictive current control, In 2022 IEEE North Karnataka Subsection Flagship International Conference (NKCon), (pp. 1–5). IEEE.

11. Magadi, Prajwal S., Anish Puradannavar, Soumya V. Kulkarni, Pramukh Manchikanti, and A. B. Raju. (2023). "Speed Control of Permanent Magnet Synchronous Motor Using Direct Torque Control-Space Vector Modulation Algorithm." In 2023 International Conference on Next Generation Electronics (NEleX), (pp. 1–6). IEEE.

12. Avinash, V. and Ratna, D. (2020). Design of PI controller for space vector modulation based direct flux and torque control of PMSM drive, In 2020 First IEEE International Conference on Measurement, Instrumentation, Control and Automation (ICMICA), (pp. 1–6). IEEE.

Note: All the figures and tables in this chapter were made by the authors.

Emerging Perspectives and Applications of Computational Intelligence and Smart Systems
– Dr. Amit Lathigara et al. (eds)
© *2026 Taylor & Francis Group, London, ISBN 978-1-041-20965-2*

10

Student Performance Prediction in Introductory Programming Courses: A Comprehensive Review of Machine Learning Techniques

Hetal Gosrani*

Reserach Scholar, School of Engineering, RK University, Rajkot
Smt. C.Z.M. Gosrani BCA College, Jamnagar

Nirav Bhatt

School of Engineering, RK University, Rajkot, Gujarat, India

■ **Abstract:** The correct assessment of student performance in programming introduction courses serves to improve both student retention and educational success. An examination of machine learning (ML) methods performing analysis on programming exercise data demonstrates evaluation of submission records and assessment of code quality along with debugging behaviours. The research reviews eleven studies that present early prediction models as well as explainable AI (XAI) frameworks for adaptive learning systems. This paper addresses three main difficulties that occur when dealing with diverse datasets, real-time system deployment and scalability issues. The use of ML interventions that rely on data has shown ability to enhance educational support and improve predictive accuracy thus indicating opportunities to optimize programming education. The review demonstrates the success levels that Random Forest algorithms together with Support Vector Machines and deep learning-based approaches achieve in recognizing students at risk. The improved interpretability of XAI lets educators see clearly how to shape effective decisions and create specific lessons for their students. Research indicates that student engagement increases when adaptive learning models work together with automatic code evaluation along with flipped classroom approaches. The main challenges pertain to limited dataset transferability and processing speed issues during application.

■ **Keywords:** Machine learning, Introductory courses, Educational data mining, Student performance prediction, Programming education, Explainable AI, Adaptive learning

1. INTRODUCTION

Introductory programming courses, often referred to as introductory, are an inevitable part of almost every computer science curriculum, but they suffer immensely due to high failure and dropout rates

*Corresponding author: hetalgsavla@gmail.com

DOI: 10.1201/9781003725046-10

(Mosquera, Suarez, and Guerrero, 2023). The earlier we are able to estimate students' performance, the more potentially helpful educators can be in helping students achieve better learning outcomes (Llanos, Bucheli, and Restrepo-Calle, 2023). Earlier, it was difficult to assess performance, but ML models are gaining more popularity because they can automatically analyse code and track changes in behavioural patterns such as submission and debugging without taking into consideration possibly skewed demographic data (Pereira et al., 2021).

Multiple ML-based prediction techniques receive investigation in this study which generates the following research questions:

RQ1: Which ML algorithms are most effective for predicting student performance in introductory courses?

RQ2: How do explainable and adaptive ML models enhance educational interventions?

RQ3: What challenges and opportunities exist in scaling ML models for wider implementation?

2. LITERATURE REVIEW

2.1 ML Techniques for Performance Prediction

Early submission data enables predictive success in student performance through the use of Random Forest and SVM models as well as additional ML models according to Islam et al. (2024). Through Explainable AI (XAI) frameworks teaching experts can uncover precaution signals which enable them to provide quick support to students who need help (Albreiki, Habuza, and Zaki, 2022). The combination of rule-based systems with ML models generates education recommendations which students easily use for their learning activities (Albreiki, Habuza, and Zaki, 2022). The implementation of AI recommendations encounters major challenges originating from imbalance in dataset information and difficulties with time-dependent system operations and scalability(Pereira et al., 2023).

2.2 Adaptive Learning and Human-AI Collaboration

Deep learning adaptive learning approaches connect with collaborative filtering methods to generate personalized learning spaces that sustain enhanced student involvement with academic material(Llanos, Bucheli, and Restrepo-Calle, 2023). Programming concept detection through ProLog2vec deep learning-based logging occurs through an analysis of programming patterns and an assessment of debugging patterns (Zhao et al., 2021) Academic achievement along with student retention rates improved notably among students due to the implementation of automated coding assessment systems in flipped learning as described by Mosquera et al., (2023). AI-driven educational interventions struggle to gain widespread adoption because of scalability difficulties alongside computational challenges and ethical issues relating to data collection privacy(Pereira et al., 2023).

3. METHODOLOGY

3.1 Hypotheses Development

The following research plan utilizes these specific hypotheses:

- Machine learning systems achieve high accuracy in forecasting students' introductory programming course outcomes through an analysis of their submission logs and debugging activities and code quality measurements.

- Predicting student performance receives enhancement through Explainable AI and adaptive learning models which generate feedback that educators along with learners can understand.

3.2 Hypothesis Testing Approach

The analytical approach for testing hypotheses in this review paper depends on evaluation of existing research. The search for relevant research in programming exercise datasets and student performance prediction retrieved documents from IEEE Xplore and Scopus as well as Google Scholar. The evaluation of selected studies demonstrated their agreement or non-agreement with each hypothesis through evaluation of intervention effectiveness and explainability together with accuracy metrics in educational contexts.

4. Results

Eleven peer-reviewed studies that evaluate different ML approach for predicting introductory student performance were studied in the research. This study examined different datasets within which information about code quality metrics as well as debugging activities and student submission records was given. The findings revealed:

4.1 Multiple ML Models

Llanos, Bucheli, and Restrepo-Calle (2023) and Van Petegem et al. (2022) confirm in their scientific studies that Random Forest and SVM offer excellent results in early predictive analysis. XGBoost and Explainable AI models represent the optimal combination for predicting vulnerable students based on research findings presented by Pereira et al. (2021). The Gradient Boosting algorithms achieve 90% success rate in predictions at every educational site according to Llanos, Bucheli, and Restrepo-Calle (2023).

H1: *Multiple ML models demonstrate strong predictive accuracy in student performance forecasting.*

4.2 XAI Frameworks and Adaptive Models

Albreiki, Habuza, and Zaki (2022) and Pereira et al. (2021) demonstrate Explainable AI tools enable educators to understand predictions which allows them to modify their teaching methods appropriately. The combination of adaptive learning models and automatic code evaluation according to Llanos, Bucheli, and Restrepo-Calle (2023) improved both student retention and engagement in the learning process. Given the current state of research,the real-time intervention scalability is not achievable with existing ML frameworks(Islam et al., 2024).

H2: *XAI frameworks and adaptive models improve both predictive explainability and student engagement, though scalability remains a challenge.*

Table 10.1 examines different machine learning models which predict student performance outcomes in introductory programming classes. The accuracy levels reached 88% because Random Forest and SVM models plus XGBoost showed the best performance when examining submission logs and debugging activities. Predicative applications of Explainable AI (XAI) by Albreiki, Habuza, and Zaki (2022) help teachers make choices while needing extra processing capabilities.

Deep learning model ProLog2vec shows good performance in detecting conceptual challenges although it needs extensive dataset collections to obtain generalization. The implementation of adaptive learning strategies which include collaborative filtering together with flipped classrooms

Table 10.1 Comparative performance of machine learning models in student performance prediction

Reference	Dataset Type	Algorithm(s)	Metric(s)	Key Findings
Albreiki, Habuza, and Zaki (2022)	Undergraduate programming data (649 students, 38 features)	XGBoost, LightGBM, SVM, Naïve Bayes, Random Forest, Explainable ML	Accuracy (86%), AUC ROC (0.96), F1-Score	Explainable ML identifies at-risk students and provides intervention insights.
Cooper (2022)	CS1 student performance data	Probabilistic Neural Network (PNN)	Accuracy (~99%)	Early alerts improve student success and reduce failure rates.
Islam et al. (2024)	Imbalanced dataset (350 CS students)	KNN, Decision Tree, Random Forest, SVC, Gaussian NB	Accuracy: SVC (96.89%), RF (91.30%), KNN (93.78%)	SVC performs best on balanced data. Feature selection improves accuracy.
Llanos, Bucheli, and Restrepo-Calle (2023)	CS1 data (grades, submissions, attempts)	GBC, Random Forest, SVC, Decision Trees, Logistic Regression, KNN, MLP	Precision, Recall, F1-Score, AUC	GBC gives the best results in key weeks. Features like attempts and time boost accuracy.
Mosquera, Suarez, and Guerrero (2023)	CS1 experimental study (82 students)	No ML used (flipped classroom study)	Test results, Grades, Mann-Whitney U Test	Flipped learning improves grades but doesn't impact submission time.
Pereira et al. (2021)	Log data from 2,056 CS1 students (first two weeks)	XGBoost, SHAP (for interpretability)	Accuracy (81.3%), AUC (0.89), F1-Score	XGBoost outperforms traditional models in early prediction. SHAP helps explainability.
Pereira et al. (2023)	Data from 2,714 students solving problems online	Deep Learning (CNN), Matrix Factorization, Cosine Similarity	Accuracy (88%), Precision, Recall, Interchangeability Rate	AI-based recommendations improve assignment quality and reduce effort.
Sunday et al. (2020)	CS1 student performance (239 students)	ID3, J48 Decision Trees	Accuracy: J48 (87.02%), ID3 (85.35%), F1-Score	J48 outperforms ID3. Attendance is key to success.
Van Petegem et al. (2022)	CS1 student data (2,080 students)	SGD, Logistic Regression, SVM, Random Forest	Balanced Accuracy (80%), F1-Score, Precision, Recall	Logistic Regression predicts success early. Accuracy remains stable across courses.
Zhao et al. (2021)	Java programming logs (207 students)	A-RNN, ProLog2vec (Deep Learning)	Accuracy: 94%, AUC ROC	Deep Learning beats traditional models in detecting struggles.
Zhidkikh et al. (2024)	CS1 submission data (trace + survey data)	Logistic Regression (yearly pass/fail, dropout prediction)	Balanced accuracy, Feature importance	High grade expectations improve success. Retaking students has a higher dropout risk.

enhances both student retention and engagement levels. The technology faces difficulties when expanding capabilities, working with diverse datasets, and providing real-time predictions.

5. DISCUSSION

The prediction performance of student success in introductory programming courses achieves high accuracy through utilization of machine learning models which include Random Forest and Support Vector Machines (SVM) and XGBoost. The models base their predictions of struggling students on examining programming exercise records combined with student submission history and their debugging activities among other interaction data. Random Forest performs best when dealing with structured data and complex decision boundaries and SVM. XGBoost along with its gradient boosting algorithm enhances predictive performance by applying feature importance analysis with regularization control.

Issues: Multiple barriers impede the wider use and practical expansion of these ML models despite their favourable outcomes. Student performance prediction models face restricted generalization capability due to limited dataset diversity because they primarily use information from unique educational institutions or learning environments. The models face barriers in adaptable implementation because of differences in curricula and assessment protocols together with demographic data between educational environments.

The deployment of educational interventions with ML components in big learning environments faces difficulties associated with achieving both real-time processing power scalability. The existing ML models experience difficulties when processing real-time student activity logs at scale with high accuracy when operating on datasets containing thousands of students. Even though these techniques show promise they require a full solution to privacy-related ethical challenges and fairness and bias concerns at scale to expand their use across educational settings.

Results: XAI and adaptive learning systems integrate to produce new educational opportunities for individualized instruction together with early response approaches. The application of XAI techniques supplies instructors with model prediction explanations to reveal the reasons for classifying particular students as at-risk. Teachers can enhance their teaching strategies and specific intervention planning through the model's transparent predictions.

The combination of balancing models which implement collaborative filtering alongside deep learning-based analytics creates personalized learning pathways through student performance data points resulting in enhanced student engagement. Automation in feedback systems along with flipped classroom teaching and real-time coding evaluation demonstrates substantial improvement of student retention rates and conceptual understanding. Future development should include finer behavioral metrics such as code structure complexity measurements and debugging efficiency analysis and collaboration pattern data to enhance prediction models and their practical applications in adaptive learning environments. By incorporating these advanced behavioural insights, educational frameworks can further refine individualized learning plans, offering real-time support tailored to students' evolving needs. Additionally, integrating explainable AI into automated assessment tools can improve transparency and trust in AI-driven recommendations, fostering a more effective and responsive learning experience.

6. PROSPECTIVE APPROACH

To improve ML models, we should integrate detailed submissions from future research regarding programming exercise. Submission resubmission patterns, plagiarism detection and behavioural log can also be combined to improve predictive accuracy. Additionally, XAI frameworks will expand XAI frameworks that increase the use of predictive models for educators.

◣ 7. CONCLUSION

The result of this study is to show the effectiveness of ML based prediction on student performance in introductory courses. This supports H1 in that ML models can successfully identify at risk students using behavioural data. The work is also validated in H2 through the benefits of interpretability with Explainable AI and engagement with adaptive learning strategies. These models will further be refined to address these dataset diversity and scalability issues, improving the educational support systems.

References

1. Albreiki, B., Habuza, T., and Zaki, N.(2022). Framework for automatically suggesting remedial actions to help students at risk based on explainable ML and rule-based models. *International Journal of Educational Technology in Higher Education* 19 (1): 1–26. doi: 10.1186/s41239-022-00354-6.

2. Cooper, Cameron I. (2022). Using Machine Learning to Identify At-risk Students in an Introductory Programming Course at a Two-year Public College. *Advances in Artificial Intelligence and Machine Learning* 2 (3): 407–421.doi:10.54364/AAIML.2022.1127.

3. Islam, A., Bukhari, F., Sattar, M. A., and Kashif, A. (2024). Determining Student's Online Academic Performance Using Machine Learning Techniques. *Informatyka, Automatyka, Pomiary w Gospodarce i Ochronie Środowiska* 14 (3): 109–117. doi:10.35784/iapgos.6173.

4. Llanos, J., Bucheli, V. A., and Restrepo-Calle, F. (2023). Early prediction of student performance in CS1 programming courses. *PeerJ Computer Science* 9:e1655. doi: 10.7717/peerj-cs.1655.

5. Mosquera, J. M. L., Suarez, C. G. H., and Guerrero, V. A. B. (2023). Effect of flipped classroom and automatic source code evaluation in a CS1 programming course according to the Kirkpatrick evaluation model. *Education and Information Technologies* 28 (10): 13235–13252. doi: 10.1007/s10639-023-11678-9.

6. Pereira, F. D., Fonseca, S. C., Oliveira, E. H. T., Cristea, A. I., Bellhäuser, H., and Rodrigues, L. (2021). Explaining Individual and Collective Programming Students' Behavior by Interpreting a Black-Box Predictive Model. *IEEE Access* 9: 117097–117119. doi: 10.1109/ACCESS.2021.3105956.

7. Pereira, F. D., Rodrigues, L., Henklain, M. H. O., Freitas, H., Oliveira, D., and Cristea, A. I.(2023). Toward Human–AI Collaboration: A Recommender System to Support CS1 Instructors to Select Problems for Assignments and Exams.*IEEE Transactions on Learning Technologies* 16 (3): 457–472. doi: 10.1109/TLT.2022.3224121.

8. Sunday, K., Ocheja, P., Hussain, S., Olyelere, S. S., Samson, B. O., and Agbo, F. J. (2020). Analyzing Student Performance in Programming Education Using Classification Techniques. *InternationalJournal of Emerging Technologies in Learning (iJET)* 15 (2): 127–144. doi: 10.3991/ijet.v15i02.11527.

9. Van Petegem, C., Deconinck, L., Mourisse, D., Maertens, R., Strijbol, N., Dhoedt, B., De Wever, B., Dawyndt, P., and Mesuere, B. (2022). Pass/Fail Prediction in Programming Courses. *Journal of Educational Computing Research* 61 (1): 68–95. doi: 10.1177/07356331221085595.

10. Zhao, H., Li, M., Lin, T., Wang, R., and Wu, Z. (2021). ProLog2vec: Detecting Novices' Difficulty in Programming Using Deep Learning. *IEEE Access* 9: 53243–53254. doi: 10.1109/ACCESS.2021.3067505.

11. Zhidkikh, D., Heilala, V., Van Petegem, C., Dawyndt, P., Järvinen, M., Viitanen, S., De Wever, B., Mesuere, B., Lappalainen, V., Kettunen, L., and Hämäläinen, R. (2024). Reproducing Predictive Learning Analytics in CS1: Toward Generalizable and Explainable Models for Enhancing Student Retention. *Journal of Learning Analytics* 11 (1): 132–150. doi: 10.18608/jla.2024.7979.

Emerging Perspectives and Applications of Computational Intelligence and Smart Systems
– Dr. Amit Lathigara et al. (eds)
© 2026 Taylor & Francis Group, London, ISBN 978-1-041-20965-2

11

Real-Time Traffic Monitoring and Request Blocking Firewall for Network Security

Jinna Kruthika*,
Gajele Manisha, Mohammad Shinaz Bhanu,
Ganjikunta Abhinaya, Kuraganti Rathnababu
Cyber Security, CMR College of Engineering & Technology,
Hyderabad, India

■ **Abstract:** The increasing complexity of Distributed Denial of Service attacks threatens network security, overwhelming resources and disrupting services. Conventional firewalls falter against these dynamic threats, requiring advanced real-time solutions. This paper presents a firewall system designed for real-time traffic inspection and request blocking to enhance network security. Combining machine learning with packet inspection and iptables-based filtering, it dynamically blocks malicious Internet Protocols. Logistic regression classifies traffic as malicious using features like packet count, with Scapy enabling sniffing. A Flask-based web interface helps administrators in monitor and manage attacks, supported by email alerts for rapid response. Experiments show it outperforms traditional firewalls against Distributed Denial of Service attacks with low false positives. Tested under heavy traffic, it proves scalable. While effective for known threats, future anomaly detection could address zero-day attacks. This scalable, automated solution bridges legacy and intelligent security frameworks, providing actionable insights.

■ **Keywords:** Anomaly detection, DDoS detection, Firewall, Iptables, Machine learning, Network security, Real-time traffic monitoring, Traffic analysis

1. INTRODUCTION

The surging trend of Distributed Denial of Service (DDoS) attacks is a serious network security threat, overwhelming systems that impair essential services such as web servers (Mahjabin et al., 2017) and financial platforms. Increasing in frequency and sophistication, these attacks overwhelm traditional firewalls reliant on static rules,exposing networks. This article proposes a real-time traffic scanning and request-blocking firewall to enhance security. By integrating machine learning,

---

*Corresponding author: jinna.kruthika.07@gmail.com

DOI: 10.1201/9781003725046-11

packet analysis, and iptables filtering, it dynamically blocks suspicious traffic. A Flask-based web interface enables real-time monitoring, while email alerts improve threat response times. Compared to conventional firewalls, this scalable solution bridges traditional and intelligent defenses, with future adaptability to zero-day threats.

2. LITERATURE REVIEW

2.1 Real-Time Traffic Monitoring and Analysis

Real-time traffic monitoring is crucial for identifying Distributed Denial of Service (DDoS) attacks through packet-level analysis. Research demonstrates that high-speed packet inspection tools can capture metrics like packet rates and protocol diversity, enabling rapid detection of anomalies (Doshi et al., 2018). This capability informs our firewall's design, which uses Scapy to outperform static systems by dynamically responding to high-traffic DDoS events.

2.2 Machine Learning for Threat Detection

Machine learning (ML) enhances threat detection by classifying traffic with high accuracy and low overhead. Studies show that supervised models like logistic regression are effective for intrusion detection systems (IDS), providing adaptability to evolving threats (Gupta & Rani, 2020). Our system adopts this model to surpass traditional firewalls, with potential for future integration of advanced ML techniques against zero-day attacks.

2.3 Rule-Based Filtering and Automation

Rule-based filtering, such as iptables-based IP blocking, provides a robust defense when automated. Research highlights its effectiveness in mitigating volumetric DDoS attacks through dynamic rule generation based on traffic analysis (Prasad et al., 2019). Our firewall integrates this with ML-driven decisions, improving scalability and response times over manual approaches.

2.4 Scalable Network Security Frameworks

Scalable frameworks ensure consistent threat detection under diverse network conditions. Studies emphasize lightweight tools and interfaces for real-time visibility and management, even during heavy traffic (Amini et al., 2016). Our system leverages SQLite logging and a Flask dashboard, supporting email alerts and maintaining performance in high-load simulations.

3. PROBLEM STATEMENT

Contemporary networks are confronted with mounting malicious traffic that floods systems, degrades services, and jeopardizes security. Traditional firewalls, bound by static rules, cannot identify changing patterns of attack in realtime, leaving networks vulnerable to rapid exploitation. Current solutions are typically sluggish and manual, lack computerized intelligence, and are handicapped by high packet rates. Scalability remains a challenge, as most systems falter under heavy traffic without sacrificing performance. Administrators also lack instant notifications and user-friendly tools, unmet by current frameworks. This study addresses these gaps with a realtime IP-blocking firewall using iptables, enhanced by a Flask-based interface and email alerts, delivering scalable, adaptive security against sophisticated threats.

4. METHODOLOGY

Developers systematically integrate Machine Learning (ML), network packet analysis, and dynamic IP control to build a real-time traffic analysis and request-blocking firewall. This section details the processes, models,and tools that achieve robust network security against Distributed Denial of Service (DDoS) attacks.

4.1 Data Generation

To train the ML model, developers created an artificial dataset simulating normal and DDoS traffic. This data set holds 100,000 rows representing traffic in an interval of time, with fields such as packet number (normal: 1–100; DDoS: 100–10,000), byte number, duration of flow, diversity of sources (IP address), use of protocols (Transmission Control Protocol [TCP], User Datagram Protocol [UDP], Internet Control Message Protocol [ICMP]), rate of packets, and rate of bytes. It has a binary label, 'is_ddos,' indicating the traffic as normal (0) or malicious (1), a representation of what happens in a real-world attack. The panda's library saves this dataset as 'dataset.csv' for model training.

4.2 Model Specifications and Training

The system employs logistic regression, a supervised ML model, for its efficiency and interpretability in binary classification. Developers split the dataset into 80% training and 20% testing using scikit-learn's train_test_split function, setting random state 42 for reproducibility. StandardScaler normalizes features like byte count and packet rate to ensure uniform scaling. The training set fits the scaler, which then processes both sets. Scikit-learn's Logistic Regression class, with default parameters (L2 regularization, 100 iterations), trains the model to predict malicious traffic probability. A confusion matrix and classification report evaluate precision, recall, and F1-score, confirming performance. Developers can save the trained model and scaler as 'lr.joblib' and 'scaler. joblib' for deployment.

4.3 Real-Time Traffic Monitoring

The Scapy library actively sniffs and analyzes traffic in realtime, filtering IP packets with its sniff function. Over a 10-second window, the system calculates statistics like packet count, byte count, source IP diversity, and protocol count. It excludes trusted IPs (e.g., 192.168.29.217) to avoid false positives, ensuring efficient monitoring that scales beyond traditional firewalls under heavy loads.

4.4 Anomaly Detection and Decision Logic

After each 10-second window, the system computes packet rate (packet number/window) and byte rate (byte number/window). A StandardScaler normalizes these statistics, feeding them into the logistic regression model to classify traffic as normal or malicious. A threshold of 500 packets prevents extreme anomalies from triggering unnecessary actions, balancing sensitivity and specificity for zero-day threat detection.

4.5 IP Blocking and Logging

The system executes iptables commands to block malicious IP addresses and logs them in an SQLite database ('firewall.db') with tables for detected attacks and blocked IPs. This hybrid ML-rules approach outperforms static firewalls, especially against evolving threats.

4.6 Notification and Management

Gmail's Simple Mail Transfer Protocol (SMTP) server sends email alerts upon threat detection, while a Flask application offers a dashboard to view logs and unblock IPs manually. This setup enhances real-time responsiveness and scalability, validated under high-traffic simulations.

5. KEY STRENGTHS

5.1 Real-Time Threat Detection

The system actively senses malicious behavior in realtime through continuous monitoring of network traffic for anomalies like floods of packets or protocol misbehaviors. It identifies possible intrusions in realtime, facilitating immediate protective measures to safeguard the network from corruption or unauthorized exploitation.

5.2 Automated Malicious IP Blocking

Once it senses malicious activity, the system automatically blocks malicious IP addresses through iptables. Automating this helps to remove human lag, decrease exposure windows, and enhance real-time protection against attacks over manual methods.

5.3 Minimized Human Intervention

The system applies Machine Learning (ML) and established threshold values to automatically detect and eliminate threats independently of human intervention. This architecture reduces administrative effort, providing constant protection with fewer operational expenses.

5.4 Experimental Validation

Testing verifies the system's capability to identify SYN flood attacks, a prevalent Distributed Denial of Service (DDoS) attack with minimal false positives, in addition to efficient logging and alerting, demonstrating its real-world practicality (Lee et al., 2018).

5.5 Scalability and Flexibility

Lightweight tools like Scapy and SQLite enable the system to efficiently handle large traffic volumes and diverse attack patterns. This scalability outperforms static systems, supporting both small and large networks under high loads.

5.6 Enhanced Threat Visibility

A Flask-based dashboard provides administrators with real-time insights into traffic patterns and blocked IPs. This visibility supports proactive management and swift decision-making against threats, enhancing control over network security.

5.7 Alerting Mechanism

The system sends real-time email alerts via Simple Mail Transfer Protocol (SMTP) when it detects threats, offering immediate visibility into breaches and enabling rapid responses to diverse network risks.

6. TECHNOLOGY USED

6.1 Packet Sniffing

Scapy captures live IP packets within a 10-second window to identify DDoS patterns, such as excessive SYN packets, ensuring precise real-time traffic analysis.

6.2 Intrusion Detection

A logistic regression classifier analyzes traffic features (e.g., packet count, rate) against a 500-packet threshold to detect SYN floods or port scans, outperforming rule-based detection alone.

6.3 Automatic Blocking

Python invokes iptables commands to automatically block malicious IPs, halting attacks instantly and enhancing defense efficiency over traditional firewalls.

6.4 Database Logging

SQLite stores attack data (IP, reason, timestamp) in tables for detected attacks and blocked IPs, enabling analysis and optimization of security measures.

6.5 Alerting

Gmail's SMTP service sends email alerts to administrators upon attack detection, facilitating immediate response to threats.

6.6 Web Dashboard

A Flask dashboard displays recent attacks and blocked IPs, offering manual unblock options for streamlined threat management.

6.7 Pattern Recognition

ML identifies traffic patterns, such as large SYN packet volumes, to optimize the detection of DDoS attacks like SYN floods or port scans, adapting to evolving threats.

7. RESULT

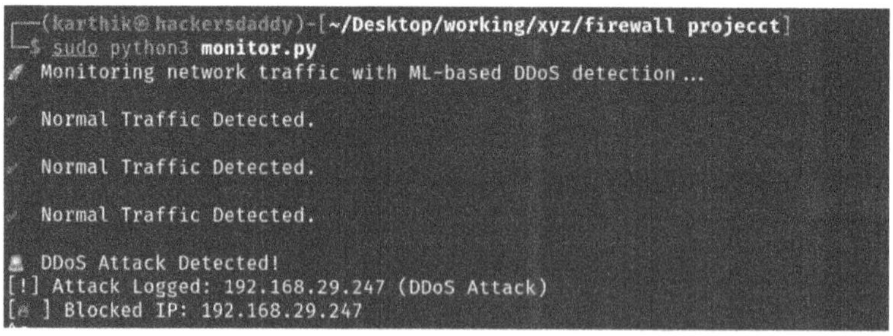

```
┌──(karthik@hackersdaddy)-[~/Desktop/working/xyz/firewall projecct]
└─$ sudo python3 monitor.py
  Monitoring network traffic with ML-based DDoS detection...

  Normal Traffic Detected.

  Normal Traffic Detected.

  Normal Traffic Detected.

  DDoS Attack Detected!
[!] Attack Logged: 192.168.29.247 (DDoS Attack)
[  ] Blocked IP: 192.168.29.247
```

Fig. 11.1 Monitoring the network traffic

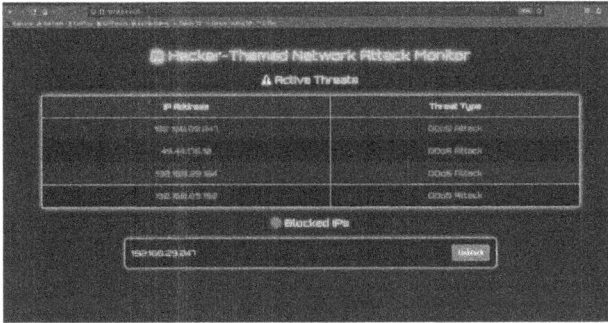

Fig. 11.2 User interface (Blocked IP page)

8. COMPARATIVE ANALYSIS

The proposed firewall outperforms Snort in DDoS detection using ML, offering higher accuracy, faster response, and better scalability under high traffic. Table 11.1: Performance Comparison of Proposed System and Snort shows these advantages.

8.1 Performance Comparison Table

Table 11.1 Performance comparison of proposed system and snort

Metric	Methodology	Snort	Notes
Detection rate	95%	85%	SYN flood simulation
False positive rate	3%	5%	Normal traffic test
Response time (s)	1.5	2.0	Time to block IP
Packet drops	2%	6%	High-traffic

Source: System simulations, 2025

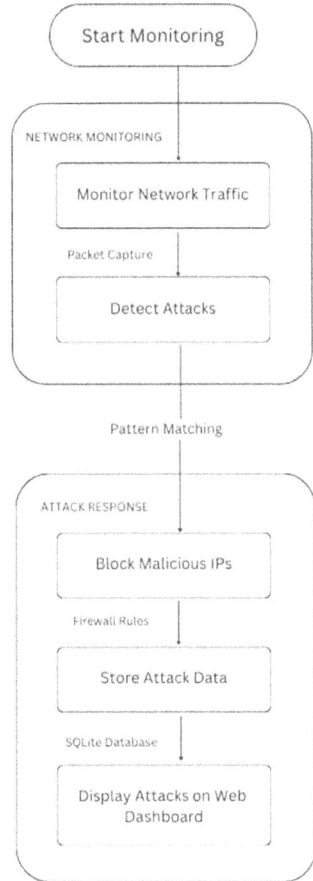

Fig. 11.3 Workflow

9. CONCLUSION

This work develops a real-time traffic analysis and request-blocking firewall that offers robust network security against Distributed Denial of Service (DDoS) attacks. By combining machine learning with packet inspection and iptables-based filtering, the system detects and counters threats in real-time, surpassing legacy firewalls. The logistic regression model, trained on a synthetic dataset, accurately classifies traffic patterns, while Scapy ensures effective packet capture. A Flask-based web portal and email alert system deliver actionable data and real-time alerts, enhancing response times. Experimental testing validates the system's low false-positive detection rate for DDoS attacks as a scalable, automated defense. This solution bridges static security and intelligent, adaptive systems, providing strong protection in dynamic threat environments.

10. FUTURE SCOPE

The proposed firewall system offers significant potential for advancing network security. Integrating a multi-model ensemble, combining logistic regression with deep learning, enhances detection accuracy for complex Distributed Denial of Service (DDoS) attacks. Implementing the system in cloud infrastructures enhances scalability for massive networks. The integration of real-time threat intelligence feeds supports proactive defense against new attack patterns, such as zero-day attacks. Enlarging protocol support and inspecting encrypted traffic solves contemporary security challenges. These enhancements increase the robustness of the system in a dynamic cyber threat situation.

References

1. Amini, M., Rezaeenour, J., & Hadavandi, E. (2016). A neural network ensemble classifier for effective intrusion detection using fuzzy clustering and radial basis function networks. *International Journal on Artificial Intelligence Tools*, *25*(02), 1550033.
2. Doshi, R., Apthorpe, N., & Feamster, N. (2018, May). Machine learning ddos detection for consumer internet of things devices. In *2018 IEEE Security and Privacy Workshops (SPW)* (pp. 29–35). IEEE.
3. Gupta, D., & Rani, R. (2020). Improving malware detection using big data and ensemble learning. *Computers & Electrical Engineering*, *86*, 106729.
4. Lee, K., Kim, J., Kwon, K. H., Han, Y., & Kim, S. (2008). DDoS attack detection method using cluster analysis. *Expert systems with applications*, *34*(3), 1659–1665.
5. Mahjabin, T., Xiao, Y., Sun, G., & Jiang, W. (2017). A survey of distributed denial-of-service attack, prevention, and mitigation techniques. *International Journal of Distributed Sensor Networks*, *13*(12), 1550147717741463.
6. Prasad, K. M., Reddy, A. R. M., & Rao, K. V. (2014). DoS and DDoS attacks: defense, detection and traceback mechanisms-a survey. *Global Journal of Computer Science and Technology*, *14*(7), 15–32.
7. Yuan, X., Li, C., & Li, X. (2017, May). DeepDefense: identifying DDoS attack via deep learning. In *2017 IEEE international conference on smart computing (SMARTCOMP)* (pp. 1–8). IEEE.

Note: All the figures and table in this chapter were made by the authors.

Emerging Perspectives and Applications of Computational Intelligence and Smart Systems
– Dr. Amit Lathigara et al. (eds)
© 2026 Taylor & Francis Group, London, ISBN 978-1-041-20965-2

12

Strategies for the Advancement of Artificial Intelligence (AI) Technologies and Their Application in Intelligence and Counter-Intelligence Operations (The Experiences of the United States and China)

David Kukhalashvili*,
Dimitri Kobiashvili, Bakari Grdzelishvili
Caucasus International University,
Tbilisi, Georgia

■ **Abstract:** Artificial intelligence (AI) has become an essential tool for intelligence agencies in today's global landscape. AI collects and analyzes a vast amount of data, including news articles and satellite imagery, which can take hours to process. By utilizing AI, users can specify parameters to generate a summary of the information they require, such as a comprehensive overview of enemy movements in the case of military operations. AI can analyze texts and images more accurately than humans. A recent article in Foreign Affairs magazine discusses how Israel has used AI to search for members of Hamas among residents of Gaza and analyze classified documents related to Iran's nuclear program. The documents include 55,000 pages of text and 55,000 additional photos and videos, and the AI's goal is to determine if Tehran is developing a nuclear weapon. It should be noted that language models based on AI can be used to rapidly spread propaganda and influence public opinion, potentially causing internal destabilization in states. This raises concerns about the impact of AI on national and cyber security.

■ **Keywords:** Artificial intelligence, Cyber security, Intelligence service, National security

1. INTRODUCTION

The aim of this research is to investigate the role of artificial intelligence (AI) in intelligence and counterintelligence operations and its significance for national security.

*Corresponding author: davit_kukhalashvili@ciu.edu.ge

DOI: 10.1201/9781003725046-12

1.1 Objective

To explore the potential of modern technologies, particularly AI, in enhancing intelligence and counter intelligence capabilities.

1.2 Subject of Study

The role of AI in intelligence and counter intelligence, including its potential applications, challenges, and implications for national security.

1.3 Hypothesis

Signs of intelligence have been documented since ancient times, providing the state with the opportunity to shape public opinion in target countries and manipulate it in accordance with its interests. Intelligence primarily operates in a covert and secret manner, legally examining target entities such as individuals and organizations to influence them towards its objectives.

To gain access to these entities, intelligence employs various techniques, including the use of classified information and other data that is essential for creating psychological profiles of individuals and their personal and private lives. Modern AI technologies are being widely adopted around the world and represent a significant tool for both intelligence and counter-intelligence operations.

The theory of techno-realism originated from offensive realism and has been applied to information technology (IT). It argues that technology is a crucial component of modern political power, and that it is easier to influence targeted outcomes through the use of technology. The theory emphasizes the role of artificial intelligence (AI) as a tool for intelligence operations.

Surveillance capitalism is another theory that describes a new surveillance-based order. It targets organizations and individuals, collecting information about them using AI technologies to market products. By analyzing this data, companies can predict the actions of their targets.

2. MAIN PART

Let us consider the stages of development of artificial intelligence and its use as an intelligence tool using the examples of the USA and China.

2.1 United States of America

The range of U.S. government research and development activities in the field of artificial intelligence includes:

The Defense Advanced Research Projects Agency (DARPA), which manages the Department of Defense's advanced research projects;

2.2 Intelligence Advanced Research Projects Activity (IARPA)

The US government's investment in AIR&D from 2015 to 2020, including only innovation and startups, was approximately 1.1 billion per year. The volume of investment from the private sector in the same direction amounted to about $1 billion. At the same time, the volume of venture capital allocated to startups increased 6-fold from 2020 to 2024.

The Strategic importance of AI R&D in the United States is reflected in the 2016 document, "National Strategic Roadmap for Artificial intelligence Research and Development," developed by the U.S.

National Science and Technology Council (NSTC). The document outlines the areas of AI R&D, development that will receive government funding. The strategy includes the following components: a federal plan for research and development in the field of **big data**; Federal Cybersecurity Research and Development Plan; National Privacy Research Strategy; National Nanotechnology Initiative Research Plan; National Computing Strategy and Initiatives; Initiative for Brain Research Using Innovative Neuro technologies; National Robotics Development Program.

The Intelligence Advanced Research projects Agency (IAPRA) was created in 2006 to conduct interdisciplinary research, identify new capabilities and develop advanced technologies in the field of intelligence. The objectives of IAPRA are based on the technical and expert knowledge available to the intelligence services, and the primary goal is to anticipate the long-term needs of the intelligence community and provide technical and research resources.

Among the important tasks of IAPRA are: deep text processing, revealing hidden meaning, highlighting key information, recognizing and responding to communicative language; Studying metaphors used in different cultures to understand the cultural norms of different people; Identifying the social goals of specific groups based on their messages; Development of methods for obtaining detailed semantic information with an emphasis on events in the form of "who-what-whom-when"; Evolution of tools for assessing the manipulation and maintenance of digital images; Analytical approach of video and audio information; Development of a satellite image recognition system. Technological advance for the automatic creation of precise three-dimensional models of objects with real physical properties from various sources of graphic data; developing automatic motion detection software for multi-camera video streaming environments; Improving the performance of facial recognition systems and anthropogenic activity detection systems; Development of high-precision matching systems extraction of biometric signatures based on low-quality data; Fulfillment of machine learning methods for processing radio frequency spectrum signals, systems for deep analytical approach of geolocation data; Development of methods for continuous analysis of electronic intelligence data, etc.

It turns out that the United States has the ability to use artificial intelligence as an important tool for both intelligence and counterintelligence activities in the interests of its own national security.

To support the above opinions, an analysis of open source information is helpful, which shows that: The National Security Agency (NSA) is the leader among American intelligence agencies in the development and implementation of AI. During surveillance, the agency collects large amounts of personal data, text messages, and the internet communications. The agency's targets are foreign governments, specific individuals, and organizations. The results obtained as a result of successful surveillance allow us to identify cyber threats and monitor networks.

2.3 People's Republic of China

The People's Republic of China deserves attention given the following data: China, along with the United States, occupies a leading position in the field of high technology and in some areas; China is a regional leader and strives for world economic, political and military primacy; China's political and economic system, which has chosen its "Chinese's path of development", differs significantly from both the socialist and capitalist systems and demonstrates advantages in some areas; The Eastern mentality leaves its mark on strategic thinking, and it is necessary to take into account the corresponding approaches.

According to open source analysis, China ranks first in the number of publications on artificial intelligence in scientific journals. In the summer of 2017, the State Council of the People's Republic of China adopted the document "Development Plan for a New Generation of Artificial Intelligence". The exact costs of AI in China are not specified, but some key Chinese indicators allow us to estimate the financial and economic costs in this area.

As of 2017, there were more that 80 million scientist in China, or more that 5.8% of the population. One hundred seventy million people in this country, or more that 12% of the population, have a higher education. In 2016, the sales volume of scientific and technical products exceeded 1 trillion US dollars. In the same year, China spent about 2.1% of its gross domestic product (GDP) – US $1.5 trillion – on scientific research (R&D) and research and development (R&D), US$216 billion more than private sector investment. The main principles of the development plan are: achieving leadership in artificial intelligence technologies, a systematic approach; market orientation; maximum openness. This document defines strategic goals and milestones for achieving them. By the end of the first stage (by 2020), the overall level of technology and application of artificial intelligence in China should meet the standards of leading countries in the world. By the end of the second stage (by 2025), the theory and technology for processing autonomous systems should be developed, specialized enterprises for the development of artificial intelligence should be created, and financial investments should be increased from 150 billion to 1 trillion dollars. **At this stage,** legislative and regulatory documents in the field of artificial intelligence should be developed. At the same time, artificial intelligence should be widely used in intelligent production, intelligent healthcare, urban planning, agriculture, defense and other areas. **By the end of the third stage (2030),** China should develop the theory and technology of artificial intelligence to such a level that it will become a major world center in this field, which will be an important factor contributing to China's global leadership. At this stage, fully independent thinking of artificial intelligence should be ensured, as well as the widespread implementation of artificial intelligence in production and everyday life, in the social sphere, the financial base of the artificial intelligence industry should be expanded form 1 trillion and 10 trillion dollars

To achieve its goals and solve its problems, China is widely and systematically exploring the prospects of developing various types of air, land, sea and underwater autonomous vehicles. An analysis of open source publications shows that the Chinese are using artificial intelligence technologies in planning and conducting cyber operations. It should be mentioned that China is actively using artificial intelligence in the military sphere, including to monitor US defense innovations. It becomes apparent that artificial intelligence is acquiring the status of an important weapon for China for solving intelligence goals and tasks.

To confirm the opinions expressed, information available in open sources will be used, according to which:

China has successfully conducted significant cyber intelligence activities against the United States aimed at:

US intellectual property (which defines the economic and national security benefits of the US)

Personal messages from senior US government and military officials.

Personal data of tens of millions of US citizens.

An analysis of open-source information shows that the Chinese government has exploited vulnerabilities in USA's aging telecommunications infrastructure to attack sensitive government systems.

3. Conclusion

The introduction of artificial intelligence technologies allows the state to strengthen its national security. Its use in intelligence and counterintelligence activities allows for damage to the political, economic, military and scientific-technical security of the target country in order to strengthen its own segments in the same direction;

The leading countries in the world in terms of the development of artificial intelligence technologies are United States and China. Accordingly, they have special conditions for their own influence and intelligence penetration, compared to the so-called third type – small states that do not have such technologies. Accordingly, in the processes of ensuring a detrimental effect on national security by small states, the development of artificial intelligence technologies by the above-mentioned countries should be considered an identified threat. The opinion is supported by the fact that there is a high probability of a detrimental effect on the national security of small states through the use of these technologies;

In identifying threats expected from artificial intelligence technologies, it is important to obtain and analyze strategic documents/information about their creation, available in open and closed sources, through intelligence and counterintelligence activities, as well as by involving the scientific and technical security sector of the country in this process;

To neutralize the harmful influences and threats of intelligence penetration from interested countries, the third type – small states – have the opportunity to carry out coordinated actions within the framework of a strategic partnership with a country that excels in artificial intelligence technologies;

Using artificial intelligence technologies, it is possible to solve the following problems: deep text processing, revealing hidden meaning, highlighting key information; recognizing and responding to the language of communication; studying metaphors used by different cultures to understand cultural norms and different peoples;

References

1. Wilson, S. (2011). *Intelligence Advanced Research Projects Activity (IARPA)-Office of Smart Collection, Great Horned Owl (GHO) Program, Proposers' Day Overview Briefing*. IARPA-BAA-11-12, August 15.
2. Wu, F., Lu, C., Zhu, M., Chen, H., Zhu, J., Yu, K., ... & Pan, Y. (2020). Towards a new generation of artificial intelligence in China. *Nature Machine Intelligence*, 2(6), 312-316.
3. Allen, G. C. (2019). Understanding China's AI strategy: Clues to Chinese strategic thinking on artificial intelligence and national security.
4. Kukhalashvili, D., & Ortanezashvili, A. (2023). Security of the Post-Soviet Space, Intelligence and Counter-Intelligence aspects. Part I: Kazakhstan. *Ukrainian Policymaker*, *12*.
5. Wang, J., Yu, X., Li, J., & Jin, X. (2018). Artificial intelligence and international norms. *Reconstructing Our Orders: Artificial Intelligence and Human Society*, 195-229.
6. Couretas, J. M. (2024). Iranian Cyber Operations
7. Kania, E. B. (2017). Battlefield singularity: Artificial intelligence, military revolution, and China's future military power.

Emerging Perspectives and Applications of Computational Intelligence and Smart Systems
– Dr. Amit Lathigara et al. (eds)
© 2026 Taylor & Francis Group, London, ISBN 978-1-041-20965-2

13

Comparative Performance of Machine Learning Algorithms for Sentiment Analysis: The Role of PCA as a Dimension Reduction Technique

Burak Borhan*, Yasin Ortakci
Karabuk University, Karabuk, Turkiye

Amit Lathigara
School of Engineering, RK University, Rajkot, Gujarat, India

■ **Abstract:** This study investigates the performance of various machine learning algorithms in the domain of sentiment analysis. It also uses a set of pre-processing techniques including tokenization, lemmatization, scaling, and Principal Component Analysis (PCA). Text review data from Amazon, IMDb, and Yelp is evaluated, and Word2Vec-based feature representations are used. Our experiments show that Support Vector Machine (SVM) often emerges as the best contender, achieving commendable results, especially when combined with PCA. However, we also observed cases where Logistic Regression or Random Forest algorithms performed comparable to SVM under certain data distributions. This finding suggests that no single algorithm is universally dominant in all scenarios. Meanwhile, the findings for real-time classification highlight that effective sentiment analysis depends not only on choosing a robust classifier, but also on using the right mix of pre-processing and dimension reduction techniques. While the combination of SVM and PCA often stands out, our results highlight the importance of adapting the approach to the dataset at hand and fine-tuning the hyper parameters. The comprehensive comparison provided in this work provides insights for researchers and practitioners aiming to optimise sentiment analysis workflows, enabling more accurate identification of positive and negative sentiments across diverse textual corpora.

■ **Keywords:** Classification, PCA, Sentiment analysis, Word2Vec

1. INTRODUCTION

Natural Language Processing (NLP), as a sub-branch of artificial intelligence that processes human language with mathematical and statistical methods, was initially applied with rule-based and

---

*Corresponding author: borhanburak@hotmail.com

DOI: 10.1201/9781003725046-13

simple statistical methods, but advances in technology have paved the way for the emergence of more advanced approaches such as deep learning. The increase in text data on social media and the internet, the proliferation of various data sources such as user comments and news texts have contributed to the prominence of sentiment analysis (Xia and Wang 2024) in the field of text mining. While e-commerce companies use text data to evaluate customer satisfaction and analyse political trends, this topic has been the subject of extensive research in the academic field.

Sentiment analysis is the process of automatically identifying positive, negative or neutral emotions expressed in texts. Sentiment analysis is of great importance in areas such as marketing, customer service, and political analysis (Kowsik et al. 2024; Zhang and Guo 2024). However, factors such as the complexity of language such as irony, metaphor, data imbalance, labelling difficulties, and polysemy (Handoyo et al. 2021) make the process difficult.

To overcome these problems, machine learning methods offer powerful solutions for classifying text data. Naive Bayes (NB) (Webb et al. 2017), Support Vector Machines (SVM) (Hearst et al.1998), Logistic Regression (LR) (LaValley 2008), Random Forest (RF) (Cutleret al. 2012) and Nearest Neighbour (KNN), while BoW, TF-IDF, or Word2Vec techniques are used to capture keyword and context clues (Das et al. 2023; Arora et al. 2021). Comparisons with different datasets and metrics reveal which model and feature extraction method is more advantageous for which data type; even small performance differences can affect decision-making (Philip et al. 2023; Devaraja et al. 2023).

This study aims to comprehensively compare the performance of SVM, NB, LR, RF and KNN algorithms in sentiment analysis and to determine the most effective approach using different evaluation metrics (accuracy, precision, recall, F1 score). Details about the used dataset and pre-processing processes will be presented in the following sections.

2. RELATED WORK

In sentiment analysis, both classical machine learning algorithms (SVM, NB, LR, RF, KNN) and deep learning based models have been examined. Ashbaugh and Zhang compared LR, NB, RF, RNN and CNN models in Amazon product reviews and found that CNN models exhibited superior performance in sentiment classification (Ashbaugh and Zhang 2024). Muqeeth evaluated SVM, RF and LR on IMDb and Twitter datasets and reported that SVM gave the best result with 84% accuracy, while LR and RF performed 81% and 77% respectively (Muqeeth et al. 2024). Similarly, Hasan and Hossan evaluated SVM, KNN, and NB classifiers and reported that SVM is a strong choice with 82.5% accuracy (Jasy et al. 2021). In the study of Shah et al. (2020), it was observed that LR was superior to other methods in processing high-dimensional data in text classification. In Iyer's study, Decision Tree, KNN, LR, NB, Perception, RF, SVM, LSTM and Bidirectional LSTM models were examined; in this study, it was emphasised that pre-processing techniques and feature selection play a critical role in model performance (Iyer 2024). Finally, Bahtiar et al. showed that LR achieved higher accuracy, precision, and recall values in the comparison of NB and LR (Bahtiar et al. 2023).

All these results show that with the right feature selection and effective pre-processing steps, classical machine learning algorithms offer powerful and competitive solutions in different data sets and scenarios.

3. MATERIAL AND METHODS

In this study, sentiment analysis was performed using text review data collected from Amazon, IMDb and Yelp. Firstly, pre-processing steps such as text cleaning, word embedding, scaling and dimensionality reduction are applied. Then, LR, SVM, NB, RF and KNN algorithms were tested with cross-validation method and evaluated with metrics such as accuracy, precision, recall and F1-score. In addition, the PCA method was used in some experiments to examine the effect of dimensionality reduction on classification performance. The best performing model was configured to make real-time predictions on sample sentences from the user.

3.1 Dataset

In this study, three separate text files with positive (1) and negative (0) labels were merged. The first file contained a review set of approximately 1000 lines (500 positive, 500 negative) from Amazon(Kotzias2015.), the second contained 748 lines (386 positive, 362 negative) from IMDb(Kotzias2015.), and the third contained 1000 lines (500 positive, 500 negative) from Yelp (Kotzias 2015.). After converting each of them to pandas Data Frame format, we merged them to form a single dataset of approximately 2748 lines (1386 positive, 1362 negative). After this merging, we randomly shuffled the lines to allow the model to generalise better on examples from different review sources. Thus, by gathering sentences and their sentiment labels from each source under one roof, we obtained a richer data distribution in the training and testing stages.

Text Cleaning and Processing

In the text cleaning phase, all words were converted to lower case and broken down into words with NLTK's word_tokenise() function. Meaningless words (stopwords) were removed, reduced to their roots with WordNetLemmatiser and punctuation marks and numbers were removed. Thus, a more consistent and meaningful data set was obtained.

3.2 Word embedding (Word2Vec)

Word2Vec method was used to convert the texts into numerical format. Using the pre-trained vectors GoogleNews-vectors-negative300, each sentence was reduced to a 300-dimensional vector with the function get_average_word2vec(). For a more balanced training process, the vectors were scaled in the range of 0-1 with MinMaxScaler.

3.3 Dimension Reduction

By applying PCA method for dimension reduction, 300 dimensional vectors were reduced to 10, 20 or 30 dimensions. Thus, the processing time is shortened and the model is more stable. Although the effect of PCA on accuracy varies depending on the model and data set, significant improvements were achieved in some cases.

3.4 Model training and evolution

LR, SVM, NB, RF and KNN models were used to classify the sentiment labelled texts. The texts were first vectorised with Word2Vec, followed by dimensionality reduction with PCA (10, 20, 30, 40 or 50 dimensions). For model training, 5-fold cross-validation was applied and each piece of data was used in training and validation phases.

Model performance was evaluated according to criteria such as accuracy, precision, recall and F1-score. Scenarios with and without dimensionality reduction (300 dimensions) were compared with PCA and the contribution of different dimensions to classification success was analysed. The results show that dimensionality reduction improves the performance of the models in some cases.

4. RESULT AND DISCUSSION

In this study, we conducted a comprehensive sentiment analysis using five machine learning classifiers (LR, SVM, NB, RF, and KNN) with a Word2Vec-based embedding approach and applied PCA in some cases. We compared two scenarios: one without PCA and another where PCA was applied at various dimensions (50, 40, 30, 20, and 10). Table 13.1 summarizes the average accuracy, precision, recall, and F1 scores across models. Notably, SVM performed best overall—especially with PCA=50, where it reached an F1 score of 0.8251, indicating that reducing the embedding space helps to filter out noise and highlight key features.

Table 13.1 Comparative performance metrics of different classifiers across various PCA dimensions

Dimensions	Metrics	LR	SVM	NB	RF	KNN
Without PCA	Acc	0.8144	0.8235	0.7591	0.7879	0.7118
	P	0.8167	0.8331	0.7345	0.8022	0.7230
	R	0.8153	0.8131	0.8196	0.7692	0.6984
	F1	0.8156	0.8228	0.7742	0.7849	0.7088
With PCA (Dim=50)	Acc	0.8100	0.8253	0.6547	0.7940	0.7449
	P	0.8159	0.8335	0.6181	0.7972	0.7306
	R	0.8052	0.8175	0.8247	0.7937	0.7836
	F1	0.8104	0.8251	0.7065	0.7953	0.7559
With PCA (Dim=40)	Acc	0.8119	0.8246	0.6714	0.8071	0.7489
	P	0.8150	0.8304	0.6323	0.8105	0.7359
	R	0.8110	0.8204	0.8333	0.8066	0.7836
	F1	0.8129	0.8250	0.7188	0.8083	0.7588
With PCA (Dim=30)	Acc	0.8079	0.8148	0.6976	0.8006	0.7482
	P	0.8118	0.8199	0.6579	0.8059	0.7310
	R	0.8059	0.8117	0.8355	0.7973	0.7922
	F1	0.8088	0.8155	0.7358	0.8013	0.7602
With PCA (Dim=20)	Acc	0.8042	0.8111	0.7384	0.7966	0.7587
	P	0.8102	0.8171	0.7008	0.8008	0.7471
	R	0.7994	0.8067	0.8420	0.7951	0.7886
	F1	0.8046	0.8115	0.7644	0.7977	0.7672
With PCA (Dim=10)	Acc	0.7937	0.7951	0.7656	0.7762	0.7606
	P	0.8005	0.8015	0.7621	0.7782	0.7499
	R	0.7879	0.7901	0.7800	0.7785	0.7886
	F1	0.7940	0.7953	0.7704	0.7782	0.7685

Acc = Accuracy, P = Precision, R = Recall, F1 = F1 Score
Source: Authors

While SVM generally led the pack, LR and RF also showed strong performance under certain conditions (e.g., LR achieved an F1 of 0.8129 at PCA=40, and RF reached 0.8083 at the same level).

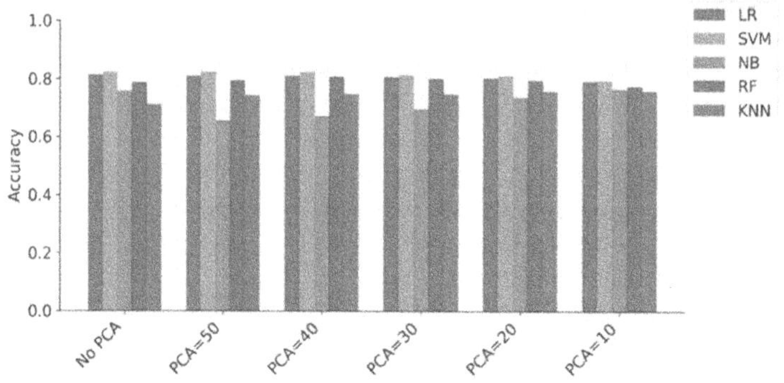

Fig. 13.1   Overall score by model & PCA dimension

Source: Authors

These variations emphasize that model success is influenced by dataset characteristics, splits, and hyper parameter settings. PCA's role was significant: reducing the 300-dimensional embedding to a lower-dimensional space often enhanced both performance and stability, though its benefits can vary with the chosen dimension (with 50 or 40 dimensions typically offering the best trade-off).

Figure 13.1 reinforces these findings, showing that the SVM model with PCA at 50 dimensions attained the highest overall accuracy score (0.8253), slightly outperforming its no-PCA counterpart (0.8235). Overall, the results underline the robustness of SVM in sentiment classification tasks, while also affirming that LR, NB, and RF remain competitive alternatives under different data conditions. Future work may involve exploring larger datasets and fine-tuning hyper parameters to determine if these alternative classifiers can match or even exceed SVM's performance.

One of the rare studies investigating the impact of dimensionality reduction on sentiment analysis (Islam et al. 2019) applied PCA to SVM, LR, NB, and RF methods, achieving a maximum accuracy of 79%. In contrast, our study combined SWM with PCA, reaching an accuracy of 83%. Additionally, as a statistical analysis, we conducted an ANOVA one-way test on the results obtained using SVM when PCA was applied with dimensions set to 50, 40, 30, 20, and 10, as well as when PCA was not applied. Since the p-value was less than 0.05, a statistically significant difference was observed among the PCA methods applied to text vectors. These findings indicate that PCA has a substantial impact on sentiment analysis.

5. Conclusion

This study has comprehensively addressed how different machine learning models (LR, SVM, NB, RF and KNN) affect sentiment analysis performance when used together with word embedding and dimension reduction techniques. It has been shown that remarkable successes can be achieved, especially with the application of PCA, when the noise reduction by preserving the important components of the data and the high accuracy potential of SVM are combined.

Thus, it has been observed that the general consistency and reliability level of the model increases both in the training and testing stages. On the other hand, models such as LR and RF have sometimes been able to stand out in certain data distributions by giving competitive results. This situation proves on cage in how critical the structure of the data set, data pre-processing preferences and model hyper parameters play in text-based tasks such as sentiment analysis. The results show that each model has its own advantages and limitations, and therefore multi-faceted evaluation should be made in real-life applications. Overall, proper integration of word embedding (Word2Vec) and PCA provides significant improvement in problems with intensive textual data such as sentiment analysis.

References

1. Arora, M., Mittal, V., & Aggarwal, P. (2021). Enactment of tf-idf and word2vec on Text Categorization. In *Proceedings of 3rd International Conference on Computing Informatics and Networks: ICCIN 2020* (pp. 199-209). Springer Singapore.
2. Ashbaugh, L., & Zhang, Y. (2024). A Comparative Study of Sentiment Analysis on Customer Reviews Using Machine Learning and Deep Learning. *Computers*, 13(12), 340.
3. Bahtiar, S. A. H., Dewa, C. K., & Luthfi, A. (2023). Comparison of Naïve Bayes and Logistic Regression in Sentiment Analysis on Marketplace Reviews Using Rating-Based Labeling. *Journal of Information Systems and Informatics*, 5(3), 915-927.
4. Cutler, A., Cutler, D. R., & Stevens, J. R. (2012). Random forests. *Ensemble machine learning: Methods and applications*, 157-175.
5. Das, M., & Alphonse, P. J. A. (2023). A comparative study on tf-idf feature weighting method and its analysis using unstructured dataset. *arXiv preprint arXiv:2308.04037*.
6. Devaraja, G., Vardhni, K., Dharshita, R., & Mahadevan, A. (2023, December). A Comparative and Analytical Study of Text Classification Models using Various Metrics and Visualizations. In *2023 OITS International Conference on Information Technology (OCIT)* (pp. 1-6). IEEE.
7. Gulam Muqeeth, Mujthaba, Associate Professor, and Research Scholar Gurughasidas Vishwavidyala. (2024). "Evaluate and Compare the Sentiment Analysis Using SVM, Random Forest and Logistic Regression." *JETIR* 11 (7): e17–24. https://www.jetir.org/view?paper=JETIR2407403.
8. Handoyo, Alif Tri, Hidayaturrahman, and Derwin Suhartono. (2021). "Sarcasm Detection in Twitter -- Performance Impact While Using Data Augmentation: Word Embeddings." *International Journal of Fuzzy Logic and Intelligent Systems* 22 (4): 401–13. https://doi.org/10.5391/IJFIS.2022.22.4.401.
9. Hearst, MA, ST Dumais, E Osuna, J Platt -Systems and their, and undefined (1998). "Support Vector Machines. *Ieeexplore.Ieee.Org*. Accessed January 18, 2025. https://ieeexplore.ieee.org/abstract/document/708428/.
10. Islam, Mazharul, Aftab Anjum, Tanveer Ahsan, and Lin Wang. (2019). "Dimensionality Reduction for
11. Sentiment Classification Using Machine Learning Classifiers." *2019 IEEE Symposium Series on Computational Intelligence, SSCI 2019*, December, 3097–3103. https://doi.org/10.1109/SSCI44817.2019.9002967.
12. Jasy, Md Deloar Hossan, Sakib Al Hasan, Md Ibrahim Khalil Sagor, Abdullah Noman, and Jiang Ming Ji. 2021. "A Performance Evaluation of Sentiment Classification Applying SVM, KNN, and Naive Bayes." *Proceedings –(2021) International Conference on Computing, Networking, Telecommunications and Engineering Sciences Applications, CoNTESA 2021*, 56–60. https://doi.org/10.1109/CONTESA52813.2021.9657115.
13. Kotzias, Dimitrios. (2015). Sentiment Labelled Sentences. UCI Machine Learning Repository. https://doi.org/10.24432/C57604.
14. Kowsik, V. V.Sai, L. Yashwanth, Srivatsan Harish, A. Kishore, Renji S, Arun Cyril Jose, and Dhanyamol M. V. (2024). "Sentiment Analysis of Twitter Data to Detect and Predict Political Leniency Using Natural Language Processing." *Journal of Intelligent Information Systems* 62 (3): 765–85. https://doi.org/10.1007/S10844-024-00842-3/METRICS.
15. LaValley, Michael P. (2008). "Logistic Regression." *Circulation* 117 (18): 2395–99. https://doi.org/10.1161/CIRCULATIONAHA.106.682658.
16. Philip, Jeethu, B. Veerasekharreddy, M. Harshini, I. V.S.L. Haritha, Shruti Patil, and Sk Khaja Shareef. (2023). "A Comparative Study of Text Classification Using Selective Machine Learning Algorithms."

Proceedings of the 7th International Conference on Intelligent Computing and Control Systems, ICICCS (2023), 482–84. https://doi.org/10.1109/ICICCS56967.2023.10142474.

17. Iyer - Authorea, 2024. "A Comparative Analysis of Sentiment Classification Models for Improved Performance Optimization." *Nhsjs.ComV IyerAuthorea Preprints, (2024)•nhsjs.Com.* https://nhsjs.com/wp-content/uploads/2024/05/A-Comparative-Analysis-of-Sentiment-Classification-Models-for-Improved-Performance-Optimization.pdf.

18. Shah, Kanish, Henil Patel, Devanshi Sanghvi, and Manan Shah. (2020). "A Comparative Analysis of Logistic Regression, Random Forest and KNN Models for the Text Classification." *Augmented Human Research (2020) 5:1* 5 (1): 1–16. https://doi.org/10.1007/S41133-020-00032-0.

19. Webb, GI, E Keogh, R Miikkulainen - Encyclopedia of machine, and undefined (2010). "Naïve Bayes." *Researchgate.Net.* Accessed January 18, 2025. https://www.researchgate.net/profile/Geoffrey-Webb/publication/306313918_Naive_Bayes/links/5cab15724585157bd32a75b6/Naive-Bayes.pdf.

20. Xia, Jingping, and Li Wang. (2024). "Social Media Text Sentiment Analysis Method Based on Comment Information Mining." *Lecture Notes of the Institute for Computer Sciences, Social-Informatics and Telecommunications Engineering, LNICST* 546 LNICST:406–21. https://doi.org/10.1007/978-3-031-51503-3_26.

21. Zhang, Xiaodong, and Chunrong Guo. (2024). "Research on Multimodal Prediction of E-Commerce

22. Customer Satisfaction Driven by Big Data." *Applied Sciences 2024, Vol. 14, Page 8181* 14 (18): 8181. https://doi.org/10.3390/APP14188181.

Emerging Perspectives and Applications of Computational Intelligence and Smart Systems
– Dr. Amit Lathigara et al. (eds)
© 2026 Taylor & Francis Group, London, ISBN 978-1-041-20965-2

14

Machine Learning-Based Attack Detection in Smart Home IoT Networks

Vinay Kumar Reddy Muthyala*

School of Engineering, RK University, Rajkot,
Gujarat, India

Amit Lathigara

School of Engineering, RK University, Rajkot,
Gujarat, India

Nur Azaliah Abu Bakar

Faculty of Artificial Intelligence, Universiti Teknologi Malaysia,
Malaysia

■ **Abstract:** The Internet of Things, or IoT, is expanding quickly and improving our lives, but it also presents serious security risks. By providing an innovative anomaly detection system that uses ensemble learning to evaluate intrusion datasets and spot novel attack patterns, this work focuses on the security of smart home IoT devices. The study emphasizes how destructive IoT assaults are becoming more frequent and how anomalies could infect devices and lead to security lapses. The results show that, especially when used with the category smart home IoT dataset, the suggested machine learning model outperforms conventional techniques in identifying anomalies. By combining machine learning and statistical analysis, the study improves the security and dependability of smart home IoT systems. This method evaluates a number of factors and uses machine learning model scores to determine how trustworthy a gadget is. The method greatly improves the security of IoT devices, as demonstrated by real-world weather conditions and a publicly accessible smart home dataset.

■ **Keywords:** IoT security, Smart home, Ensemble learning, IoT networks

1. Introduction

Lighting, temperature, climate control, doors, and windows are just a few of the home appliances that may be managed and observed with smart home automation. These days, people can use a smartphone to manage and control all of these household appliances. Convenience and improved

---

*Corresponding author: muthyalavinayreddy@gmail.com

DOI: 10.1201/9781003725046-14

control over household appliances are provided by this technology. However, there are serious security issues because of the dynamic and varied character of smart homes as well as their internet access (Chandak et al., 2023). Strict cybersecurity protections are required since these environments hold a significant amount of private and sensitive user data. Affordable, low-power sensor technologies, ubiquitous high-speed connectivity, rising cloud usage, and improved data processing and analytics capabilities are some of the main drivers of growth (Alduailij et al., 2022). It is projected that this rise will be further accelerated by the introduction of 5G technology, the spread of smart city projects worldwide, and the increase in connected devices, which will present attractive prospects for IoT suppliers.

Smart homes have numerous advantages, but they also come with more risk and complexity. The vast attack surface created by the networked networks exposes user information and device functioning to possible dangers. Device manipulation, data breaches, and illegal access are the main security issues with smart homes. False alarms, interruptions to regular routines, damaged privacy, and monetary losses are all possible outcomes of these vulnerabilities (Al-Shareeda et al., 2023).

The dynamic and varied nature of IoT environments exposes them to a wide range of security risks. Information exchange is required due to the vast array of IoT devices, increasing the architecture's susceptibility to assaults (Chi et al., 2022). These gadgets are more vulnerable due to their interconnectedness and variety, which makes them desirable targets for bad actors. Intrusion Detection Systems (IDS) are frequently used to improve security. Attacks can be detected by network-based IDS without the need for device-specific software (Jiang et al., 2022). Even though these systems work well in many situations, they can't identify every hazard.

2. PROPOSED FRAMEWORK

Figure 14.1 shows proposed framework, which provides a thorough method for identifying attacks on Smart Home systems. It covers the phases of feature engineering, preprocessing, data collecting, and machine learning model deployment. In order to protect Smart Home environments, this method entails painstaking steps meant to find weaknesses, increase the importance of features, and create reliable predictive models.

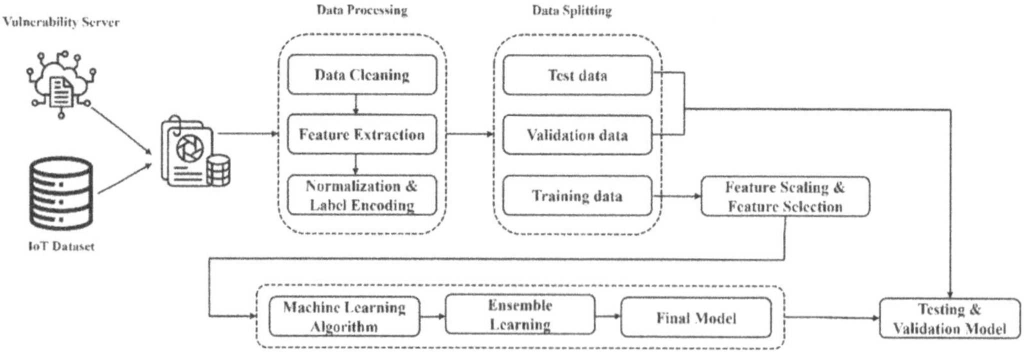

Fig. 14.1 Proposed design of smart home attack detection and prevention system

Data Processing stage entails compiling information from a variety of sources, such as the CICIoT2023 dataset (Malhotra, M., 2023), the Common Vulnerabilities and Exposures (CVE), and the National Vulnerability Database (NVD) (Nakagawa et al., 2021). The objective is to create an extensive database of security flaws and situations relevant to smart homes. By carefully cleaning the data, this crucial stage guarantees the completeness and quality of the information gathered. The procedure preserves the dataset's integrity by using exact replacement approaches to deal with duplicates, null values, and missing data.

Feature scaling is the next stage after dataset splitting. The popular Standard Scalar technique is typically used to achieve feature scaling. This technique modifies the features so that the distribution's mean is zero and its standard deviation is one. To prevent the training and test data from being scaled using the same mean value, it is essential to split the data prior to the feature scaling procedure. This guarantees that the scaling process is impartial and stops data leaks (Soe et al., 2020).

2.1 Ensemble Learning for Model Training

Several Machine learning algorithms, including K-Nearest Neighbors, Decision Trees, Logistic Regression, and Random Forests, are used in our system's training phase.

Training Dataset: The training dataset consists of labeled instances with matching outputs assigned to particular categories. In order to identify patterns in the data and gain the ability to correctly classify new cases, these algorithms go through a learning process.

Our approach gains from the advantages of each machine learning technique by using a variety of machine learning algorithms in ensemble learning, which enhances model performance and resilience in identifying Smart Home system threats.

3. EVALUATION

The IoT Intrusion dataset (Malhotra, M., 2023) is the one used in this paper. For this dataset, IoT device traffic data was generated. These devices were designed to operate with other peripherals linked to the same networks, and data was then collected to capture the interactions and activities inside this IoT ecosystem. There is total 9 different IoT devices utilized for this purpose and generated traffic are in mostly equal ratio. Each of these IoT smart devices generates attack traffic. A total of 115 distinct features are employed in the assessment process. Each feature is listed along with a description. Several attack classes to group all attacks into distinct categories. 11 distinct attack classifications exist in all. Convolutional, pooling, flattening, and dense layers are among the layers that make up this model. After passing through multiple convolutional layers with max pooling layers strewn throughout, the input layer receives data with the shape (115, 1). The model is probably made for a multi-class classification problem because it has four dense layers, the last of which produces eleven classes. The network's intricacy and capacity to recognize complicated patterns in the data are demonstrated by the model's 216,395 trainable parameters.

A graph displaying the model accuracy over 17 training epochs, displaying both training and validation accuracy, is shown in Fig. 14.2. The training accuracy is shown by the blue line, while the validation accuracy is shown by the orange line. The graph shows that over the first few epochs, both the training and validation accuracies rise significantly before leveling off at epoch 4. Over the course of the remaining epochs, the training accuracy continues to increase steadily, reaching about 94%.

Both training and validation losses decline significantly in the first epochs, with the biggest drop occurring between epochs 1 and 3, as can be seen from the graph in Fig. 14.3. Following this initial drop, the training loss steadily decreases and settles around a low value near 0.015. The validation

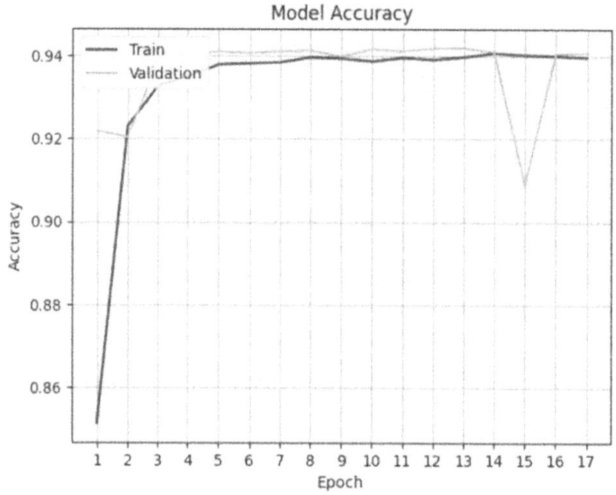

Fig. 14.2 Model accuracy graph

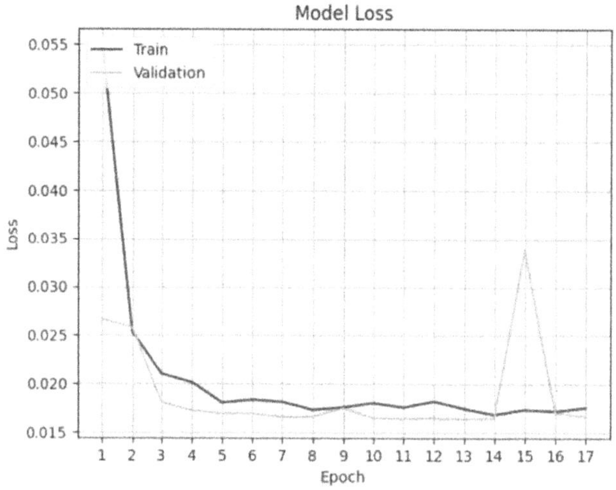

Fig. 14.3 Model loss graph

loss exhibits a comparable pattern, beginning with a steep decline and stabilizing at approximately epoch 4.

4. CONCLUSION

The framework provides a methodical way to secure environments in smart homes. Through the integration of multiple critical processes, including data collection, preprocessing, feature engineering, dataset splitting, model training, and evaluation, this structured technique improves the security of Smart Home IoT devices. The framework guarantees strong and efficient security safeguards for Smart Home systems with this comprehensive approach. Preprocessing and dataset construction place a strong emphasis on the value of cleanliness and diversity. Extensive data purification guarantees completeness and correctness, providing a strong basis for additional research and analysis. We apply traditional machine learning models to an IoT dataset using an ensemble learning approach in order to address the research issue. The goal of this approach is to find abnormalities and attacks. We employ label encoding and feature extraction techniques for data processing, making sure the data is ready for precise detection. Following a thorough examination, we are also quite pleased with the outcomes we obtained.

References

1. Alduailij, M., Khan, Q. W., Tahir, M., Sardaraz, M., Alduailij, M., & Malik, F. (2022). Machine-learning-based DDoS attack detection using mutual information and random forest feature importance method. Symmetry, 14(6), 1095.
2. Al-Shareeda, M. A., Manickam, S., & Ali, M. (2023). DDoS attacks detection using machine learning and deep learning techniques: Analysis and comparison. Bulletin of Electrical Engineering and Informatics, 12(2), 930–939. https://ssrn.com/abstract=4515135
3. Araya, J., et al. (2023). Anomaly-based cyberattacks detection for smart homes: A systematic literature review. Internet of Things, 22. Elsevier.
4. Chandak, A. V., et al. (2023). DDoS attack detection in smart home applications. Journal of Software Practices and Experience. Wiley.
5. Chi, H., Fu, C., Zeng, Q., & Du, X. (2022). Delay wreaks havoc on your smart home: Delay-based automation interference attacks. IEEE Symposium on Security and Privacy (S&P), 1575. IEEE Computer Society.
6. Churcher, A., Ullah, R., Ahmad, J., urRehman, S., Masood, F., Gogate, M., Alqahtani, F., Nour, B., & Buchanan, W. J. (2021). An experimental analysis of attack classification using machine learning in IoT networks. Sensors, 21(2), 446. https://doi.org/10.3390/s21020446
7. Jiang, C., et al. (2022). Effective anomaly detection in smart home by integrating event time intervals. Procedia Computer Science, 210, 53–60. Elsevier.
8. Malhotra, M. (2023). UNB-CIC IoT dataset [Data set]. Kaggle. Retrieved from https://www.kaggle.com/datasets/madhavmalhotra/unb-cic-iot-dataset
9. Mishra, A., Cheng, A. M. K., & Zhang, Y. (2020). Intrusion detection using principal component analysis and support vector machines. 2020 IEEE 16th International Conference on Control Automation (ICCA), 907–912. IEEE.
10. Nakagawa, H. Y. F., et al. (2021). Attack detection in smart home IoT networks using CluStream and Page-Hinkley test. IEEE Latin-American Conference on Communications (LATINCOM).
11. Patel, A. A., & Soni, S. J. (2015). A novel proposal for defending against vampire attack in WSN. 2015 Fifth International Conference on Communication Systems and Network Technologies (CSNT), 624–627. https://doi.org/10.1109/CSNT.2015.94
12. Soe, Y. N., Feng, Y., Santosa, P. I., Hartanto, R., & Sakurai, K. (2020). Machine learning-based IoT-botnet attack detection with sequential architecture. Sensors, 20(16), 4372. https://doi.org/10.3390/s20164372
13. Soni, S. J., & Nayak, S. D. (2013). Enhancing security features and performance of AODV protocol under attack for MANET. 2013 International Conference on Intelligent Systems and Signal Processing (ISSP), 325–328. https://doi.org/10.1109/ISSP.2013.6526928
14. Tuan, T. A., Long, H. V., & Son, L. H. (2020). Performance evaluation of botnet DDoS attack detection using machine learning. Evolutionary Intelligence, 13, 283–294. https://doi.org/10.1007/s12065-019-00310-w

Note: All the figures in this chapter were made by the authors.

Emerging Perspectives and Applications of Computational Intelligence and Smart Systems
– Dr. Amit Lathigara et al. (eds)
© 2026 Taylor & Francis Group, London, ISBN 978-1-041-20965-2

15

Development of Knee Joint Implant using VAT Polymerization and Investment Casting

Ratnadeepsinh M. Jadeja*
PhD Scholar, Department of Mechanical Engineering,
School of Engineering, RK University, Rajkot, Gujarat, India

Marmik M. Dave
Assistant Manager, R&D Department,
Atul Greentech Pvt. Ltd.

■ **Abstract:** The development of additive manufacturing has greatly benefited medical applications, especially in the field of orthopaedic implants. The VAT polymerization method - an innovative 3D printing process that allows great precision - is used to manufacture a knee joint implant. To maximize the performance of knee implants, the study looks into material choices, design factors, fabrication methods, and mechanical qualities. To build a customized implant, information was gathered from X-ray and scan images unique to each patient. Vat polymerization was used to 3D print patient-specific implants, and then the generated implant was used in investment casting as a pattern to create a metal component out of biocompatible material. According to the findings, VAT polymerization offers a promising substitute for conventional manufacturing techniques in the production of knee implants with superior surface polish, structural stability, and patient-specific fit.

■ **Keywords:** 3D printing, Additive manufacturing, CAD model, Investment casting, Knee joint implant, Vat polymerization

1. INTRODUCTION

For increased mobility and pain alleviation, knee implants are required for conditions affecting the knee joint, such as osteoarthritis and severe traumas. Surface smoothness and customization are two areas where traditional manufacturing methods, such as casting and machining, sometimes fall short. With photopolymerizable polymers, vat polymerization, an additive manufacturing technique, provides accurate fabrication capabilities. The viability of developing knee implants using VAT polymerization and using investment casting to produce a long-lasting metal implant

---

*Corresponding author: ratnadeepsinh.jadeja@rku.ac.in

DOI: 10.1201/9781003725046-15

is covered in this study. Figure 15.1 and Fig. 15.2 show the difference between manufacturing methods, where in one vat polymerization used where scope of modification in the design as per the requirement of patient, and without vat polymerization is suitable for mass production and no scope for modification.

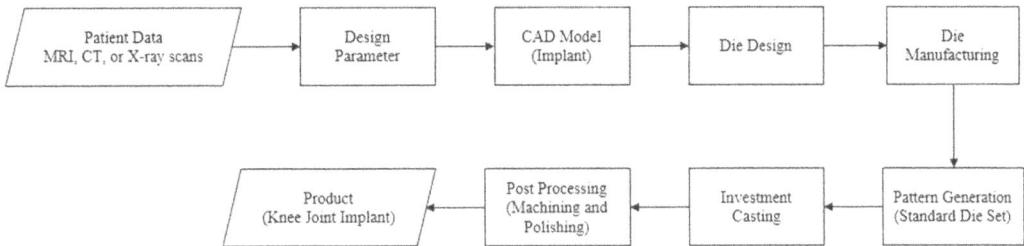

Fig. 15.1 Flowchart of implant manufacturing method without vat polymerization

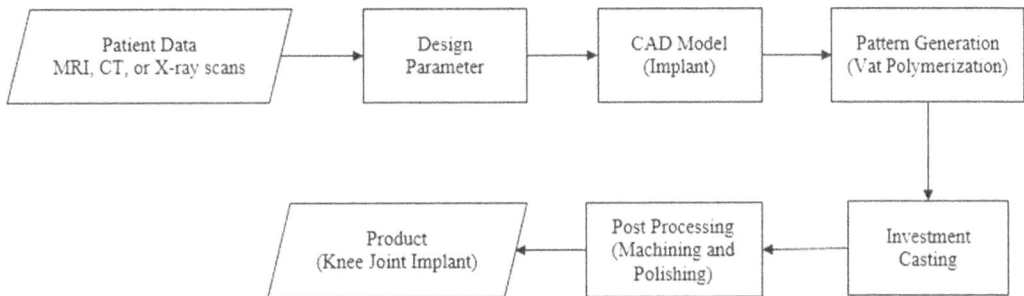

Fig. 15.2 Flowchart of implant manufacturing method with vat polymerization

2. METHODOLOGY

A methodical, multi-step process is used to develop a patient-specific knee joint implant, employing Vat-polymerization to guarantee accuracy, biocompatibility, and mechanical integrity for the ideal fit.

Data Collection: One of the most important steps in creating a knee joint implant customized for each patient is gathering data. First, high-resolution medical imaging data is obtained from patients, usually by MRI, CT, or X-ray scans (Aziliz et al., 2024). These imaging methods offer comprehensive details regarding the knee joint, such as joint alignment, cartilage thickness, and bone structure. Using specialist medical imaging software, the gathered scan data is then transformed into a 3D digital model, guaranteeing an exact depiction of the knee's anatomy. A personalized implant that precisely fits the patient's joint structure is designed using this 3D model as the basis (Balwan and Shinde 2019). The effectiveness of the implant, the reduction of postoperative problems, and the overall functioning and lifetime of the knee joint replacement are all directly impacted by the precision of the data-gathering phase (Ramavath et. al., 2023).

Design and CAD Modeling: For the generation of an exact, patient-specific knee joint implant, the design and CAD modeling is crucial. A 3D digital model of the patient's knee joint is created utilizing sophisticated CAD (Computer-Aided Design) software based on the gathered medical imaging data.

To guarantee excellent functionality and anatomical accuracy. Biomechanical compatibility, stress distribution (Harrysson et. al., 2017), and articulation with adjacent tissues are important design factors(Aziliz et al., 2024). Different implant geometries are assessed to improve stability and reduce wear over time. Furthermore, stress distribution is examined using modeling techniques to make sure the implant can sustain physiological demands (Harrysson et. al., 2017). The completed CAD model is then made compatible with the Vat Polymerization process by optimizing it for 3D printing. This method lowers surgical risks and improves patient outcomes by enabling the manufacturing of highly personalized knee implants. For the CAD Modelling, Fusion 360 is used and Fig. 15.3 shows the Femoral and Tibial components.

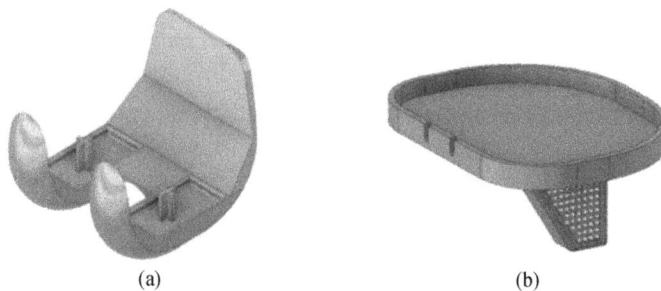

(a)   (b)

Fig. 15.3   3D model (a) Femoral component (b) Tibial component

Pattern Generation: A high-precision additive manufacturing method called vat polymerization is employed to create patterns for Investment Casting. Using a light source, such as a laser in stereolithography (SLA) or a digital projector in digital light processing (DLP), this procedure selectively cures photopolymer resins layer by layer. The process makes it possible to create intricately detailed, smooth-surfaced items with intricate geometries - difficult to archive using traditional manufacturing methods (Pagac et. al., 2021). The primary benefit is its remarkable accuracy in implant production. Resins can also be utilized to make sacrificial or temporary patterns for investment casting, which enables the creation of long-lasting metal implants (Sameni et. al., 2022). A promising method in contemporary orthopedic applications, the knee joint implant production technique uses Vat polymerization to offer higher surface quality, precise dimensional control, and greater customization. Figure 15.4 shows the printed pattern for investment casting.

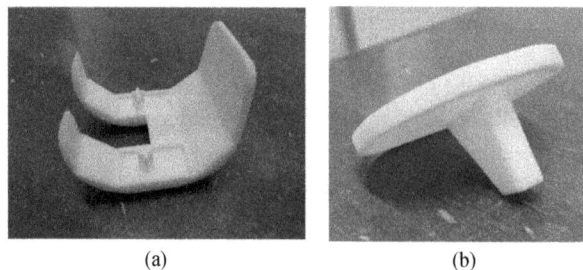

(a)   (b)

Fig. 15.4   Pattern for investment casting (a) Femoral component (b) Tibial component

Investment Casting: The process of turning the 3D printed knee joint implant pattern into a robust, biocompatible metal implant requires investment casting. Initially, a 3D printed pattern (Fig. 15.4)

acts as a sacrificial pattern in this procedure. In order to make a mold, the printed pattern is covered with a ceramic slurry (Fig. 15.5). The mold is then heated to burn out the polymer, leaving behind a perfect cavity that precisely mimics the geometry of the implant (Fig. 15.6). To create the final implant structure, molten biocompatible metal is poured into the mold (Sarojrani et. al., 2012). After solidification, the mold is removed, and post-processing methods like heat treatment, polishing, and machining are used to improve the implant's surface finish and mechanical strength. Investment casting is the perfect technique for producing orthopedic implants that are suited to the demands of specific patients because it guarantees the creation of intricate, personalized implants with superior dimensional precision, strength, and biocompatibility.

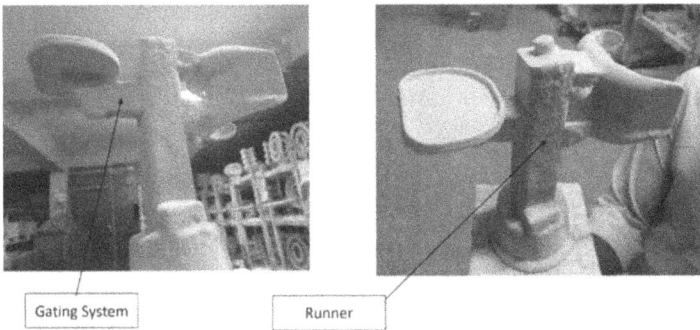

Fig. 15.5 Mold preparation

Fig. 15.6 Burnt out of the polymer   **Fig. 15.7** Metal pouring

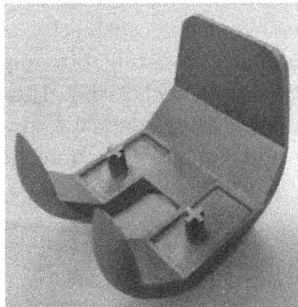

Fig. 15.8 Femoral component after machining work

3. RESULTS AND DISCUSSION

The fabricated knee implants exhibit:

High Dimensional Accuracy: Vat Polymerization ensures minimal deviations from the CAD model.

Improved Surface Finish: The layer-by-layer curing process results in smoother surfaces compared to other additive manufacturing techniques.

Adequate Mechanical Properties: The implants demonstrate high strength and durability suitable for load-bearing applications.

Customization Potential: Patient-specific implants enhance fit and functionality, reducing post-surgery complications.

Enhanced Durability: The investment casting process allows for the production of durable metal implants with high biocompatibility.

4. CLINICAL IMPLICATIONS

Orthopedic surgery is significantly impacted by the use of investment casting and Vat polymerization in knee implants. These implants reduce the amount of pain caused by fitting issues and lower rejection rates by utilizing patient-specific designs. Better joint function and longevity are the results of the implant's more natural fit, which is guaranteed by the great precision of Vat polymerization. Furthermore, this approach reduces manufacturing lead times and costs, making personalized orthopedic solutions. Future advancements in this technology could further refine implant performance, paving the way for more effective and long-lasting knee joint replacements.

5. FUTURE WORK

While the combination of Vat Polymerization and investment casting presents a promising approach, further studies are needed. Future research should focus on:

- Long-term clinical trials to assess patient outcomes.
- Development of hybrid material compositions for enhanced performance.
- Integration of smart technology for better experience after implant surgery.

6. CONCLUSION

Vat Polymerization emerges as a promising technique for the fabrication of patient-specific knee joint implants due to its precision, customization capabilities, and material adaptability. Moreover, polymer-based patterns can be turned into long-lasting, biocompatible metal implants with superior mechanical qualities through the use of investment casting. This strategy could be further optimized by upcoming developments in material science, clinical testing, and smart implant technologies, increasing the efficacy, accessibility, and durability of tailored joint replacements.

References

1. Aziliz Guezou-Philippe, Arnaud Clavé, Ehouarn Maguet, Ludivine Maintier, Charles Garraud (2024). Fully automated workflow for the design of patient-specific orthopaedic implants: application to total knee arthroplasty. hal-04519267

2. Balwan, A.R., and V.D. Shinde (2019). "Development of Patient Specific Knee Joint Implant." Materials Today: Proceedings (27) 288–293.
3. Ramavath D, Yeole SN, Kode JP, Pothula N, Devana SR (2023). Development of patient-specific 3D printed implants for total knee arthroplasty. Explor Med. (4) 1033–1047.
4. Harrysson OL, Hosni YA, Nayfeh JF (2007). Custom-designed orthopedic implants evaluated using finite element analysis of patient-specific computed tomography data: femoral-component case study. BMC MusculoskeletDisord:8:91
5. Pagac M, Hajnys J, Ma QP, Jancar L, Jansa J, Stefek P, Mesicek J. (2021). A Review of Vat Photopolymerization Technology: Materials, Applications, Challenges, and Future Trends of 3D Printing. Polymers (Basel). 13(4):598.
6. Sameni F, Ozkan B, Karmel S, Engstrøm DS, Sabet E. (2022). Large Scale Vat-Photopolymerization of Investment Casting Master Patterns: The Total Solution. Polymers 14(21):4593.
7. Sarojrani Pattnaik, D. Benny Karunakar, P.K. Jha (2012). Developments in investment casting process - A review, Journal of Materials Processing Technology 212(11) 2332–2348.

Note: All the figures in this chapter were made by the authors.

Emerging Perspectives and Applications of Computational Intelligence and Smart Systems
– Dr. Amit Lathigara et al. (eds)
© 2026 Taylor & Francis Group, London, ISBN 978-1-041-20965-2

16

A Comparative Review of a Single-Stage and Two-Stage Grid-Connected PV Systems

Vishal G. Jotangiya*

PhD. Scholar, Department of Electrical Engineering,
Faculty of Technology, RK University,
Rajkot, Gujarat, India

Electrical Engineering Department, Lukhdhirji Engineering College,
Morbi, Gujarat, India

Riaz Israni,
Chiragkumar Parekh

Department of Electrical Engineering, School of Engineering,
RK University, Rajkot, Gujarat, India

■ **Abstract:** Power conversion technologies have advanced to increase efficiency, dependability, and power quality as a result of the growing use of photovoltaic (PV) systems in grid-connected applications. Among the prevalent architectures, single-stage and two-stage PV systems have emerged as key contenders. In this paper, the topology, efficiency, cost, power quality, control complexity, and suitability for different environmental conditions of these systems are compared. Single-stage PV systems offer a smaller component count, lower cost, and a more compact design by combining grid interfacing and Maximum Power Point Tracking (MPPT) into a single power conversion stage. However, they pose challenges in achieving good efficiency across a multi range of operating conditions and ensuring optimal grid compliance. On the other hand, two-stage systems separate the MPPT function from grid interfacing, typically using a DC-DC converter followed by a DC-AC inverter. This decoupling enhances control flexibility, power quality, and performance in dynamic weather conditions but increases overall cost, size, and control complexity. The trade-offs between the two methods in terms of lifetime cost, total harmonic distortion (THD), and efficiency is highlighted by a comparison of simulation and experimental results from the literature.

■ **Keywords:** Control flexibility, DC-DC converter, MPPT, Photovoltaic (PV) systems

*Corresponding author: vgjotangiya@gmail.com

DOI: 10.1201/9781003725046-16

1. INTRODUCTION

Global apprehensions concerning climate change and the swift exhaustion of fossil fuel reserves necessitate prompt measures to accelerate the generation of clean energy utilizing existing renewable energy resources (RER). According to their operational and technical specifications, PV systems have the benefit of operating in both grid-integrated and islanding/standalone modes (Aisyah Mohd Yusof and Ali 2019; Murillo-Yarce et al. 2020).The worldwide expansion of photovoltaic (PV), wind, fuel cell technologies, and other RER, along with their benefits such as reduced operating expenses and improved energy security, has led to a marked increase in the incorporation of RER-based power generation into utility grids(Karuppiah, Natarajan et al. 2024; Parekh and Israni 2024). PV systems can be divided into two primary categories: grid-connected (on-grid) PV systems and standalone PV systems(ElNozahy and Salama 2013).Due to the researcher's persistent efforts, the modest standalone PV system has been converted into a grid connected PV system(Panigrahi, Mishra, and Srivastava 2018).Consequently, the optimization of interface inverter efficiency is of utmost importance for the enhancement of overall system performance. A variety of inverter topologies have been proposed and implemented across photovoltaic panels, wind energy systems, and other renewable energy sources, each exhibiting distinct advantages and disadvantages(Israni et al. 2023).

Most PV systems must use MPPT under various operating conditions, such as changes in load conditions, irradiation, and ambient temperature, in order to operate as efficiently and profitably as possible. Many iterations of MPPT algorithms have been developed(Zhu, Yao, and Wu 2011; Maurya and Saket 2023). (Haddadi, Farhangi, and Blaabjerg 2019) presents a novel three-phase Single-Stage Multi-Port Inverter (SSMPI) that combines the advantageous aspects of multi-string inverters with the previously mentioned benefits of the current fed topologies.

2. GRID TIED PV INVERTER TOPOLOGIES

Inverters can be divided into various groups according to their configuration and component types. These classifications are based on a number of variables, including the number of power conversion stages –single and multi-stage, with transformer and transformer less configuration. The following is a brief discussion and description of each category:

2.1 Inverters Based on Number of Power Conversion Stages

Depending on power processing stages, the power converter can be split into single-stage and two-stage configurations(Kumar, Gupta, and Gupta 2017; Zeb et al. 2018).

Two-Stage Converter

The electrical output generated by PV system is characterized by lower direct current (DC) levels, thereby necessitating the incorporation of a DC–DC converter for effective connection with electrical grids. DC to DC converter functions as voltage amplifier and additionally employed for the purpose of MPPT. In certain instances, these converters also provide the functionality of galvanic isolation functionality (Xiao et al. 2016).

The two-stage topology is shown in Fig. 16.1. The operational efficiency of dual-stage converters is diminished as a consequence of the incorporation of an additional converter. However, the diminished conversion efficacy is partially compensated by enhanced extraction efficiency.

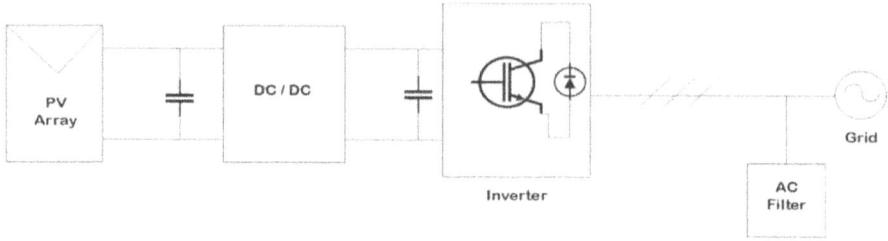

Fig. 16.1 Two-stage converter topology

Source: Authors

One-Stage Converter Topology

In addition to controlling injected grid currents, the single stage inverter also regulates voltage amplification and tracks the maximum power point. In one-stage, one DC–AC inverter is used as a connecting unit shown in Fig. 16.2. High conversion efficiency is the goal of the straightforward single-stage power interface. Additionally, it incorporates precise MPPT control, enhances power factor, maintains a steady switching frequency and lowers current harmonics (Zeb et al. 2018).

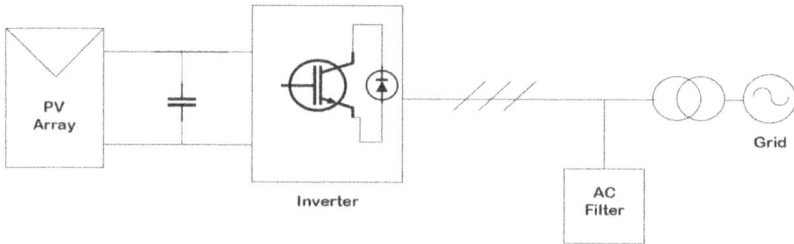

Fig. 16.2 One- stage converter topology

Source: Authors

Single-stage inverters, on the other hand, can withstand low DC input voltage range and poor power quality. As a result, in order to create a panel, numerous solar cells must be connected either in series or shunt, increasing the DC voltage at the solar side and inevitably raising equipment costs. Table 16.1 below provides a comparison of the one-stage and two-stage topologies.

Table 16.1 Discrimination of different grid-connected solar system topologies (Gawhade et al. 2021)

Sr. No.	Parameters	Single-stage conversion	Two-stage conversion
1	Size	Small	Large
2	Cost	Low	High
3	Efficiency	High	Low
4	Control algorithm	Complex	Simple
5	Reliability	High	Low
6	Voltage Total Harmonic Distortion (THD)	Low	High
7	DC voltage stability	Low	High

Table 16.2 Performance correlation between the two topologies (Zhu, Yao, and Wu 2011)

	Single + Perturb and Observe (P&O)	Single + Incremental conductance (Inc)	Two + P&O	Two + Inc
Power efficiency	98.48%	99.49%	98.11%	97.52%
THD (Phase Voltage)	4.5%	2.7%	0.27%	0.28%
MPPT accuracy (ideal : 4000w)	3977w	3856.6w	3910.6w	3917.3w

To (Zhu, Yao, and Wu 2011)emphasizes lack of comprehensive studies comparing the design details and performance of these two topologies, particularly in relation to MPPT methods. Table 16.2 delineates a comparative analysis of performance metrics derived from three distinct indices: power efficiency, phase voltage total harmonic distortion (THD), and maximum power point tracking (MPPT) accuracy, which correspondingly signify economic viability, power quality, and the efficacy of the controller. As anticipated, a single stage has higher power efficiency than two stages. The THD item indicates that, as a result of a steadier DC level voltage control effect, the AC side of two stages has fewer non-characteristic and lower order harmonics than the single stage. It shows that the MPPT accuracy of the two topologies differs not significantly.

2.2 Inverters with and Without Transformers

Compared to transformer topologies, transformer-less inverter are more efficient and less expensive; however, they require additional circuitry to address the issue of DC current injection. Transformer-less topology also have the drawback of lacking galvanic isolation between the PV array and the utility grid, which can lead to voltage variations between the PV module and the ground depending on the inverter circuit. Table 16.3 lists the distinctions between with transformer and without transformer.

Table 16.3 Distinctions between with transformer and without transformer inverters (Zeb et al. 2018)

Inverter	Based on Line-frequency transformer	Based on High-frequency transformer	Transformer less
Advantages	Secure due to galvanic isolation, high reliability, simpler design	High efficiency	Compact design, low weight, and high efficiency
Disadvantages	High weight and volume, low efficiency	Complex and Costly technology	Additional safety measures essential

One of the primary drawbacks of one-stage PV topology is that, especially in lower power applications (such as AC-module inverters), their output voltage range for PV panels and strings are restricted, which will lower overall efficiency.

3. Conclusion

A study concludes that the single-stage PV system offers slightly higher overall efficiency compared to the double-stage system, primarily due to its simpler design and reduced component count. The single-stage configuration is identified as more cost-effective, as it requires fewer components and lower initial investment. However, this comes at the expense of a larger DC bus capacitor, which may affect the overall system size. The two-stage system maintains a more stable DC bus voltage due to the boost converter's role, which helps in reducing oscillations and improving power quality.

This stability is crucial for maintaining consistent performance under varying load conditions. Additionally, the choice of transformer has a significant impact on inverter design.

References

1. Aisyah Mohd Yusof, Nur, and Zaipatimah Ali. (2019). "Review of Active Synchronization for Renewable Powered Microgrid." International Journal of Engineering & Technology 8 (1.7 SE-Articles): 14–21. https://doi.org/10.14419/ijet.v8i1.7.25950.
2. ElNozahy, M S, and M M A Salama. (2013). "Technical Impacts of Grid-Connected Photovoltaic Systems on Electrical Networks—A Review." Journal of Renewable and Sustainable Energy 5 (3).
3. Gawhade, Pragya, and Amit Ojha. (2021). "Recent Advances in Synchronization Techniques for Grid-Tied PV System: A Review." Energy Reports 7:6581–99.
4. Haddadi, Amir Mousa, Shahrokh Farhangi, and Frede Blaabjerg. (2019). "A Reliable Three-Phase Single-Stage Multiport Inverter for Grid-Connected Photovoltaic Applications." IEEE Journal of Emerging and Selected Topics in Power Electronics 7 (4): 2384–93. https://doi.org/10.1109/JESTPE.2018.2872618.
5. Israni, Riaz K, Renu Yadav, Rahul Singh, and B Rajagopal Reddy. (2023). "Power Quality Enhancement in Wind-Hydro Based Hybrid Renewable Energy System by Interlocking of UPQC." In 2023 3rd International Conference on Energy, Power and Electrical Engineering (EPEE), 112–23. https://doi.org/10.1109/EPEE59859.2023.10352036.
6. Karuppiah, Natarajan, Mounica, Patil, Bhanutej, J.N., Saravanan, S., Reddy, Rakshith, and Israni, Riaz. (2024). "Revolutionizing Renewable Energy Integration: The Innovative Gravity Energy Storage Solution." E3S Web of Conf. 547:3028. https://doi.org/10.1051/e3sconf/202454703028.
7. Kumar, Ajay, Dr Gupta, and Vikas Gupta. (2017). "A Comprehensive Review on Grid-TiedSolar Photovoltaic System." Journal of Green Engineering 7:213–54. https://doi.org/10.13052/jge1904-4720.71210.
8. Maurya, Jyoti, and R K Saket. (2023). "Performance Analysis of Single-Stage and Two-Stage VSI-Fed IM Drive for Solar Pump Irrigation Systems." In 2023 IEEE IAS Global Conference on Renewable Energy and Hydrogen Technologies (GlobConHT), 1–6. https://doi.org/10.1109/GlobConHT56829.2023.10087514.
9. Murillo-Yarce, Duberney, José Alarcón-Alarcón, Marco Rivera, Carlos Restrepo, Javier Muñoz, Carlos Baier, and Patrick Wheeler. (2020). "A Review of Control Techniques in Photovoltaic Systems." Sustainability 12 (24). https://doi.org/10.3390/su122410598.
10. Panigrahi, Ramanuja, Santanu K Mishra, and S C Srivastava. 2018. "Grid Integration of Small-Scale Photovoltaic Systems-A Review." In 2018 IEEE Industry Applications Society Annual Meeting (IAS), 1–8. https://doi.org/10.1109/IAS.2018.8544503.
11. Parekh, Chiragkumar, and Riaz Israni. (2024). "Power Quality Enhancement in Hydroelectric-Solar PV-Based Hybrid System by Exploiting CPD." International Journal of Smart Grid and Green Communications 2:231–48. https://doi.org/10.1504/IJSGGC.2024.10065799.
12. Xiao, Weidong, Mohamed S El Moursi, Omair Khan, and David Infield. (2016). "Review of Grid-Tied Converter Topologies Used in Photovoltaic Systems." IET Renewable Power Generation 10 (10): 1543–51. https://doi.org/https://doi.org/10.1049/iet-rpg.2015.0521.
13. Zeb, Kamran, Waqar Uddin, Muhammad Adil Khan, Zunaib Ali, Muhammad Umair Ali, Nicholas Christofides, and H J Kim. (2018). "A Comprehensive Review on Inverter Topologies and Control Strategies for Grid Connected Photovoltaic System." Renewable and Sustainable Energy Reviews 94:1120–41. https://doi.org/https://doi.org/10.1016/j.rser.2018.06.053.
14. Zhu, YongLi, JianGuo Yao, and Di Wu. (2011). "Comparative Study of Two Stages and Single Stage Topologies for Grid-Tie Photovoltaic Generation by PSCAD/EMTDC." In 2011 International Conference on Advanced Power System Automation and Protection, 2:1304–9. https://doi.org/10.1109/APAP.2011.6180580.

Emerging Perspectives and Applications of Computational Intelligence and Smart Systems
– Dr. Amit Lathigara et al. (eds)
© 2026 Taylor & Francis Group, London, ISBN 978-1-041-20965-2

17

Advancements in Automatic Load Frequency Control with Integration of Renewable Energy Sources: A Comprehensive Review

Parin H. Chauhan*

Ph.D. Scholar, Department of Electrical Engineering,
Faculty of Technology, R K University,
Rajkot, Gujarat, India
Lukhdhirji Engineering College, Morbi,
Gujarat, India

Riaz Kurbanali Israni, Chiragkumar Parekh

Department of Electrical Engineering, Faculty of Technology,
R K University, Rajkot, Gujarat, India

■ **Abstract:** In a large power system, the role of Automatic load frequency control (ALFC) is to ensure stable operation of power system with controlling system frequency against the change in load. With higher load demand and more integration of Renewable Energy Sources (RES) in to the grid, frequency control faces more complex challenges which further affect the system stability. As ALFC is a dynamic control, the intermittent nature of RES causes more frequency deviations and hence correct assessment of frequency deviations becomes important. The main representation in this paper is a detailed review for the advancement in ALFC, considering conventional, current, as well as advanced electric power system with all constraints, the use of semi-conductor, electrical storage system (ESSs) for ALFC methodology has been considered with various grids. The literature review addresses various challenges attached with RES inclusion and provides future directives for ALFC in power system that bridge the research gap associated with ALFC control with RES integration. Various control methodologies including virtual inertia, adaptive control, artificial intelligence (AI)-based approaches, and predictive control techniques, are explored to enhance system performance in terms of frequency deviations.

■ **Keywords:** Automatic load frequency control, Renewable energy sources

*Corresponding author: chauhanparin@gmail.com

DOI: 10.1201/9781003725046-17

1. INTRODUCTION

An interconnected power system deals with main two types of controls: Voltage control and Frequency control. First is controlled locally as the voltage level is not the same for generation, transmission or distribution and hence it is called local control whereas the other is controlled globally as the frequency remains same throughout the system. As the load demand is increasing day by day the inclusion RES with Automatic Load Frequency Control (ALFC) is important for having stable, reliable, and sustainable operation of advanced interconnected power systems (Shankar et al. 2017). However, with the advancement in high inclusion of RES, like PV and wind, frequency control becomes more complex due to variable nature of these RES. Conventional power plant like thermal has its own inertia which helps in controlling the frequency but these RES do not have such inertial support which causes problems in ALFC. To eliminate such issues, modern ALFC strategies includes advanced control methods such as artificial intelligence (AI), machine learning (ML), fuzzy logic, and adaptive predictive control to dynamically adjust generation and maintain system balance(Shankar et al. 2017).

2. AUTOMATIC LOAD FREQUENCY CONTROL (ALFC)

The basic diagram shown here for ALFC consists of various components works in coherence to maintain power system stability in terms of frequency deviations against the change in load demand. When load changes the frequency of the generator will change and is detected by the speed governor. This error signal will be processed by the ALFC controller. This controller sends corrective signals to the speed-governor via linkage mechanisms to restore the balance by adjusting the turbine power output (Shahgholian, Shafaghi, and Mahdavi-Nasab 2010). Prime-mover in ALFC can be of two types: Tandem Compound and Cross Compound. Most of the ALFC analysis are carried out by researchers have taken the approximated prime-mover models for ALFC which will cause inaccurate assessment of frequency deviations which further results in to unstable operation of the system (Chauhan et al. 2013).

Table 17.1 Comparison of advanced and approximated prime-mover model (Shahgholian, Shafaghi, and Mahdavi-Nasab 2010)

Features	ALFC with Advanced Prime-Mover Model	ALFC with Approximated Prime-Mover Model
Response Time	Slower response	Faster response
Efficiency	Higher efficiency due to reheating	Lower efficiency
Governor Action	More complex	Simpler
Frequency Control Performance	Slower initial response but more stable operation over long durations.	Quick response but frequency deviations under load changes.
Suitability	Best suited for largepower plants	Suitable for peak-load plants
Load Following Capability	Moderate	High
Power System Stability	Better long-term stability	May require additional damping techniques

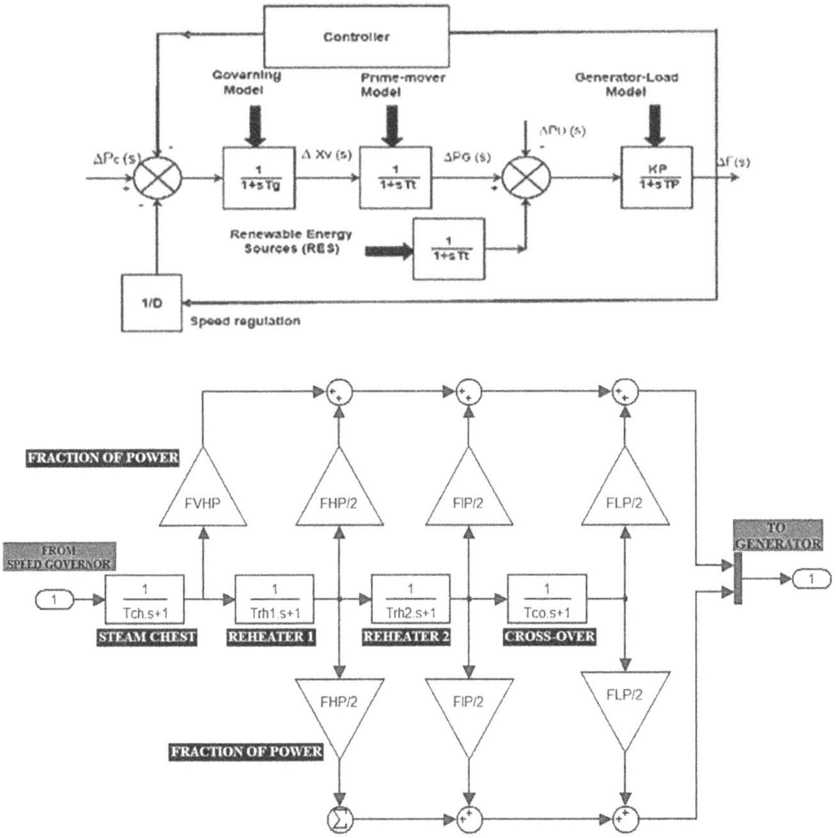

Fig. 17.1 Block diagram of ALFC & generic model of prime-mover for ALFC

Source: Authors

3. CHALLENGES AND LIMITATIONS OF RES INCLUSION IN ALFC

With rapidly increasing load demand it becomes almost compulsion to integrate all available RES in the system (Israni et al. 2023) which do not have inertial support. Hence frequency deviations will be more in such cases which further affects the system stability. Total generation is considered by taking into the account the power from thermal as well as PV (Parekh and Israni 2024). The error signal is given to generator and load model which will generate the change in frequency

Table 17.2 RES vs conventional generators (Brahmendra Kumar et al. 2019)

Factors to be considered	RES (Solar)	RES (Wind)	Conventional Generator
Variations	High	Less	Minimum
Expense	High	Moderate	Less
Maintenance requirements	Almost Zero	Very much	Medium
Inertia	Zero	Less	Higher
Power Capacity	Lesser	Very Less	Higher

signal(Kerdphol et al. 2018). But more integration of RESs can cause severe and serious frequency control challenges. RES needs battery storage for effective integration into Automatic Load Frequency Control (ALFC) for controlling the frequency. Following is the comparison of RES with synchronous generators(Brahmendra Kumar et al. 2019):

Key challenges observed in (Brahmendra Kumar et al. 2019) are as follow: In ALFC initial few seconds are the time when inertial response comes into the picture and supports frequency control but with RES this control is difficult to achieve naturally. When load disturbance occurs, the frequency starts to change(Karuppiah et al. 2024). Secondly, with inclusion of RES in the system the grid voltage will start to decrease which may cause system instability. In other research (Sharma and Saikia 2015) ALFC with three area system and thermal with solar in one area is considered with single reheat turbine. The PI & PID are compared with and without considering both solar and thermal power plants. In this research grey wolf optimizer technique for the control is used in ALFC (Sharma and Saikia 2015).

◢ 4. POSSIBLE SOLUTIONS FOR CHALLENGES & LIMITATIONS OF RES INCLUSION IN ALFC

The virtual inertia delivers the power that is required during inertial response of ALFC. It will mimic the nature of conventional generators. Various algorithms have been tried in (Skiparev et al. 2021) for virtual inertia with different control strategies like PI, PID, Fuzzy, Robust H infinity and other AI controls algorithms. In the analysis of (Faragalla et al. 2022) virtual inertia control is advanced by using WHALE algorithm which is optimized with different parameters and the uncertainties of the rotational inertia of the generation. Research made by author shows the importance of proposed technique with various disturbance applied in the ALFC (Faragalla et al. 2022). This can be summarize in following table. In (Hajiakbari Fini and Hamedani Golshan 2018) the researcher

Table 17.3 Comparison of virtual inertia control algorithms (Faragalla et al. 2022; Kerdphol, Rahman, and Mitani 2018; Skiparev et al. 2021)

Algorithm	Applications	Advantage	Drawbacks	Complexity	Robustness
Classical	H - ∞ (Robust)	More sound frequency control& Solid Peak Reduction	Significant peak during connection disturbance & order reduction needed	Medium	High
	Diagram (Coefficient) Method	Very strong robustness & order reduction doesn't required	Limited robustness	Medium	High
Advanced	Fuzzy Method	Flexible reaction	Limited adaptation & long computational time	High	High
	AI Learning (Reinforcement Method)	Advanced feedback system & high robustness	More sample data needed	Very high	Very high
Hybrid	PI/PID & PSO	Numerically Simple analysis	Slow Convergence& limited robustness	Low	Low
	Model predictive control	More sound & fast predication and optimization	Complex optimization	High	High

has analysed an improvement in the frequency stability of micro grid having low inertia system. In the analysis they have tuned the inertia with ALFC and proposed more advanced ALFC with RES integration where the overall inertia of the system is very less (Mokhtar et al. 2022; Masikana, Sharma, and Sharma 2024).

5. Conclusion

Integration of renewable energy sources (RES) leads to reduced system inertia which makes the automatic load frequency control (ALFC) more difficult and complex. To have correct and accurate assessment of frequency deviations with and without the integration of RES, approximation in the prime-mover analysis can be eliminated. Furthermore, the inertial support can be obtain using virtual inertia techniques with advanced controllers during the RES integration in order to support the frequency control. These advancement in ALFC helps the power system to have stable and reliable operations.

References

1. Brahmendra Kumar, G. V., Ratnam Kamala Sarojini, K. Palanisamy, Sanjeevikumar Padmanaban, and Jens Bo Holm-Nielsen. (2019). "Large Scale Renewable Energy Integration: Issues and Solutions." Energies 12 (10). https://doi.org/10.3390/en12101996.
2. Chauhan, Parin, Viren Pandya, Jaypalsinh Chauhan, and Rahul Karangia. (2013). "Simulation & Analysis of ALFC with Higher Order Prime-Mover Models for Single Control Area." (2013) International Conference on Energy Efficient Technologies for Sustainability, ICEETS 2013, 1084–89. https://doi.org/10.1109/ICEETS.2013.6533538.
3. Faragalla, Asmaa, Omar Abdel-Rahim, Mohamed Orabi, and Esam H. Abdelhameed. (2022). "Enhanced Virtual Inertia Control for Microgrids with High-Penetration Renewables Based on Whale Optimization." Energies 15 (23): 1–18. https://doi.org/10.3390/en15239254.
4. Hajiakbari Fini, Masoud, and Mohamad Esmail Hamedani Golshan. (2018). "Determining Optimal Virtual Inertia and Frequency Control Parameters to Preserve the Frequency Stability in Islanded Microgrids with High Penetration of Renewables" Electric Power Systems Research 154:13–22. https://doi.org/10.1016/j.epsr.2017.08.007.
5. Israni, Riaz, Renu Yadav, Rahul Singh, and B Reddy. (2023). "Power Quality Enhancement in Wind-Hydro Based Hybrid Renewable Energy System by Interlocking of UPQC." In , 112–23. https://doi.org/10.1109/EPEE59859.2023.10352036.
6. Karuppiah, Natarajan, Patil Mounica, J. N. Bhanutej, S. Saravanan, Rakshith Reddy, and Riaz Israni. (2024). "Revolutionizing Renewable Energy Integration: The Innovative Gravity Energy Storage Solution." E3S Web of Conferences 547. https://doi.org/10.1051/e3sconf/202454703028.
7. Kerdphol, Thongchart, Fathin Saifur Rahman, and Yasunori Mitani. (2018). "Virtual Inertia Control Application to Enhance Frequency Stability of Interconnected Power Systems with High Renewable Energy Penetration." Energies 11 (4). https://doi.org/10.3390/en11040981.
8. Kerdphol, Thongchart, Fathin Saifur Rahman, Yasunori Mitani, Masayuki Watanabe, and Sinan Kufeoglu. (2018). "Robust Virtual Inertia Control of an Islanded Microgrid Considering High Penetration of Renewable Energy." IEEE Access 6:625–36. https://doi.org/10.1109/ACCESS.2017.2773486.
9. Masikana, S. B., Gulshan Sharma, and Sachin Sharma. (2024). "Renewable Energy Sources Integrated Load Frequency Control of Power System: A Review." E-Prime - Advances in Electrical Engineering, Electronics and Energy 8 (May): 100605. https://doi.org/10.1016/j.prime.2024.100605.
10. Mokhtar, Mohamed, Mostafa I. Marei, Mariam A. Sameh, and Mahmoud A. Attia. (2022). "An Adaptive Load Frequency Control for Power Systems with Renewable Energy Sources" Energies 15 (2): 1–22. https://doi.org/10.3390/en15020573.

11. Parekh, Chiragkumar, and Riaz Israni. (2024). "Power Quality Enhancement in Hydroelectric-Solar PV-Based Hybrid System by Exploiting CPD" International Journal of Smart Grid and Green Communications 2:231–48 https://doi.org/10.1504/IJSGGC.2024.10065799.

12. Shahgholian, Ghazanfar, Pegah Shafaghi, and Homayoun Mahdavi-Nasab. (2010). "A Comparative Analysis and Simulation of ALFC in Single Area Power System for Different Turbines " ICECT 2010 - Proceedings of the 2010 2nd International Conference on Electronic Computer Technology, 50–54. https://doi.org/10.1109/ICECTECH.2010.5479992.

13. Shankar, Ravi, S. R. Pradhan, Kalyan Chatterjee, and Rajasi Mandal. (2017). "A Comprehensive State of the Art Literature Survey on LFC Mechanism for Power System." Renewable and Sustainable Energy Reviews 76 (February): 1185–1207. https://doi.org/10.1016/j.rser.2017.02.064.

14. Sharma, Yatin, and Lalit Chandra Saikia. (2015). "Automatic Generation Control of a Multi-Area ST - Thermal Power System Using Grey Wolf Optimizer Algorithm Based Classical Controllers." International Journal of Electrical Power and Energy Systems 73:853–62. https://doi.org/10.1016/j.ijepes.2015.06.005.

15. Skiparev, Vjatseslav, Ram Machlev, Nilanjan Roy Chowdhury, Yoash Levron, Eduard Petlenkov, and Juri Belikov. (2021). "Virtual Inertia Control Methods in Islanded Micro-grids" Energies 14 (6): 1–20. https://doi.org/10.3390/en14061562.

Emerging Perspectives and Applications of Computational Intelligence and Smart Systems
– Dr. Amit Lathigara et al. (eds)
© *2026 Taylor & Francis Group, London, ISBN 978-1-041-20965-2*

18

A Feasibility Study on The Use of Los Angeles Machine to Grind Fly Ash

Mehul Rangani*,
Sanjay Joshi, Ajaysinh Vaghela
Department of Civil Engineering, Faculty of Technology,
RK University, Rajkot, Gujarat, India

Danuta Barnat-Hunek
Faculty of Civil Engineering and Architecture,
Lublin University of Technology, 40 Nadbystrzycka St.,
20-618 Lublin, Poland

■ **Abstract:** Over the last decade, fly ash has been used more and more in concrete as a supplementary cementitious material, which has helped improve durability, sustainability, and cost-effectiveness in construction. Fly ash is a by-product of coal burning in thermal power plants and consists of fine particles mainly made up of silica, alumina, and iron oxide. Although it has its benefits, its performance in actual applications is usually constrained by particle size, morphology, and fineness that influence its pozzolanic reactivity. The existence of coarser and irregular particles limits the effectiveness of fly ash as an SCM, requiring additional processing to maximize its performance. The work examines the performance of the Los Angeles (LA) machine as a grinding device for grinding fly ash particles finer and more uniformly distributed, thus enhancing their surface area and reactivity. The Los Angeles machine, normally employed for testing aggregate abrasion, was modified in this work to determine if it can be used to grind fly ash to a finer particle size for incorporation into concrete. The grinding operation serves to reduce coarser fly ash particles into finer ones, so they can effectively engage in pozzolanic reactions when combined with cementitious materials. Experimental runs were carried out with different parameters, such as grinding time, speed of rotation, and the amount of grinding media employed in the Los Angeles machine. Fly ash samples were ground under varying grinding conditions and their resulting particle size distribution, specific surface area, and morphology were studied. The ground fly ash was subsequently compared with unground fly ash to assess the enhancement in its fineness and reactivity. Some advanced characterization methods like laser diffraction particle size analysis and X-ray diffraction (XRD) were used to investigate the impact of grinding on the physical and chemical characteristics of fly ash. The findings of the research show that the Los Angeles machine can efficiently grind fly ash into smaller particles, greatly expanding its surface area and its pozzolanic reactivity. The ground fly ash exhibited enhanced performance as a supplementary cementitious material by encouraging

*Corresponding author: mehul.rangani@rku.ac.in

DOI: 10.1201/9781003725046-18

improved particle packing, minimizing voids in the concrete matrix, and increasing the overall strength and durability of concrete. The enhancement of fineness was responsible for a greater proportion of hydration reactions upon mixing with cement, which resulted in better early-age and long-term mechanical characteristics.

■ **Keywords:** Ball milling, Surface area, Particle size

1. INTRODUCTION

A by-product of burning coal, fly ash is obtained in huge amounts from thermal power plants. Because of its pozzolanic qualities and capacity to increase the workability and durability of concrete, it has been extensively utilized as an additional cementitious ingredient in concrete. However, the shape and size of the fly ash particles frequently limit how effective they are in concrete. By expanding its surface area and decreasing its particle size, fly ash grinding can enhance concrete's reactivity and performance.

2. LITERATURE REVIEW

The impact of grinding on the characteristics of fly ash has been the subject of numerous investigations. The literature review that follows discusses the use of several ball milling machines to grind fly ash and other waste materials to create cement mortar composite.

R. Szabo et al. (2019) checked the effect of the grinding process of fly ash on geo polymer foam concrete, in which raw fly ash was grinded in a ball mill for different time durationsof 5,10,20,30,60 and 120 minutes. The results of foam characterization and the rheology of the polymer paste show the change in cell structure and physical characteristics of geopolymer foam. Hitesh, Reeta Wattal, et al. (2018) examined the feasibility of law-energy ball milling for the development and characterization of coal fly ash and found the method economical, reliable, low maintenance, and easy to operate. L. Krishnaraj and P.T. Ravichandran (2019) the effectiveness of fly ash as a substitute waste material for manufacturing cement and found that because of the ultrafine nature of ground particles, the end concrete shows great mechanical properties as compared to ordinary. The replacement percentages were taken as high as 30% of cementitious material with a change in surface area because grinding was from $727 \, cm^2/g$ to almost $3500 \, cm^2/g$ specific surface area. S.K. Nath and Sanjay Kumar in 2020 identified the effect of particle fineness on microstructure and mechanical properties of geopolymer concrete. To check the same, they grinded the samples for 30 and 60 minutes in ball milling. The results show improved microstructure and mechanical properties of geopolymer concrete. Again in 2021. L. Krishnaraj and P.T. Ravichandran checked the feasibility of the grinded fly ash for masonry construction work and found that the grinding process significantly changed the specific surface area and particle size of fly ash particles without changing chemical composition. Also, they found improved microstructure of mortar with fewer micro cracks. Bao Liu et al. (2022) highlighted the nonlinear relation between specific surface area and fineness and their effect on mechanical and durability properties by using iron tailing powder. They also highlighted how higher specific surface area and fineness level increase pozzolanic activities within the concrete and change the microstructure of the composite. Yingbin Wang and Yang Li et al. (2022) further elaborated the study by considering different grinding methods for utilizing waste glass powder in cement composites.

They checked the samples with three conditions i.e., ethyl alcohol condition, dry condition, and wet condition. They found the relation between grinding condition and milling efficiency. At last, in 2023 YU Xuan Liew et al. identified the feasibility of using the LOS Angeles abrasion machine as an alternative to ball milling and for manufacturing of nano palm oil fuel ash. Also, they identified the effect of adding the same in mortar mixture. They identified that with a slower speed of grinding, the LA abrasion machine is grinding the particles sufficiently with higher rotations.

3. EXPERIMENTAL INVESTIGATION

3.1 Material Preparation by Using Los Angeles Machine

The experimental investigation was carried out using a Los Angeles machine, which is a type of abrasive grinding machine. The machine consists of a steel drum with a series of steel balls and a grinding charge. The fly ash was grided for different durations, ranging from 1 to 5 hours, then intervals of 5 hours from 5 to 20 Hours. In this research work the influence of grinding time over particle size and surface area of fly ash. The sample has been collected with the time of specified interval and it is tested by the advanced testing and the results are concluded. The few samples result in 1-hour griding and 5 hours griding are indicated below in Fig. 18.1 & Table 18.1 and Fig. 18.2 & Table 18.2 and other grinding hours results are indicated in Table 18.3. To measure the particle size, the samples of varying griding times were analyzed and tested using a Master-Sizer 3000 particle size analyzer. The particle size was tested using the dry technique.

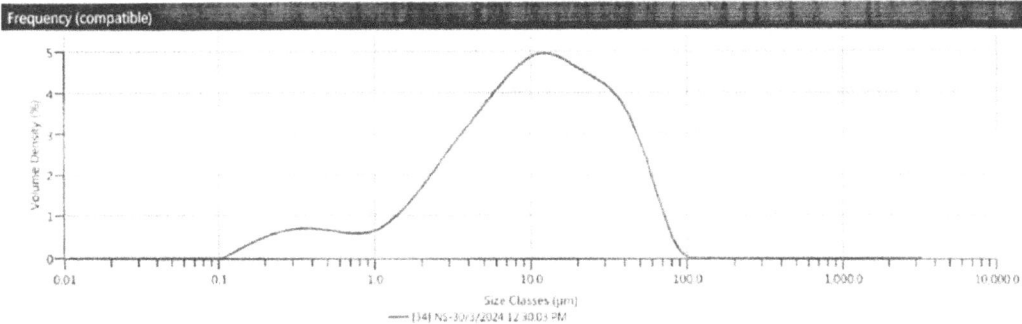

Fig. 18.1 PSD histogram of 1 hours grided industrial ash

Source: Sprint testing laboratory result

Table 18.1 Result of particle size distribution of 1 hours grided industrial ash

Size (µm)	% Volume Under
0.01 to 0.113	0
0.128 to 0.675	0.1 to 6.33
0.767 to 5.92	6.82 to 33.26
6.72 to 51.8	36.74 to 95.34
58.9 to 86.4	97.29 to 99.91
98.1 to 3500	100

Source: Sprint testing laboratory result

Figure 18.1 displays the outcome of the above histogram: Size of particle: Dv(10)- 1.47 μm toDv(100)- 97.6 μm; surface area = 2247 (m²/kg).

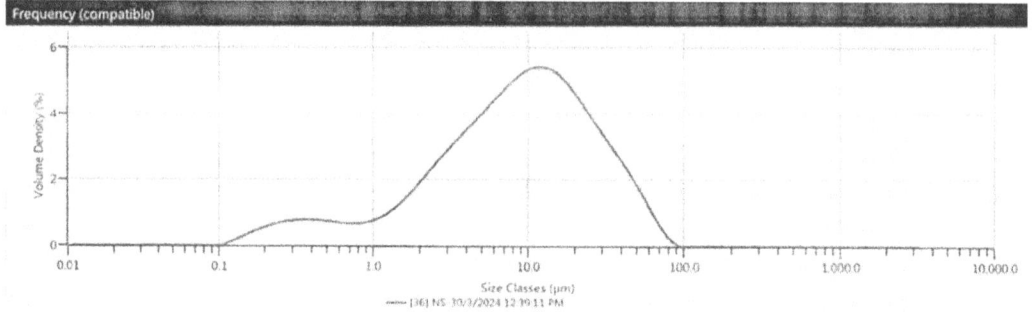

Fig. 18.2 PSD histogram of 5 hours grided industrial Ash

Source: Sprint testing laboratory result

Table 18.2 Result of particle size distribution of 5 hours grided industrial ash

Size (μm)	% Volume Under
0.01 to 0.113	0
0.128 to 0.675	0.11 to 6.99
0.767 to 5.92	7.55 to 36.83
6.72 to 51.8	40.59 to 97.26
58.9 to 76	98.5 to 99.81
86.4 to 3500	100

Source: Sprint testing laboratory result

Figure 18.2 displays the outcome of the above histogram: Size of particles: Dv(10)-1.25 μm toDv(100)-86.3 μm; surface area = 2468 (m²/kg).

4. RESULTS AND DISCUSSION

Table 18.3 displays the result of the experimental study and illustrates how grinding time affects fly ash's particle size and surface area.

Table 18.3 Physical properties of various grinding times of industrial ash

	Griding Time (Hours)							
Properties	1	2	3	4	5	10	15	20
Specific surface area (m²/kg)	2247	2281	2296	2392	2468	3106	3647	4201
Particle Size(μm) Dv(10) to Dv(100)	1.47 to 97.6	1.45 to 92.2	1.44 to 86	1.36 to 86.20	1.25 to 86.3	0.776 to 35.3	0.583 to 31	0.385 to 26.2

Source: Sprint testing laboratory result

The findings demonstrate that the Los Angeles machine significantly reduces particle size and increases surface area when fly ash is ground. Fly-ash surface area increases from 2247 m²/kg to 4201 m²/kg, but its particle size decreases from 1.47 to 97.6 µm to 0.385 to 26.2 µm.

5. CONCLUSION

The feasibility of grinding fly ash with the Los Angeles machine is well explored within this study. The results present that the particle size of the fly ash was effectively reduced with the machine, and its surface area was greatly amplified, thereby activating its pozzolanic potential. The particle size of fly ash is reduced from 1.47 to 97.6 µm to 0.385 to 26.2 µm, while the surface area is increased from 2247 m²/kg to 4201 m²/kg. This enhancement in fineness directly leads to improved performance in concrete application through increased particle packing, increased possibilities of cementitious reaction, and enhanced overall strength and durability. The study also offers insights into how to optimize grinding parameters like rotation rate, grinding time, and media type to yield the most effective processing. These findings form a basis for more studies to further the use of fly ash in sustainable building materials.

References

1. Yu Xuan Liew et.all (2023), "Suitability of Using LA Abrasion Machine for the Nano Manufacturing of Palm Oil Fuel Ash and Incorporating in Mortar Mixture", The Open Civil Engineering Journal.
2. Yingbin Wang, Yang Li et.all (2022), "Preparation of waste glass powder by different grinding methods and its utilization in cement-based materials, Advanced Powder Technology 33.
3. Bao Liu, Haining Meng et.all (2022)," Relationship between the fineness and specific surface area of iron tailing powder and its effect on compressive strength and drying shrinkage of cement composites.
4. Shubham Sharma, Vikas Patyal,(2022) "Mechanical, morphological, and fracture deformation behavior of MWCNTs-reinforced (Al–Cu–Mg–T351) alloy cast nanocomposites fabricated by optimized mechanical milling and powder metallurgy techniques", Nanotechnology Reviews.
5. Qiaoyi Han, Peng Zhang et.all,(2022)"Comprehensive review of the properties of fly ash-based geopolymer with additive of nano-SiO2", Nanotechnology Reviews.
6. L. Krishnaraj, P.T. Ravichandran (2021), "Characterization of ultra-fine fly ash as a sustainable cementitious material for masonry construction", Ain Shams Engineering Journal 12.
7. Hitesh, Reeta Wattal et.all, 2021" Development and characterization of coal fly ash through low-energy ball milling", Materials Today: Proceedings.Volume 47, Part 11, Pages 2970–2975.
8. S.K. Nath, Sanjay Kumar (2020), "Role of particle fineness on engineering properties and microstructure of fly ash derived geopolymer", Construction and Building Materials 233.
9. L. Krishnaraj, P.T. Ravichandran (2019)," Investigation on the grinding impact of fly ash particles and its characterization analysis in cement mortar composites", Ain Shams Engineering Journal.
10. Fei Cheng, Ying Feng et.all, (2019),"Practical strategy to produce ultrafine ceramic glaze: Introducing a polycarboxylate grinding aid to the grinding process", Advanced Powder Technology.
11. Hussein M Hamada1, Gul Ahmed Jokhioet.all.(2018)," Applications of Nano palm oil fuel ash and Nano fly ash in concrete", IOP Conference Series: Materials Science and Engineering.
12. R. SZABÓ et.all (2017), "Effect of Grinding Fineness of Flyash on the properties of Geopolymer Foam", Archives of Metallurgy and Materials.

Emerging Perspectives and Applications of Computational Intelligence and Smart Systems
– Dr. Amit Lathigara et al. (eds)
© *2026 Taylor & Francis Group, London, ISBN 978-1-041-20965-2*

19

Effective Analysis of Mental Health Using AI and Machine Learning in Social Media Analytics

M. Lavanya
Department of Computer Science and Engineering,
Sri Venkateswara College of Engineering & Technology,
Chittoor, Andhra Pradesh, India

Parvez Belim*, Nirav Bhatt
School of Engineering, RK University, Rajkot, Gujarat, India

Rani Sahu
Department of Computer Science, IES College of Technology,
Bhopal, Madhya Pradesh, India

Neha Chauhan
School of Engineering, RK University, Rajkot, Gujarat, India

■ **Abstract:** More than 300 million people worldwide suffer from depression, necessitating advanced AI-driven approaches for early detection. Social media enables mental health analysis by extracting emotional patterns from digital content. This study applies machine learning (ML) to identify depressive tendencies in social media text. A curated data set labeled with emotional categories—angry, neutral, surprised, happy, and sad—supports a predictive framework integrating sentiment analysis, NLP, and multi-model fusion. The proposed model, combining Recurrent Neural Networks (RNN) and Long Short-Term Memory (LSTM) with fully connected layers, achieves 99.0% accuracy, minimizing misclassification. Comparative analysis demonstrates its superiority over existing deep learning models. Additionally, this research synthesizes contemporary depression detection techniques, datasets, and classification methodologies. The findings emphasize AI's potential in scalable, non-invasive mental health assessments while addressing ethical concerns in digital health surveillance and AI-driven psychological interventions. This study highlights AI's transformative role in mental health analytics, enabling early diagnosis and timely intervention.

■ **Keywords:** Mental health, Depression, Social media, Emotional

*Corresponding author: parvez.belim@rku.ac.in

DOI: 10.1201/9781003725046-19

1. INTRODUCTION

Mental health issues, including depression, anxiety, and stress, are global challenges that significantly impact individuals' quality of life. Traditional diagnostic and therapeutic approaches often face barriers such as stigma, accessibility, and the subjective nature of self-reporting. With the ubiquity of social media platforms like Twitter, Reddit, and Instagram, a new avenue for understanding mental health has emerged. Users' posts, comments, and interactions often contain valuable signals indicative of their emotional well-being.

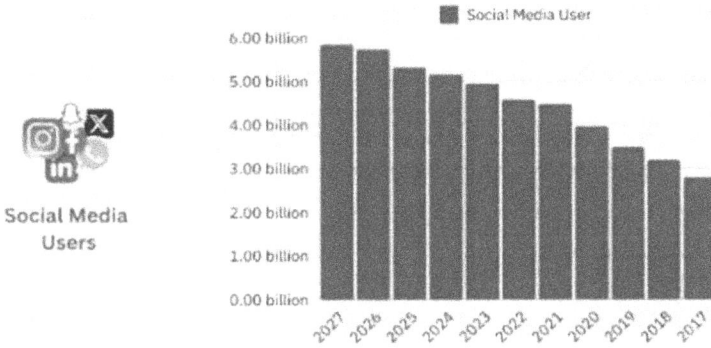

Fig. 19.1   Social media users (in billion)

Source: https://prioridata.com/data/social-media-usage/

2. LITERATURE REVIEW

Artificial Intelligence (AI) has emerged as a transformative tool in mental health analysis, offering promising solutions for early detection, monitoring, and intervention. This literature review explores key studies that have employed AI models for mental health assessment, leveraging diverse datasets and machine learning techniques.

Table 19.1   Literature review

Title	Authors	ModelUsed	Dataset	Accuracy
[1]	Amanatetal. (2022)	LSTM,RNN	Tweets-ScrapedDepression Dataset	98%
[2]	Nusratetal.(2024)	BERT	TwitterDataset	96%
[3]	Baueretal.(2024)	(LLMs), BERT	RedditPosts	N/A
[4]	Rafiquletal.(2018)	DT,SVM, KNN	Facebookusercomments	85%
[5]	Safaetal. (2023)	ML,DL	Twitter,Reddit,Facebook	N/A
[6]	LinandZhang (2021)	LR,KNN,RF	sociodemographicdata	89%
[7]	Kannanetal. (2024)	SVM,DT, LR	Behavioralassessments	N/A
[8]	Mansooretal(2024)	BERT, LSTM	Twitter,Reddit,Facebook	89.3%
[9]	Shenetal.(2017)	MDL	Twitter	88%
[10]	Kasannenietetal. (2024)	BERT,CNN	Reddit,Twitter,andothers	96.4%

Source: Authors

3. PROPOSED METHODOLOGY

A social media dataset from Kaggle was refined, balanced, and preprocessed using segmentation, stemming, and lemmatization. Word-scoring techniques were applied for text analysis before training machine learning models to classify tweets as depressed or non-depressed.

The data set was split into training and evaluation subset stone sure accurate learning and validation. The methodology follows five key stages: Data Acquisition, Preprocessing, Feature Engineering, Model Training, and Evaluation.

The Fig. 19.2 proposed model outlines a pipeline for analyzing mental heal through social media using machine learning. It starts with data extraction, followed by preprocessing steps such as cleaning, annotation, and text normalization. Feature engineering involves tokenization, word embeddings, stemming, and encoding. The dataset is then partitioned for training and testing, with classification performed using models like SVM, Naïve Bayes, Random Forest, CNNs, LSTM, and BERT. Finally, model performance is assessed using key evaluation metrics to generate mental health predictions.

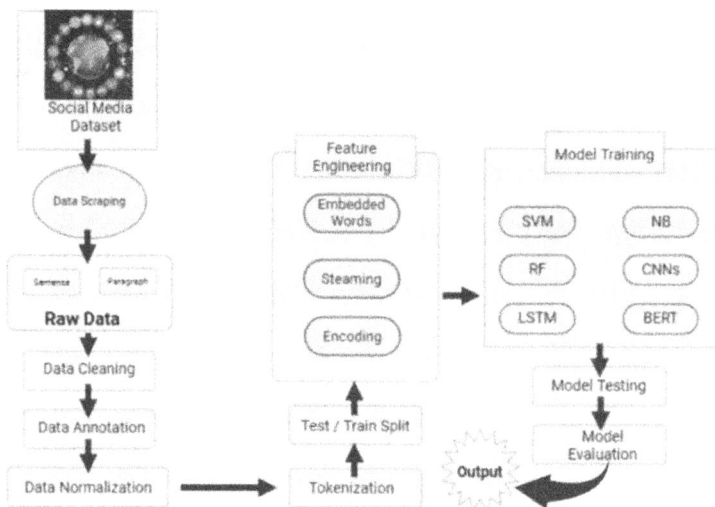

Fig. 19.2 Proposed methodology

Source: Authors

Examples of Depressive Sentences

1. "If also empty inside, and nothing seems to help."
2. "Noon cares about me; I' mall alone."

Examples of Depressive Tweets

1. "I'm so tired of pretending I'm ok when I'm falling apart inside."
2. "Everyday feels like a battle I'm losing. I don't know how much longer I can do this."

4. PRINCIPAL COMPONENT ANALYSIS FOR DATA VISUALIZATION

PCA is dimensionality reduction technique that we used for feature visualization. In order to optimize inter-class scattering and reduce inner scattering of samples, PC A uses the covariance matrix to conduct decomposition and create eigenvalues. The one that the dispersion of matrices inside and across classes is depicted in equations (1) and (2).

$$F_D = \frac{1}{n-1} \sum_{i=1}^{n} \left(X_i - \bar{X}_j \right) \left(X_i - \bar{X}_j \right)^T \tag{1}$$

$$F_E = \frac{1}{n-1} \sum_{i=1}^{n} \sum_{x_k \in C_i} \left(X_i - \bar{X}_j \right) \left(X_i - \bar{X}_j \right)^T \tag{2}$$

Figure 19.3 presents training vs. validation accuracy (left) and training vs. validation loss (right) over epochs. The increasing training accuracy and decreasing loss indicate effective model learning. Validation accuracy initially improves but later declines, while validation loss rises, suggesting overfitting—strong performance on training data but poor generalization to new data. For mental health analysis in social media, this highlights limited generalizability. Strategies like early stopping, regularization, and data augmentation can enhance model robustness.

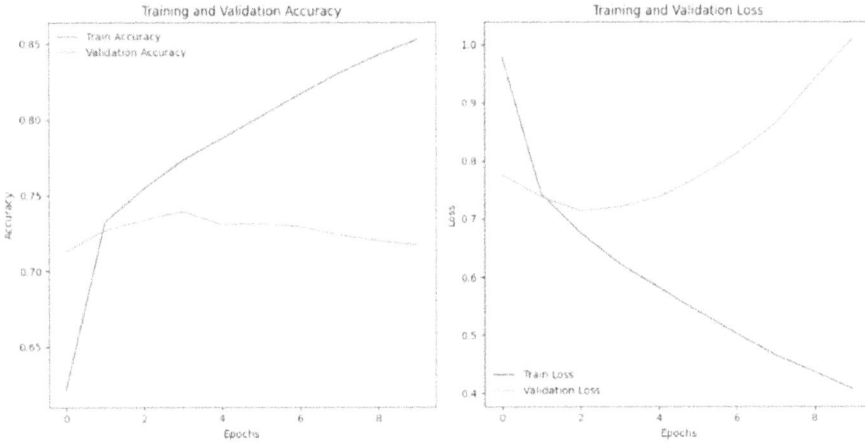

Fig. 19.3 Training and validation performance of model

Source: Authors

Figure 19.4 is a 3D surface plot illustrating the classification of mental health states (Depressed vs. Non-Depressed) using PCA. The three axes represent principal components from social media analytics data, while the colored surface indicates feature variations. Labels highlight regions where AI models categorize data points as Depressed or Non- Depressed.

Figure 19.5 highlights strong diagonal elements, indicating high prediction accuracy for each class (e.g., "Neutral" with 3,646 correct predictions, "Happy" with3, 273). Some Misclassifications occur, such as "Neutral" mislabeled as "Happy" (521 times) and Angry" as "Neutral" (201 times). Notably, the model never predicted the "Surprise" class, suggesting possible class imbalance or mislabeling.

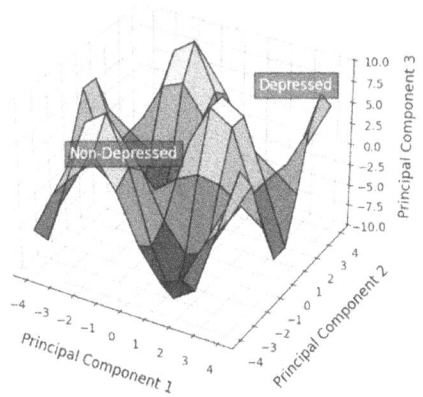

Fig. 19.4 PCA on a rich emotional state feature, a 3-D plot

Source: Authors

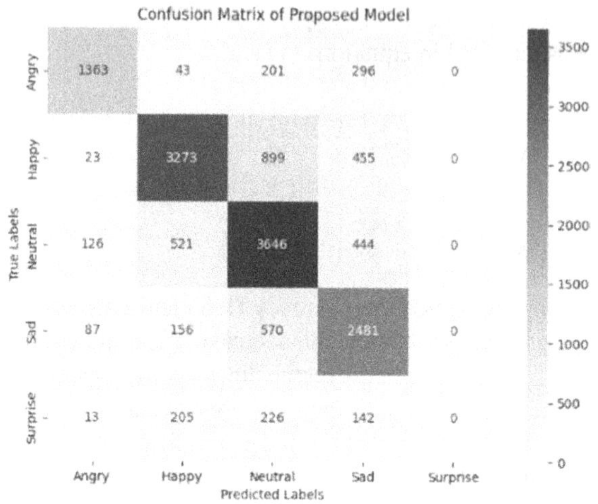

Fig. 19.5   Confusion matrix of proposed model

Source: Authors

5. Conclusion

This research presents a deep learning framework using an RNN-LSTM architecture for automated depression detection from text. The data set underwent preprocessing, including stemming, lemmatization, and PCA for feature extraction. Trained on a large Kaggle dataset, the model achieved 99% accuracy while minimizing false positives. Comparative analysis with Naïve Bayes, SVM, CNN, and Decision Trees showed significant improvements in accuracy, precision, recall, and F1-score. The findings highlight deep learning's potential for scalable mental health assessment. Future research will integrate hybrid recurrent networks and expand datasets to enhance emotional pattern recognition.

References

1. Amanat, A., Rizwan, M., Javed, A.R., Abdelhaq, M., Alsaqour, R., Pandya,S., & Uddin, M. (2022). Deep learning for depression detection from textual data. *Electronics*, *11*(5), 676.

2. Nusrat, M. O., Shahzad, W., & Jamal, S. A. (2024). Multi Class Depression Detection Through Tweets using Artificial Intelligence. *arXiv preprint arXiv:2404.13104*.

3. Bauer, B., Norel, R., Leow, A., Rached, Z. A., Wen, B., & Cecchi, G. (2024). Using Large Language Models to Understand Suicidalityina Social Media–Based Taxonomy of Mental Health Disorders: Linguistic Analysis of Reddit Posts. *JMIR mental health*, *11*, e57234. de Souza Filho, E. M., Rey, H. C. V., Frajtag, R. M., Cook, D. M. A., de Carvalho, L. N. D., Ribeiro, A. L. P., & Amaral, J. (2021). Can machine learning be useful as a screening tool for depression in primary care? *Journal of Psychiatric Research*, *132*, 1–6.

4. Islam, M. R., Kabir, M. A., Ahmed, A., Kamal, A. R. M., Wang, H., & Ulhaq, A. (2018). Depression detection from social network data using machine learning techniques. *Health information science and systems*, *6*, 1–12.

5. Safa, R., Edalatpanah, S. A., & Sorourkhah, A. (2023). Predicting mental health using social media: A roadmap for future development. In *Deep learning in personalized healthcare and decision support* (pp. 285–303). Academic Press.

6. Kannan, K. D., Jagatheesaperumal, S. K., Kandala, R. N., Lotfaliany, M., Alizadehsanid, R., & Mohebbi, M. (2024). Advancements in machine learning and deep learning forearly detection and management of mental health disorder. *arXiv preprint arXiv:2412.06147*. Mansoor, M.A., & Ansari, K. H. (2024). Early Detection of Mental Health Crises through Artifical-Intelligence-Powered Social Media Analysis: A Prospective Observational Study. *Journal of Personalized Medicine*, *14*(9), 958.
7. Shen, G., Jia, J., Nie, L., Feng, F., Zhang, C., Hu, T., ... & Zhu, W. (2017, August). Depression detection via harvesting social media: A multimodal dictionary learning solution. In *IJCAI* (pp. 3838–3844).
8. Kasanneni, Y., Duggal, A., Sathyaraj, R., & Raja, S. P. (2024). Effective Analysis of Machine and Deep Learning Methods for Diagnosing Mental Health Using Social Media Conversations. *IEEE Transactions on Computational Social Systems*.

Emerging Perspectives and Applications of Computational Intelligence and Smart Systems
– Dr. Amit Lathigara et al. (eds)
© 2026 Taylor & Francis Group, London, ISBN 978-1-041-20965-2

20

Enhancing Multi-Modal MRI Brain Tumor Detection using Hybrid Convolutional Based Deep Learning Model

Nirav Ranpara*

Research Scholar, School of Engineering,
RK University, Rajkot, Gujarat, India

Amit Lathigara, Chetan Shingadiya

School of Engineering, RK University, Rajkot,
Gujarat, India

Róża Dzierżak

Department of Electronics and Information Technology,
Lublin University of Technology, ul. Nadbystrzycka 38A,
20-618 Lublin, Poland

■ **Abstract:** A critical yet difficult job in medical imaging is the detection and segmentation of brain cancers from multi-modal magnetic resonance imaging (MRI), primarily because brain tumors are complicated and heterogeneous. Clinical practice is greatly limited by the time-consuming, heavily reliant on radiologist expertise, and inter-observer variability of traditional manual segmentation procedures. The low generalizability of models across varied datasets, the challenge of segmenting small tumor sub-regions, and the underutilization of multi-modal MRI data are some of the major problems that still exist despite recent advances in deep learning showing promise in automating brain tumor segmentation. In order to overcome these inadequacies, this study presents a new deep learning framework that maximizes segmentation accuracy and model robustness by merging a transformer-based architecture with a hybrid Convolutional neural network (CNN). The recommended model works remarkably well in segmenting tumor locations across a range of MRI sequences, obtaining a significant accuracy of 98.85% and a low loss of 0.02. This work's key innovation is its capacity to use hybrid deep learning architectures to harness multi-modal MRI data. This is a considerable improvement over standard CNN-based models, which frequently fall short of properly exploiting the rich, complimentary information supplied by diverse imaging modalities.

■ **Keywords:** Deep learning, Brain tumor, Data augmentation, CNN, Image processing

*Corresponding author: research4004@gmail.com

DOI: 10.1201/9781003725046-20

1. INTRODUCTION

Due to their complexity and heterogeneity, brain tumors represent some of the most challenging medical conditions to manage and can significantly affect patient outcomes. In multi-modal magnetic resonance imaging (MRI), reliable and exact tumor site identification is critical for prognosis, therapy planning, and diagnosis. Multi-modal MRI offers supplementary information on various tumor features, typically comprising T1-weighted (T1), T1-weighted contrast-enhanced (T1c), T2-weighted (T2), and fluid-attenuated inversion recovery (FLAIR) sequences. However, segmenting these photos by hand takes a lot of time, is sensitive to inter-observer variability, and relies largely on the experience of radiologists (Bakas et al., 2018; Menze et al., 2015).These challenges have necessitated the development of automated techniques for brain tumor segmentation.

In the past decade, the application of deep learning techniques for brain tumor segmentation has progressed markedly. The ability of Convolutional neural networks (CNNs) to immediately acquire hierarchical information from unprocessed images has established them as the preeminent framework. Recent advancements have introduced intricate designs such as U-Net (Ronneberger et al., 2015) and its variants to enhance segmentation precision, whereas earlier approaches employed traditional CNN structures for voxel-wise classification (Havaei et al., 2017; Tadhani & Vekariya, 2024). Moreover, by focusing on relevant tumor sites and incorporating contextual data, attention mechanisms (Bhalodiya et al., n.d.; Oktay et al., 2018) and multi-scale feature extraction methods have enhanced the efficacy of deep learning models significantly. Even with these advancements, a number of challenges still need to be handled.

2. PROPOSED APPROACH

In the first stage, brain MRI data is acquired from clinical and publically accessible datasets, including BRATS. Pre-processing is important to standardize the data because varied sources have variable image quality and acquisition methods. Data augmentation solutions are used to solve the problem of small datasets and minimize over fitting. By producing new training instances, these improvements improve the generalizability and durability of the model. Among the augmentation approaches are:

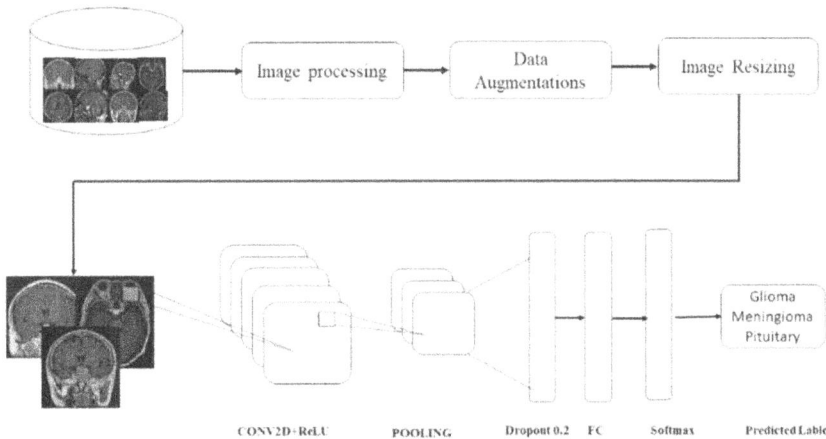

Fig. 20.1 Proposed brain tumor detection approach

Convolutional neural networks (CNNs), which are meant to automatically extract high-level information from multi-modal MRI scans, provide the foundation of the recommended system. A number of critical parts make up the CNN design, and each one increases the model's potential to recognize complicated tumor patterns. Essential picture features including borders, textures, and tumor boundaries are recognized by Convolutional layers activated by ReLU. These layers hone and extract increasingly abstract patterns that are necessary for precision tumor identification as the network becomes deeper.

3. IMPLEMENTATION

A publicly accessible brain MRI dataset containing 7,023 images(*Brain Tumor MRI Dataset*, n.d.) that have been rigorously sorted into four different classes—glioma, meningioma, pituitary tumor, and no tumor—is used to test the suggested methodology. Every photograph in the collection captures distinctive morphological abnormalities associated to different medical diseases, presenting a distinct cross-sectional view of the human brain. In order to assure extensive coverage of all tumor forms and to enable effective training, validation, and testing of the suggested deep learning model, the dataset is meticulously labelled, offering a balanced distribution across the four categories. The model's capacity to generalize over a range of inputs and increase diagnostic reliability is made feasible by the variety of imaging situations and tumor characteristics that provide a realistic description of clinical settings.

As mentioned in Fig. 20.2; by expanding the number and diversity of datasets, data augmentation improves generalization and strengthens deep learning models. To prevent the introduction of artefacts, effective augmentation must be in line with the problem and the properties of the data. Geometric changes (e.g., rotation, scaling) and intensity adjustments (e.g., brightness, contrast) are common methods in medical imaging. Carefully implementing these techniques guarantees that the augmented data will always be representative and pertinent. Accurate performance evaluation, reproducibility, and transparency all depend on proper documentation of augmentation strategies.

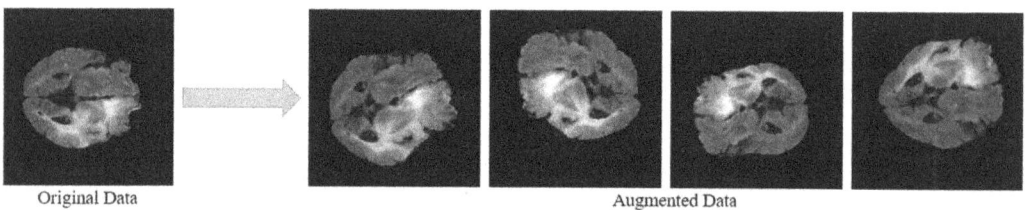

Original Data   Augmented Data

Fig. 20.2 Random sample of data augmentation

The CNN-based architecture of the recommended deep learning model is intended for the detection of cyber attacks. In order to maintain spatial information, it processes input data of the shape (240, 240, 3), expanding it to (244, 244, 3), via a Zero Padding layer. The first phase in feature extraction is the Conv2D layer (32 filters), which is followed by an Activation layer for non-linearity and Batch Normalization for stable training. In order to minimize computation and preserve critical features, two Max Pooling layers sequentially reduce the spatial dimensions to (14, 14, 32).In order to input feature maps into a Dense layer (128 neurons), a Flatten layer turns them into a 1D vector (6,272). The last Dense layer (1 neuron) handles classification, most likely employing a sigmoid activation, while a Dropout layer prevents against over fitting. The 807,937 parameters (807,873 trainable) model strikes a balance between accuracy and efficiency. Its CNN-based architecture leverages deep learning techniques to identify cyber attacks effectively.

4. RESULT AND DISCUSSION

As mentioned in Fig. 20.3, with 25 epochs, the recommended model reduced the loss to 0.002 and obtained an accuracy of 98.8%, exhibiting exceptional performance. The ROC curve and the training and testing accuracies, as presented in the figures below, support the model's good generalization to unknown data. It's amazing learning abilities and precise forecasts are demonstrated in its high accuracy and minimum loss. Furthermore, the model showed a recall of 0.99, which implies that 99% of genuine positives were discovered, and a precision of 0.98, which shows that 98% of forecast positives were accurate. These metrics demonstrate that the model is good at decreasing false negatives and discovering pertinent positives. The model has been proved to be a dependable and efficient tool with high precision, recall, and overall accuracy, making it suited for real-world applications needing outstanding classification performance.

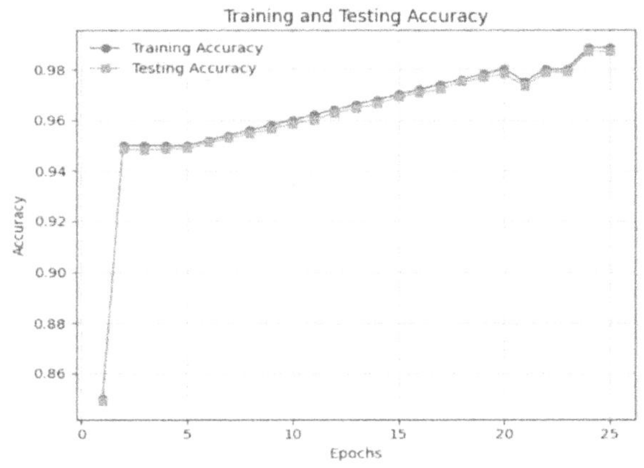

Fig. 20.3 Training and testing accuracy

5. CONCLUSION

To sum up, the suggested deep learning model for brain tumor identification has exhibited excellent performance, reaching a loss of 0.02 and an accuracy of 98.85%. These data illustrate the model's capacity to correctly categorize images of brain tumors, indicating its potential as a trustworthy diagnostic tool in medical imaging. The model's resilience was considerably strengthened by the addition of data augmentation approaches, which enabled it to generalize across a variety of tumor forms, imaging settings, and likely causes of data fluctuation. The model's capacity to learn essential features and minimize errors is evidenced by its outstanding accuracy and little loss, and its resistance to over fitting underlines its promise for practical clinical applications. The results also show that the model can be of considerable benefit to healthcare practitioners by delivering accurate and effective tumor detection, which is vital for boosting patient outcomes and diagnostic precision. All things considered, this deep learning model not only demonstrates cutting-edge performance but also adds to the burgeoning corpus of studies on the application of AI for medical diagnostics, especially with regard to the diagnosis of brain cancers.

References

1. Amin, J., Anjum, M. A., Sharif, M., Jabeen, S., Kadry, S., & Moreno Ger, P. (2022). A New Model for Brain Tumor Detection Using Ensemble Transfer Learning and Quantum Variational Classifier. *Computational Intelligence and Neuroscience, 2022*. https://doi.org/10.1155/2022/3236305
2. Arabahmadi, M., & Farahbakhsh, R. (2022). Deep Learning for Smart Healthcare: A Survey on Brain Tumor. *Sensors, 22*, 1–27.
3. Arif, M., Ajesh, F., Shamsudheen, S., Geman, O., Izdrui, D., & Vicoveanu, D. (2022). Brain Tumor Detection and Classification by MRI Using Biologically Inspired Orthogonal Wavelet Transform and Deep Learning Techniques. *Journal of Healthcare Engineering, 2022*. https://doi.org/10.1155/2022/2693621
4. Bhalodiya, D., Tadhani, J., & Computer, R. D.-I. J. of. (n.d.). Mining Recurring Patterns in Time Series. *Researchgate.Net*. Retrieved February 6, 2025, from https://www.researchgate.net/profile/Dharmesh-Bhalodiya/publication/333117438_Mining_Recurring_Patterns_in_Time_Series/links/5d036599a6fdccd130993dc1/Mining-Recurring-Patterns-in-Time-Series.pdf
5. *Brain Tumor Classification (MRI)*. (n.d.). Retrieved February 6, 2025, from https://www.kaggle.com/datasets/sartajbhuvaji/brain-tumor-classification-mri
6. *Brain Tumor MRI Dataset*. (n.d.). Retrieved February 6, 2025, from https://www.kaggle.com/datasets/masoudnickparvar/brain-tumor-mri-dataset
7. Khairandish, M. O., Sharma, M., Jain, V., Chatterjee, J. M., & Jhanjhi, N. Z. (2022). A Hybrid CNN-SVM Threshold Segmentation Approach for Tumor Detection and Classification of MRI Brain Images. *Irbm, 43*(4), 290–299. https://doi.org/10.1016/j.irbm.2021.06.003
8. Menze, B. H., Jakab, A., Bauer, S., Kalpathy-Cramer, J., Farahani, K., Kirby, J., Burren, Y., Porz, N., Slotboom, J., Wiest, R., Lanczi, L., Gerstner, E., Weber, M. A., Arbel, T., Avants, B. B., Ayache, N., Buendia, P., Collins, D. L., Cordier, N., … Van Leemput, K. (2015). The Multimodal Brain Tumor Image Segmentation Benchmark (BRATS). *IEEE Transactions on Medical Imaging, 34*(10), 1993–2024. https://doi.org/10.1109/TMI.2014.2377694
9. Tadhani, J. R., Vekariya, V., Sorathiya, V., Alshathri, S., & El-Shafai, W. (2024). Securing web applications against XSS and SQLi attacks using a novel deep learning approach. *Scientific Reports 2024 14:1, 14*(1), 1–17. https://doi.org/10.1038/s41598-023-48845-4
10. Musallam, A. S., Sherif, A. S., & Hussein, M. K. (2022). A New Convolutional Neural Network Architecture for Automatic Detection of Brain Tumors in Magnetic Resonance Imaging Images. *IEEE Access, 10*, 2775–2782. https://doi.org/10.1109/ACCESS.2022.3140289
11. Qureshi, S. A., Ahmed Raza, S. E., Hussain, L., Malibari, A. A., Nour, M. K., Rehman, A. U., Al-Wesabi, F. N., & Hilal, A. M. (2022). Intelligent Ultra-Light Deep Learning Model for Multi-Class Brain Tumor Detection. *Applied Sciences (Switzerland), 12*(8). https://doi.org/10.3390/app12083715
12. Tadhani, J. R., & Vekariya, V. (2024). A Survey of Deep Learning Models, Datasets, and Applications for Cyber Attack Detection. *AIP Conference Proceedings, 3107*(1). https://doi.org/10.1063/5.0208470/3287908

Note: All the figures in this chapter were made by the authors.

Emerging Perspectives and Applications of Computational Intelligence and Smart Systems
– Dr. Amit Lathigara et al. (eds)
© *2026 Taylor & Francis Group, London, ISBN 978-1-041-20965-2*

21

Real-Time Adaptive Decision-Making in Unstructured Agricultural Scenarios Based on WSN and Deep Learning

Suresh Kumar K.*

Rajalakshmi Engineering College, Tamil Nadu, India

**Malathi M., Pavithra P.,
Shanmugapriya K., Deepika S.**

PG Scholar, Panimalar Engineering College,
Tamil Nadu, India

■ **Abstract:** Autonomous precision agriculture has been proposed in this research as an integrated framework of WSN and Deep Learning and it promises revolutionizing decision-making in farming through real-time intervention. However current systems suffer adaptability, energy efficiency, and even on-the-go processing. Continuous monitoring of dose patterns for soil condition, climate, and crop health by WSN nodes is associated with deep learning algorithms to deliver a real-time result from comparative analysis of learned patterns. In this regard, the system will use self-learning for continuous adaptability to changes in the environment. It is lightweight in model size and deployments in low-power edge devices as opposed to cloud-based computing hence increasing efficiency in remote agriculture settings. This research proposes an intelligent and sustainable farming ecosystem by augmenting farms with the ability to autonomously optimize resource consumption, improve yields, and reduce risks for increased productivity. Real-time decision-making result is the energy-efficient system for increased agricultural efficacy, less waste, and climate-resilient farming.

■ **Keywords:** Deep learning, Cloud based computing, Decision making, Resource consumption, WSN

1. INTRODUCTION

Agriculture is very important in the provision of food security and economic stability globally. Nevertheless, intensive farming is exposed to serious challenges like climate change, resource depletion, and growing global food demand. Conventional farming practices depend a lot on human labor and decision-making through intuition, and this results in inefficiencies in irrigation, pest

*Corresponding author: sureshkumar.k@rajalakshmi.edu.in

DOI: 10.1201/9781003725046-21

management, and yield prediction. To tackle these challenges, new technologies like Wireless Sensor Networks (WSN) and Deep Learning have been incorporated into smart farming, facilitating real-time monitoring and automated decision-making.

Wireless Sensor Networks (WSN) are decentralized sensor nodes distributed across the field, which monitor real-time information on important agricultural factors like soil moisture, temperature, humidity, and crop conditions. WSN enables precision agriculture by delivering timely and precise information for improved resource utilization. Yet, the enormous data generated by WSN must be intelligently processed and analyzed, and this is where deep learning can step in.

2. BACKGROUND STUDY

2.1 Unstructured and Dynamic Agricultural Environments

Recent advancements in deep learning and the Internet of Things (IoT) have significantly transformed precision agriculture with real-time data collection and intelligent decision-making. A number of studies have examined wireless sensor network (WSN) integration with machine learning algorithms to enhance crop monitoring and disease detection. For example, (Rahaman, M. M., at al. 2022) designed a crop-diagnosis platform that uses deep learning models to inspect nutrient content in a natural environment, enhancing precision agriculture efficiency.

2.2 Limited Real-Time Adaptability

In spite of these developments, current smart farming solutions continue to have some key challenges. One of the main limitations is the dynamic and unstructured nature of agricultural settings, where soil type, climate, and crop types affect the applicability of deep learning models. In addition, environmental factors like terrain and weather conditions affecting connectivity in WSN nodes make real-time data transmission and processing difficult (Jeong, Y., at al. 2018)

To overcome these challenges, researchers are investigating energy-efficient edge processing methods that support real-time decision-making and minimize reliance on cloud-based solutions. A suggested a deep convolutional neural network (DCNN) to optimize irrigation scheduling using soil moisture estimation for agricultural water efficiency improvement (Shaikh, A., at al. 2023)

3. METHODOLOGY

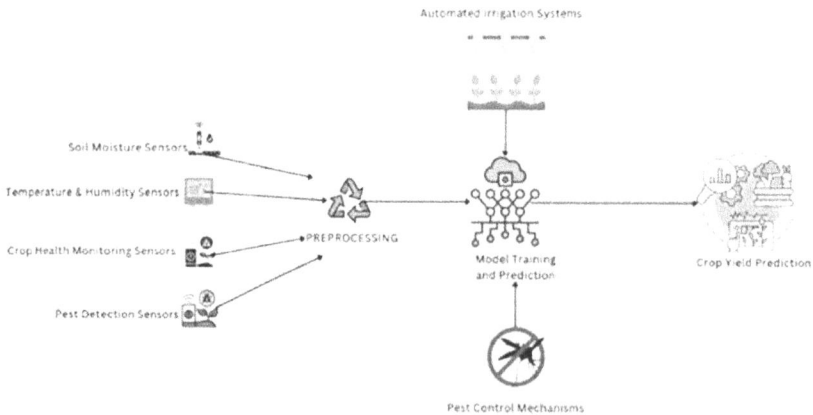

Fig. 21.1 Proposed system architecture

3.1 Data Collections and Preprocessing

The data collection process in the suggested precision agriculture system is to collect real-time environmental and soil parameters through Wireless Sensor Networks (WSN) and combine historical crop data from Kaggle datasets. The WSN nodes, with the facilities of soil nutrient sensors (N, P, K), temperature, humidity, pH, and rain sensors, are always sensing the field conditions and reporting the data to a central processing unit. Real-time sensor data is needed to make autonomous decisions about irrigation and pest control. For the purpose of providing high-quality data for predictive modeling, the dataset collected is subjected to preprocessing and feature engineering.

3.2 Model Training and Prediction

High-quality data for predictive modeling is ensured by the dataset collected after preprocessing and feature engineering. Missing values are handled, sensor reading values are normalized, and the categorical variable is encoded if required. Input variables to the model are carefully chosen with features like N, P, K (nutrients in soil), temperature, humidity, pH, and rainfall. Statistical treatments are applied to outliers to avoid biased predictions.

3.3 Autonomous Decision-Making

Irrigation Control: In case soil nitrogen content is low and moisture is lacking, the system initiates irrigation.

$$Irrigation = \begin{cases} \text{Start Irrigation, } P > 50 \\ \text{"No Irrigation Needed", } otherwise \end{cases}$$

By examining historical trends in data, future yield forecasts can be improved, resulting in ongoing improvements in agricultural efficiency.

$$Yield\ Improvement = \frac{Predicted\ Yield - Existing\ Yield}{Existing\ Yield} \times 100\%$$

4. EXPERIMENTAL RESULT ANALYSIS

The model utilized environmental parameters like Nitrogen (N), Phosphorus (P), Potassium (K), temperature, humidity, pH, and rainfall to forecast crop yield through a Linear Regression model.

Table 21.1   Dataset

N	P	K	Temperature	Humidity	ph	Rainfall	Label
90	42	43	20.87974	82.00274	6.502985	202.9355	rice
85	58	41	21.77046	80.31964	7.038096	226.6555	rice
60	55	44	23.00446	82.32076	7.840207	263.9642	rice
74	35	40	26.4911	80.15836	6.980401	242.864	rice
78	42	42	20.13017	81.60487	7.628473	262.7173	rice

4.1 Model Performance

A Linear Regression was applied to model crop yield depending on the input agricultural features provided. The available dataset was employed to train the model, while the performance was assessed using scores like (MAE) 2.2564, Mean Squared Error (MSE) is 6.8259 and R-squared (R^2) scores is 0.0032. The model possessed a fair ability to explain soil characteristics, climate, and the crop yield.

4.2 Real-Time WSN Data Prediction

For facilitating real-time decision-making, WSN data was simulated from the dataset to obtain dynamic prediction of crop yield depending on the environmental conditions at the moment. The model incorporated real-time sensor readings, such as soil nutrient content, temperature, humidity, and rainfall, to offer real-time predictions during runtime. Autonomous decision-making was incorporated, which facilitated activities such as irrigation control and pesticide advice without human interaction.

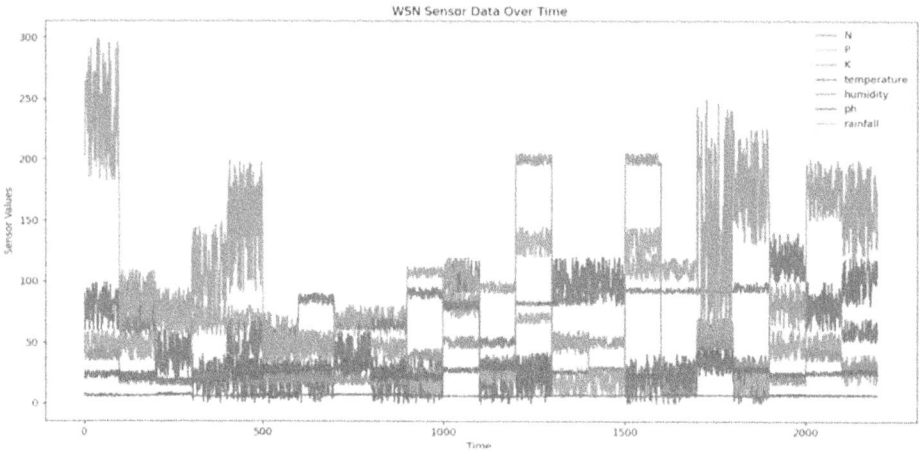

Fig. 21.2   Real time WSN data prediction

4.3 Results Visualization

Visualization methods were employed to examine sensor data and contrast forecasted crop yield with available values. Line charts were employed to illustrate trends in soil nutrients, temperature, and humidity over time, enabling the identification of patterns and changes in environmental conditions.

5. CONCLUSION

The suggested framework effectively enhanced the prediction of crop yield and decision-making using WSN and machine learning. Real-time autonomous decisions in farming enabled efficiency and sustainability in farming

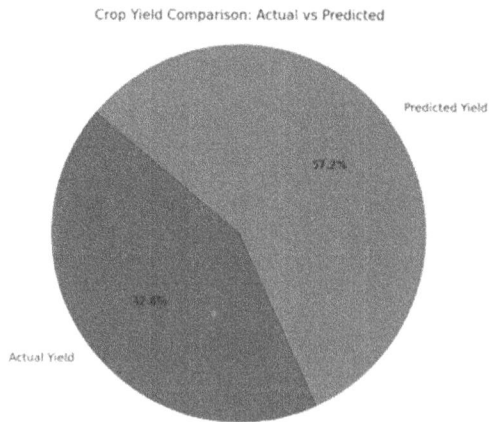

Fig. 21.3   Actual vs predicted crop yield

operations. For enhancement of accuracy and flexibility, the future will explore the integration of sophisticated deep models like LSTMs and CNNs. Moreover, improving the system with edge computing optimization will allow processing in real time on IoT devices, minimizing the reliance on cloud computing and efficiency in general. All these innovations will make precision agriculture more effective, robust, data-based, and sustainable.

References

1. Rahaman, M. M., & Azharuddin, M. (2022). Wireless sensor networks in agriculture through machine learning: A survey. *Computers and Electronics in Agriculture*, *197*, 106928.
2. Jeong, Y., Son, S., Lee, S., & Lee, B. (2018). A total crop-diagnosis platform based on deep learning models in a natural nutrient environment. *Applied Sciences*, *8*(10), 1992.
3. Velásquez, D., Sánchez, A., Sarmiento, S., Toro, M., Maiza, M., & Sierra, B. (2020). A method for detecting coffee leaf rust through wireless sensor neworks, remote sensing, and deep learning: Case study of the Caturra variety in Colombia. *Applied Sciences*, *10*(2), 697.
4. Lynda, D., Brahim, F., Hamid, S., & Hamadoun, C. (2023). Towards a semantic structure for classifying IoT agriculture sensor datasets: An approach based on machine learning and web semantic technologies. *Journal of King Saud University-Computer and Information Sciences*, *35*(8), 101700.
5. Isranil, R. K., Nedunchezhian, T., Tanna, P., & Soni, S. (2025) AI-endorsed techniques for smart energy utilization: A holistic review. *Digital Transformation and Sustainability of Business*, 271–274.
6. Shaikh, A., Ranjan, N. M., Rai, S., Rao, P. A., & Shinde, G. (2023, May). Incorporating LoRa Based Wireless Sensor Network and Machine Learning Technologies to Improve Precision Agriculture. In *International Conference on Information Science and Applications* (pp. 397-412). Singapore: Springer Nature Singapore.
7. Math, RajinderKumar Mallayya, and Nagaraj V. Dharwadkar. "An intelligent irrigation scheduling and monitoring system for precision agriculture application." *International Journal of Agricultural and Environmental Information Systems (IJAEIS)* 11.4 (2020): 1–24.
8. Kumar, P., Udayakumar, A., Anbarasa Kumar, A., Senthamarai Kannan, K., & Krishnan, N. (2023). Multiparameter optimization system with DCNN in precision agriculture for advanced irrigation planning and scheduling based on soil moisture estimation. *Environmental Monitoring and Assessment*, *195*(1), 13.
9. Abougreen, A. N., & Chakraborty, C. (2021). Applications of machine learning and internet of things in agriculture. *Green Technological Innovation for Sustainable Smart Societies: Post Pandemic Era*, 257–279.
10. Liu, J., Shu, L., Lu, X., & Liu, Y. (2023). Survey of intelligent agricultural IoT based on 5G. *Electronics*, *12*(10), 2336.
11. Reyana, A., Kautish, S., Karthik, P. S., Al-Baltah, I. A., Jasser, M. B., & Mohamed, A. W. (2023). Accelerating crop yield: multisensor data fusion and machine learning for agriculture text classification. *IEEE Access*, *11*, 20795–20805.

Note: All the figures and table in this chapter were made by the authors.

Emerging Perspectives and Applications of Computational Intelligence and Smart Systems
– Dr. Amit Lathigara et al. (eds)
© 2026 Taylor & Francis Group, London, ISBN 978-1-041-20965-2

22

Assessing Global Water Scarcity Using the WEAP Model: Implications for Sustainable Development

Kajal Dudhatra*
Research Scholar,
School of Engineering, RK University, Rajkot,
Gujarat, India

Sajaykumar Joshi
School of Engineering, RK University, Rajkot,
Gujarat, India

■ **Abstract:** Water resource management is a growing global concern due to population growth, industrial expansion, and climate change. The Water Evaluation and Planning (WEAP) model has been widely applied to analyze and forecast water demand and supply scenarios. This paper reviews multiple studies conducted in India, Algeria, Ethiopia, Morocco, and Iraq, focusing on their methodologies, key findings, and proposed strategies. The review highlights how WEAP integrates hydrological, climatic, and socio-economic parameters to simulate future water demands. The findings emphasize increasing demand, climate-driven variability, and the necessity for sustainable water management strategies. Recommendations include integrating remote sensing data and machine learning for enhanced predictive capabilities.

■ **Keywords:** WEAP model, Water resource management, Climate change, Hydrological modelling, Sustainable water planning

1. Introduction

With population growth, climate change, and rising freshwater demand, water scarcity has become an increasingly critical global issue. Many regions are experiencing severe water stress, leading to conflicts over water allocation, declining agricultural productivity, and environmental degradation. As a result, the need for sustainable and integrated water management strategies has become more critical than ever. Effective water resource management requires a holistic approach that

---

*Corresponding author: kajal.thumar@rku.ac.in

DOI: 10.1201/9781003725046-22

considers hydrological, climatic, and socio-economic factors to ensure equitable and efficient water distribution.

One popular tool for assessing patterns of water supply and demand is the WEAP model (Trivedi & Rathod, 2024). WEAP provides a comprehensive framework for evaluating different water management scenarios by integrating multiple variables, including climate data, land use patterns, population growth, and policy interventions. Its user-friendly interface and ability to model both surface and groundwater resources make it a valuable tool for decision-makers, researchers, and policymakers. By simulating various management strategies, WEAP helps in identifying potential solutions to mitigate water scarcity and improve resource allocation.

Over the years, numerous studies have applied WEAP in diverse geographical settings, from arid and semi-arid regions to humid and temperate climates. Researchers have utilized model to evaluate how climate change affects the availability of water, evaluate water allocation policies, and explore the effectiveness of conservation strategies. By compiling and analyzing these studies, this review aims to provide a comprehensive assessment of WEAP's effectiveness in addressing water scarcity challenges. The findings will offer insights into the model's strengths, limitations, and potential improvements for future applications in water resource planning and management.

2. LITERATURE REVIEW

2.1 Application of WEAP in Water Demand and Supply Assessment

Several studies have applied WEAP to evaluate current and future water demand at the city and basin levels. In Bhavnagar City, Gujarat, India, WEAP was used to simulate water demand scenarios from 2025 to 2050, accounting for population growth and industrial expansion. The findings indicated a 25% increase in total demand under a reference scenario, with a 40% rise in a high-growth scenario, posing risks of unsustainable water consumption. Strategies such as groundwater recharge augmentation and desalination plant integration were suggested to mitigate seasonal water shortages.

A similar approach was applied in Annaba Province, Northeastern Algeria, where baseline (2020) and future demand (2050) scenarios were simulated under three socio-economic pathways: Business-As-Usual (BAU), High Population Growth (HPG), and Water Conservation (WC). Results showed that BAU would lead to a 17% deficit by 2050, increasing to 28% under HPG, while the WC scenario reduced demand by 15% through improved irrigation efficiency.

2.2 Impact of Climate Change on Water Resources

Climate change-induced alterations in hydrology have also been extensively analyzed using WEAP. In the Katha Basin, downscaled CMIP5 (Tun et al., 2024). Climate data was integrated into WEAP, revealing streamflow reductions of 12–18% by 2060 under RCP 8.5. The study incorporated the Soil Moisture Method to estimate evapotranspiration and percolation rates, highlighting a 40% increase in unmet demand by 2040 under a business-as-usual scenario.

Similarly, in Chennai Basin, India, WEAP, combined with GIS-based high-resolution spatial demand mapping, projected that urban expansion could increase demand by 55% by 2050 (Razi Sadath et al., 2023), causing a 30% reduction in reservoir storage during drought years. The study recommended rainwater harvesting mandates, artificial recharge zones, and wastewater reuse to improve water sustainability.

In the Subarnarekha River Basin, Monte Carlo simulations were utilized to assess supply risks under different climate scenarios. Under a 2°C temperature rise, irrigation water demand increased by 22%, while groundwater recharge declined by 18%(Kumar et al., 2021). Optimized reservoir releases through WEAP's Catchment Delineation function reduced unmet demand by 14% annually.

2.3 Agricultural and Hydrological Modelling Using WEAP

The WEAP model has also been used to assess agricultural water demand and hydrological processes. In the Veda River Basin, Madhya Pradesh, crop-specific evapotranspiration data was integrated using the FAO56 Penman-Monteith model. The study projected a 30% increase in agricultural water demand due to a shift from rainfed to irrigated cropping systems (Carpenter & Choudhary, 2022). The adoption of deficit irrigation techniques showed potential water savings of 12–18%.

In the Ur River Watershed, Madhya Pradesh, demand scenarios were simulated under Low Growth, BAU, and High Growth, alongside RCP 4.5 and 8.5 climate projections. The worst-case scenario suggested that 50% of demand would remain unmet by 2050, emphasizing the need for reservoir dredging and groundwater recharge zones to enhance supply (Nivesh et al., 2023).Future agricultural expansion could increase withdrawal by 35%, leading to groundwater overexploitation, necessitating policy-driven abstraction limits and regulated conjunctive water use strategies.

2.4 WEAP Model for Subbasin Hydrology and Water Allocation

Several studies have evaluated WEAP's hydrological modelling capabilities by comparing it with other hydrological models. In Ethiopia's Central Rift Valley Basin, WEAP's stream flow simulation had an NSE of 0.75, though SWAT (Soil and Water Assessment Tool) outperformed WEAP in groundwater recharge estimation (Abera & Ayenew, 2021).

Another study in the Didessa Sub-Basin, West Ethiopia, used WEAP to model multi-sector demand projections. Findings indicated that urban and industrial demand could increase by 40% by 2050, with supply deficits expected by 2035 if storage expansion is not implemented.

Refining previous studies on the Katha Basin, another research integrated ensemble climate projections, confirming that streamflow would decline by 10–15% under RCP 4.5 and up to 20% under RCP 8.5.

2.5 Water Management and Adaptation Strategies

In the Tensift River Basin, Morocco, WEAP's Catchment Delineation and Allocation tool was used to optimize reservoir storage regulation, demonstrating that climate resilience strategies could reduce annual shortages by 25% (Khebiza et al., 2009).

In the Upper Indus Basin, WEAP-SWAT modelling projected a 35% increase in irrigation demand under extreme climate scenarios. The study also modelled transboundary water agreements to optimize allocation among India, Pakistan, and China (Ahmad et al., 2025).

Lastly, in Baghdad City, Iraq, WEAP was used to simulate urban water constraints, with findings indicating that 50% of demand would remain unmet by 2045 unless alternative sources like treated wastewater reuse were implemented (Al-Mukhtar & Mutar, 2021).

3. Conclusion

The reviewed studies highlight the growing challenge of water scarcity driven by rising demand, climate change, and inefficient water use. WEAP model simulations across various regions, including India, Algeria, Ethiopia, Iraq, and Morocco, predict significant increases in water demand (25–50%) by 2050, coupled with declining streamflow (10–20%) and groundwater depletion (Trivedi & Rathod, 2024; Nivesh et al., 2023; Abera &Ayenew, 2021; Ahmad et al., 2025). Climate change projections (e.g., RCP 4.5 & 8.5) indicate that higher temperatures and altered precipitation patterns will exacerbate water stress. Studies emphasize strategies like efficient irrigation, wastewater reuse, and conservation policies, as well as supply-side interventions like rainwater harvesting and reservoir optimization (Khebiza et al., 2009; Kumar et al., 2021). WEAP's scenario-based analysis aids in assessing future risks, though integrated approaches such as WEAP-SWAT or WEAP-MODFLOW may enhance performance in groundwater modelling. In conclusion, proactive integrated water resource management (IWRM) is essential for long-term water sustainability.

References

1. Trivedi, A. A., & Rathod, D. (2024). Assessment of Future Water Demand Utilizing the WEAP Model in Bhavnagar City, Gujarat, India. International Journal for Research in Applied Science & Engineering Technology, 12(III).
2. Berredjem, Abdel Fatah, Ahlem Boumaiza, and Azzedine Hani. (2023) "Simulation of current and future water demands using the WEAP model in the Annaba province, Northeastern Algeria: a case study." AQUA—Water Infrastructure, Ecosystems and Society 72, no. 9: 1815–1824.
3. Tun, Win Lwin, Cho Cho Thin Kyi, and Yin Yin Htwe. (2024) "Evaluation of Stream Flow and Water Demand due to Climate Change in the Katha Basin Using Water Evaluation and Planning (WEAP) Model." The Indonesian Journal of Computer Science 13, no. 5.
4. RaziSadath, PuthanVeettil, Mariappan RinishaKartheeshwari, and Lakshmanan Elango. (2023) "WEAP Model Based Evaluation of Future Scenarios and Strategies for Sustainable Water Management in the Chennai Basin, India." AQUA—Water Infrastructure, Ecosystems and Society 72, no. 11: 2062–2080.
5. Kumar, Randhir, Pratibha Kumari, P. K. Parhi, V. K. Tripathi°, and Ajai Singh (2021). "Evaluating water supply risk in the middle reaches of Subarnarekha river basin by using WEAP model." Indian Journal of Environmental Protection 4, no. 8.
6. Carpenter, Abhay, and Mahendra Kumar Choudhary. (2022) "Water demand and supply analysis using WEAP model for Veda river basin Madhya Pradesh (Nimar region), India." Trends in sciences 19, no. 6: 3050–3050.
7. Agarwal, Sunny, Jyoti P. Patil, V. C. Goyal, and Ajai Singh. (2019) "Assessment of water supply–demand using water evaluation and planning (WEAP) model for Ur River watershed, Madhya Pradesh, India." Journal of The Institution of Engineers (India): Series A 100: 21–32.
8. Nivesh, Shreya, Jyoti Parasharam Patil, Vikas Chandra Goyal, Bhagwat Saran, Ajay Kumar Singh, Anurag Raizada, Anurag Malik, and Alban Kuriqi. (2023) "Assessment of future water demand and supply using WEAP model in Dhasan River Basin, Madhya Pradesh, India." Environmental Science and Pollution Research 30, no. 10: 27289–27302.
9. Abera Abdi, Debele, and Tenalem Ayenew. (2021) "Evaluation of the WEAP model in simulating subbasin hydrology in the Central Rift Valley basin, Ethiopia." Ecological processes 10, no. 1: 41.
10. Adgolign, Tena Bekele, GVR Srinivasa Rao, and YerramsettyAbbulu. (2016) "WEAP modeling of surface water resources allocation in Didessa Sub-Basin, West Ethiopia." Sustainable Water Resources Management 2: 55–70.

11. Tun, Win Lwin, Cho Cho Thin Kyi, and Yin Yin Htwe. (2024) "Evaluation of Stream Flow and Water Demand due to Climate Change in the Katha Basin Using Water Evaluation and Planning (WEAP) Model." *The Indonesian Journal of Computer Science* 13, no. 5.

12. Khebiza, Mohammed Yacoubi, M. Messouli, F. Hammadi, and B. Ghallabi. (2009)"Adaptation actions to reduce water system vulnerability to climate change in Tensift River basin (Morocco)." In *IOP Conference Series. Earth and Environmental Science*, vol. 6, no. 29. IOP Publishing,

13. Ahmad, Sareer, Muhammad Waseem, Hira Wahab, Abdul Qadeer Khan, Zulqarnain Jehan, Izhar Ahmad, and Megersa Kebede Leta. (2025) "Assessing water demand and supply in the Upper Indus Basin using integrated hydrological modeling under varied socioeconomic scenarios." *Applied Water Science* 15, no. 1: 5.

14. Al-Mukhtar, Mustafa M., and Ghasaq S. Mutar. (2021)" Modelling of future water use scenarios using WEAP model: a case study in Baghdad City, Iraq." *Eng. Technol. J* 39: 488–503.

Emerging Perspectives and Applications of Computational Intelligence and Smart Systems
– Dr. Amit Lathigara et al. (eds)
© 2026 Taylor & Francis Group, London, ISBN 978-1-041-20965-2

23

IoT-Driven Batch Tracking System for Pharmaceutical Quality and Compliance in Industry 4.0

Abhishek Tripathi[1], B. Hanuma Kumar
Department of Computer Science and Engineering,
Kalasalingam Academy of Research and Education,
Tamil Nadu, India

Shubham Anjankar[2]
Dept of Electronics Engineering,
Shri Ramdeobaba College of Engineering and Management,
Nagpur, India

Suresh Balpande
Department of Information Technology and security,
Ramdeobaba University Nagpur, India

**Saral Kumar Gupta,
Chandra Mohan S. Negi**
Banasthali Vidyapith, Banasthali, Rajasthan,
India

■ **Abstract:** Aiming at the goals of responsible consumption and production (SDG 12), this work presents an IoT-enabled environmental monitoring and batch tracking system intended to improve the sustainability and quality assurance of pharmaceutical manufacture. Combining BMP180 and DHT11 sensors to monitor critical environmental parameters including atmospheric pressure, temperature and humidity enables the system guarantee optimal conditions all through the production process. The Blynk App and computer system real-time display and analysis capacity are made possible by an ESP8266 microcontroller transferring data gathered by these sensors to the cloud. Through automatic deviations-based warnings, this system reduces operating costs, removes manual intervention, and advances proactive decision-making. The strategy provides tracking, best use of resources, sustainability in drugs, and promotion of compliance.

■ **Keywords:** IoT, Pharmaceuticals, Sensors, Monitoring and Sustainability

Corresponding author: [1]tripathi.abhishek.5@gmail.com, [2]anjankarsc@rknec.edu

DOI: 10.1201/9781003725046-23

1. INTRODUCTION

Pharmaceutical manufacturing depends on accurate environmental monitoring to keep (Marques et al., 2020) regulatory compliance, stability, and quality of commodities. Three main elements that greatly influence the chemical stability and efficacy of therapeutic drugs are temperature, humidity, and air pressure (Chen et al., 2020). Real-time data integration made feasible by IoT enabled automation lets many different industries change their operations (Tripathi et al., 2024). Usually depending on manual procedures, conventional monitoring systems are prone to mistakes and inefficiencies that could compromise the consistency of production settings and so cause likely risks. This work presents an IoT-enabled environmental monitoring and batch tracking system that automates and maximizes the manufacturing process (Rane et al., 2023) in order to solve these difficulties. Moreover, track and trace solutions in Internet of Things architecture let pharmaceutical firms increase supply chain traceability, thereby boosting inventory control and assisting the fight against counterfeit medications. From production to distribution, these systems give total control that allows faster and more focused better decisions to be made (Sardjono et al., 2023). Pharmacy 4.0 projects illustrate that IoT-driven solutions offer responsibility and openness, so combining modern quality and safety requirements with production procedures guarantees (Sardjono et al., 2023).

This change in favour of IoT acceptance marks a major first towards guaranteeing competitive advantage and environmentally friendly manufacturing techniques in the pharmaceutical industry. Combining DHT11 and BMP180 sensors, the device real-time monitors humidity, temperature, and air pressure. These sensors interfaced a central processing unit compiling and passing data to a cloud platform from an ESP8266 microcontroller. Although the Blynk App lets production managers track and assess this data so they could remotely verify conditions and get alerts when deviations show up, the automated character of the system removes manual involvement and operational inefficiencies even while it offers quick corrective actions. By so reducing resource waste and increasing operational efficiency, constant and precise monitoring helps the system to support values of sustainable development. This method shows how IoT technology could enhance dependability, traceability, and quality control in pharmaceutical production, therefore providing a basis for more effective and strong manufacturing lines. The paper of organization as follows: Section 2 looks at essential stuff. Section 3 runs over the advised approach. Section 4 connects in the system's application. Section 5 offers lots of analysis and outcomes. Section 6 finishes the research; Section 7 addresses future scope and developments.

2. LITERATURE REVIEW

More and more in the current era of digital transformation and the Fourth Industrial Revolution, the pharmaceutical industry makes use of innovative technologies to increase operational efficiency, product quality, and regulatory compliance. Including IoT into pharmaceutical manufacturing offers excellent chances to guarantee quality control, lower risks, and streamline processes. By enabling real-time monitoring of critical parameters—qualities required to maintain optimum circumstances during pharmaceutical manufacture (Rane et al., 2023), IoT aids to produce outstanding data collecting and analysis. Including sensors such as BMP180 and DHT11 into manufacturing systems allows manufacturers to monitor air pressure, temperature, and humidity, therefore confirming that ambient conditions meet high quality criteria (Sharma et al., 2020). Apart from reducing machine wear and tear, this link allows predictive maintenance, hence cutting production costs and downtime

(Ullagaddi, 2024). Including IoT technologies into pharmaceutical manufacturing and supply chain management has tremendously enhanced process dependability, quality, and efficiency. IoT solutions offer connected networks of devices with individual identities seamless real-time monitoring and data transfer. These systems maximize unit operations during pharmaceutical manufacture by tracking critical components including temperature, humidity, and pressure so ensuring constant production efficiency and quality control (Sharma et al., 2020). As sensor networks and mobile apps control environmental aspects such temperature swings and humidity, IoT has also changed the way medications and vaccinations are given. These systems ensure the integrity of certain drugs by use of MQTT and safe data transfer techniques, therefore obtaining a verified dependability rate of above 89% (Bhatti et al., 2024). Low-cost wireless devices employing ESP8266 have industrial applications with great promise. By use of a master-slave communication topology, these systems minimize SCADA-like solutions for pharmaceutical production, thereby enabling real-time parameter monitoring, data logging, and control in industrial environments (Albayari et al., 2023). By means of real-time updates and alarms concerning medicine storage, IoT-based intelligent medicine management systems also use Raspberry Pi and VNC technologies to improve security and surveillance in storage facilities, so ensuring the safety and availability of pharmaceutical products (Jesudoss et al., 2019).

3. METHODOLOGY

Starting with turning on the ESP8266 microcontroller, initializing all components—including the DHT11 and BMP180 sensors—for environmental monitoring, the presented flowchart of batch tracking system in Fig. 23.1 starts. Measuring temperature, humidity, and atmospheric pressure lets the sensors guarantee important values kept for pharmaceutical manufacturing. Should a sensor fail, the system retries till successful? Once turned on, the sensors compile environmental data in real time. This information guides the ESP8266 microcontroller in cloud transmission preparation. Should mistakes in data collecting or processing surface, the system loops back to fix them.

The ESP8266 moves processed data to the cloud using WiFi connection. Acting as a centralized storage and analysis platform, the cloud assures flawless communication by security. The Blynk App shows consumers real-time data that acts as a visual dashboard for remote monitoring manufacturing conditions. Should data move fail, the system retries till successful. Every batch makes use of a unique ID that connects particular production cycles

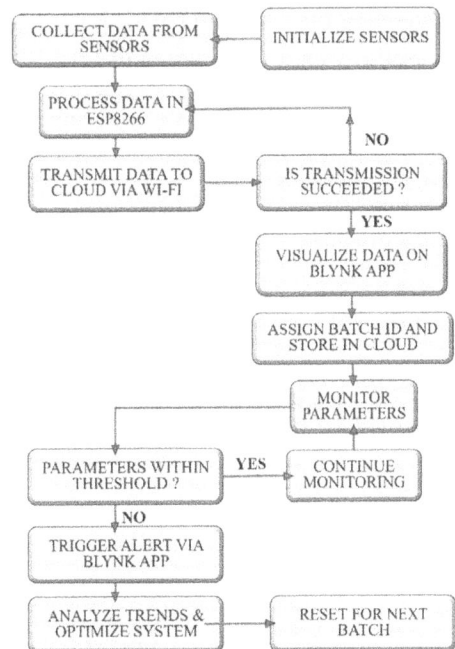

Fig. 23.1 Batch tracking process flowchart for pharmaceutical monitoring

to environmental data. The system tracks variables against set thresholds all the time. Should differences be found, the Blynk App instantly warns the user for quick corrections. Monitoring starts once things have level off. If no deviations occur, the system logs data in real-time and analyzes historical trends to optimize future production cycles. At the end of the batch, the system either resets for the next batch or concludes the process. This structured and responsive methodology ensures precision, efficiency, and stability in pharmaceutical production.

4. SYSTEM IMPLEMENTATION

Combining hardware components, software platforms, and cloud connectivity, the IoT-enabled batch tracking system provides pharmaceutical manufacture real-time monitoring and traceability. BMP180 and DHT11 sensors linked to the ESP8266 microcontroller—the main processing and communication unit—make up the hardware setup. The DHT11 sensor senses temperature and humidity; the BMP180 sensor detects air pressure. Offering low latency and perfect communication, the ESP8266 analyzes the data and transfers it to the cloud using its built-in Wi-Fi module.

The processed data is housed on a cloud platform; consolidated source of all batch-related environmental data. Showing on a user-friendly dashboard temperature, humidity, and pressure information, the Blynk App is the main interface for real-time data display. Any value higher than the designated thresholds triggers the program to create alarms ensuring quick notice and action. The technology maximizes forthcoming production cycles by allowing data to be continuously tracked and recorded, so enabling historical trend analysis. By means of accurate and automatic monitoring, efficient traceability using QR codes, robust hardware, cloud, and mobile interface connection, this completely integrated solution guarantees minimum possible manual intervention optimization of the pharmaceutical production process. Connected to the ESP8266 microcontroller, Fig. 23.2 (a) shows how DHT11 and BMP180 sensors are batch monitoring box integrated. Emphasizing cloud-based monitoring and connected hardware, a laptop displays running scripts for real-time data movement and viewing. Every manufacturing batch has a QR code with batch ID, formulation guidelines, and manufacture dates. Originally intended at the beginning of the batch, these QR codes link the obtained industrial environmental data. This method guarantees batch-specific traceability and provides stakeholders with quick access data via QR code scanning, therefore boosting responsibility and allowing timely, necessary remedial action. The QR code module is

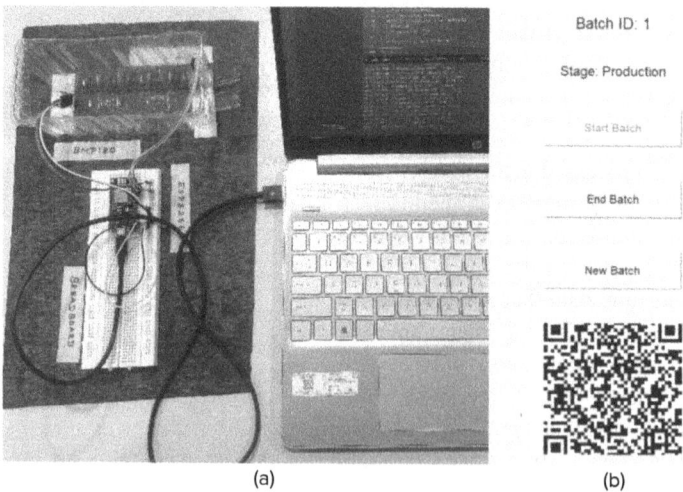

(a) (b)

Fig. 23.2 (a) Integrated hardware setup for batch monitoring and IoT connectivity, (b) Batch tracker application interface with QR code generation and IoT connectivity

used to generate QR codes and PIL is used to show them in the Batch Tracker Application as shown in Fig. 23.2 (b), which was created with Python's Tkinter toolkit for the GUI. By allowing users to initiate, process, and finish batches, the tool promotes effective management of pharmaceutical batches. Through the request library, it logs batch data to ThingSpeak, guaranteeing precise and easy data tracking.

5. RESULTS AND DISCUSSION

Figure 23.3 represents the environmental parameters recorded for the IoT-enabled batch tracking system in pharmaceutical production, with data collected using hardware setups tested at four locations in Tamil Nadu- Chennai (November 3, 2024), Krishnankoil (September 20, 2024), Madurai (October 12, 2024), and Virudhunagar (September 22, 2024). Temperature, humidity, and atmospheric pressure were monitored using DHT11 and BMP180 sensors, ensuring optimal conditions for drug formulation and maintaining product quality. The variations in these parameters demonstrate the system's capability to accurately monitor real-time data under different environmental conditions, supporting effective production oversight.

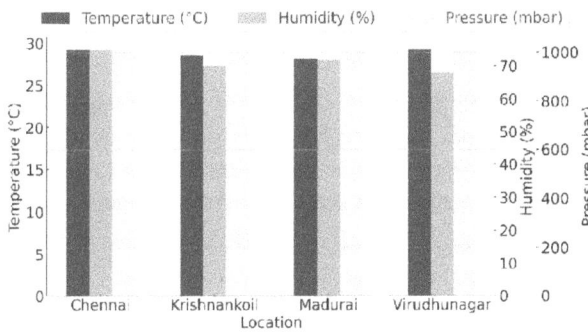

Fig. 23.3 Environmental parameters for IoT-enabled batch monitoring system across locations

Figure 23.4 illustrates the variation in environmental parameters—temperature, humidity, and pressure for Krishnan coil location on a single day from 8 AM to 8 PM. The temperature ranges from 22°C to 28.5°C, peaking at 2 PM, while humidity increases steadily, reaching 75% at 2 PM and stabilizing thereafter. Atmospheric pressure shows minor fluctuations, with a range of 1005

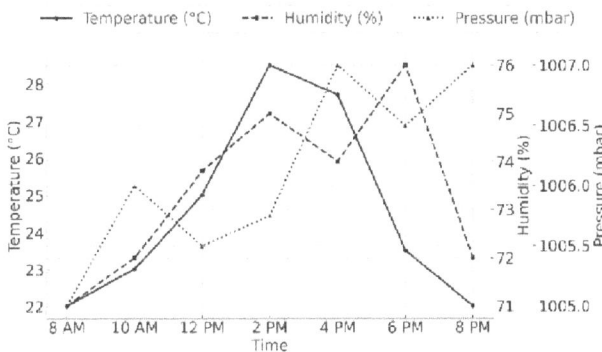

Fig. 23.4 Environmental parameters for throughout the day

mbar to 1007 mbar, highlighting the system's ability to monitor real-time environmental conditions. These parameters are critical for maintaining optimal conditions in pharmaceutical batch tracking systems used.

6. Conclusion

Combining BMP180 and DHT11 sensors with ESP8266 and cloud-based visualization via the Blynk App, the IoT-enabled batch tracking solution for pharmaceutical manufacturing combines Above all, this gadget provides real-time monitoring of vital environmental parameters including temperature, humidity, and air pressure, thereby preserving perfect conditions in drug development processes. By automating data collecting and processing, the system will be able to reduce hand-off rates, operating errors, and improve quality control. Constant logging and trend analysis include features that provide expected insights improving regulatory compliance and sustainable production techniques. Different climatic factors expose IoT's success as well as its dependability in increasing industrial efficiency and openness. This system offers a scalable and reasonably priced solution for modern pharmaceutical organizations wishing to add innovative technology into their operations for continuous quality and operational excellence.

7. Future Work

Future research aims to integrate additional sensors for monitoring elements including particle matter and chemical concentrations, hence enhancing data accuracy and environmental insights. Machine learning approaches will be part of predictive analytics, so supporting proactive decision-making. Planned improvements also assure scalability for more general industrial use and boost cloud capability.

References

1. Marques, C. M., Moniz, S., de Sousa, J. P., Barbosa-Povoa, A. P., &Reklaitis, G. (2020). Decision-support challenges in the chemical-pharmaceutical industry: Findings and future research directions. *Computers & Chemical Engineering, 134*, 106672.
2. Chen, Yingjie, Ou Yang, Chaitanya Sampat, Pooja Bhalode, Rohit Ramachandran, and Marianthi Ierapetritou. "Digital twins in pharmaceutical and biopharmaceutical manufacturing: a literature review." *Processes* 8, no. 9 (2020): 1088.
3. Tripathi, Abhishek, K. Manvith, G. Manju Srinath, N. Kamal Yuvan, and U. Tharun Reddy. "Implementation of an Alert System Integrated into Smart Wireless Helmets Utilizing IoT Sensors." In *2024 2nd International Conference on Networking and Communications (ICNWC)*, pp. 1–5. IEEE, 2024.
4. Rane, Santosh B., Sandesh Wavhal, and Prathamesh R. Potdar. "Integration of Lean Six Sigma with Internet of Things (IoT) for productivity improvement: a case study of contactor manufacturing industry." *International Journal of System Assurance Engineering and Management* 14, no. 5 (2023): 1990–2018.
5. Sardjono, Wahyu, Indah Ristya Mustika Arum, Aninda Rahmasari, and Erma Lusia. "Impact of Track and Trace (T&T) in Industrial Revolution 4.0 of the Pharmaceutical Industry (Pharma 4.0)." In *E3S Web of Conferences*, vol. 388, p. 03014. EDP Sciences, 2023.
6. Rane, Santosh B., Sandesh Wavhal, and Prathamesh R. Potdar. "Integration of Lean Six Sigma with Internet of Things (IoT) for productivity improvement: a case study of contactor manufacturing industry." *International Journal of System Assurance Engineering and Management*, 14, no. 5 (2023): 1990–2018.
7. Sharma, Deepak Kumar, Saakshi Bhargava, and Kartik Singhal. "Internet of Things applications in the pharmaceutical industry." In *An Industrial IoT Approach for Pharmaceutical Industry Growth*, pp. 153–190. Academic Press, 2020.

8. Ullagaddi, P. "Leveraging digital transformation for enhanced risk mitigation and compliance in pharma manufacturing." *Journal of Advances in Medical and Pharmaceutical Sciences*, 26, no. 6 (2024): 75–86.

9. Sharma, Apoorva, Jaswinder Kaur, and Inderbir Singh. "Internet of things (IoT) in pharmaceutical manufacturing, warehousing, and supply chain management." *SN Computer Science*, 1, no. 4 (2020): 232.

10. Bhatti, David Samuel, Muhammad Mueed Hussain, Beomkyu Suh, Zulfiqar Ali, IsmatovAkobir, and Ki-Il Kim. "IoT-enhanced transport and monitoring of medicine using sensors, MQTT, and secure short message service." *IEEE Access* (2024).

11. Albayari, Osama, Amin Yousef, Ramzi Albawab, Bahaa Jibrini, and Mohammad Salah. "Low-Cost Wireless Monitoring and Control: A Case Study for Industrial Implementation." In *2023 IEEE Jordan International Joint Conference on Electrical Engineering and Information Technology (JEEIT)*, pp. 171–176. IEEE, 2023.

12. Jesudoss, A., Daniel, M. J., & Richard, J. J. (2019, October). Intelligent medicine management system and surveillance in IoT environment. In *IOP Conference Series: Materials Science and Engineering* (Vol. 590, No. 1, p. 012005). IOP Publishing.

Note: All the figures in this chapter were made by the authors.

Emerging Perspectives and Applications of Computational Intelligence and Smart Systems
– Dr. Amit Lathigara et al. (eds)
© 2026 Taylor & Francis Group, London, ISBN 978-1-041-20965-2

24

AI-Driven Load Balancing in Cloud Computing: A Deep Reinforcement Learning Approach

Paresh Tanna, Nirav Bhatt

School of Engineering, RK University, Rajkot,
Gujarat, India

Sunil Soni*, Jaydeep Tadhani

IT department, Government Polytechnic, Rajkot,
Gujarat, India

Riaz Israni

School of Engineering, RK University, Rajkot,
Gujarat, India

■ **Abstract:** Because cloud computing offers scalable, on-demand resources, it has completely transformed contemporary IT infrastructure. However, maintaining optimal resource utilization and guaranteeing Quality of Service (QoS) is severely hampered by dynamic workload variations and unpredictable traffic patterns. This research suggests a Deep Reinforcement Learning (DRL)-based load balancing system that uses Proximal Policy Optimization (PPO) and Deep Q-Networks (DQN) for intelligent resource allocation in order to overcome these constraints. By interacting with the cloud environment and maximizing critical performance metrics like response time, throughput, and energy efficiency, the suggested model dynamically learns the best workload distribution strategies. We do comprehensive experiments using CloudSim and real-world cloud workload traces (Google Cloud dataset) to assess the efficacy of the DRL-based strategy. Our strategy achieves shorter response times, fewer SLA violations, and more resource efficiency when compared to traditional and heuristic-based load balancing techniques. The findings demonstrate DRL's ability to manage cloud workloads independently while cutting down on computational overhead. On order to create more intelligent and energy-efficient cloud computing infrastructures, this study lays the groundwork for future research on self-adaptive, AI-driven load balancing strategies.

■ **Keywords:** Cloud computing, Load balancing, Deep reinforcement learning, Deep Q-networks (DQN), Proximal policy optimization (PPO)

---

*Corresponding author: profsjsoni@gmail.com

DOI: 10.1201/9781003725046-24

1. INTRODUCTION

A key component of contemporary IT architecture, cloud computing allows businesses to offer scalable, on-demand computational capabilities at low operating costs. Effective load balancing has emerged as a key difficulty in managing workloads across dispersed cloud resources as cloud services continue to expand. By distributing tasks evenly among virtual machines (VMs), load balancing enhances system performance, response time, and resource usage. Performance bottlenecks and ineffective resource allocation result from the static and heuristic-based solutions provided by traditional load balancing algorithms like Round Robin, Least Connection, and Weighted Load Balancing, which frequently are unable to adjust to dynamic cloud environments (Alhilali & Montazerolghaem, 2023).

Approaches powered by artificial intelligence (AI) have drawn a lot of interest in optimizing load balancing through adaptive decision-making and predictive analytics (Chen et al., 2023; Wang et al., 2023). For real-time task distribution in cloud systems, Deep Reinforcement Learning (DRL) in particular has shown promise (Haris & Zubair, 2024). DRL models make wise decisions based on observed workload patterns and are constantly learning from system interactions. In contrast to conventional rule-based techniques, DRL optimizes key performance metrics like latency, throughput, SLA (Service Level Agreement) adherence, and energy usage to dynamically modify resource allocation tactics (Bolanowski et al., 2023).

In order to intelligently manage cloud workloads, this study suggests a load balancing framework based on Deep Q-Network (DQN) and Proximal Policy Optimization (PPO). By engaging with the cloud environment and optimizing long-term incentives, the DRL agent discovers the best job allocation policy. The performance of the suggested approach is compared with that of conventional and heuristic-based load balancing strategies using CloudSim and actual Google Cloud workload traces. According to experimental findings, the DRL-based strategy improves resource efficiency, decreases SLA violations, and speeds up response times, making it a good fit for contemporary cloud computing systems.

2. PROPOSED APPROACH

The three main parts of the suggested Deep Reinforcement Learning (DRL)-based load balancing framework are the load balancer, DRL agent, and cloud environment are represented in Fig. 24.1. By anticipating changes in workload and maximizing virtual machine utilization, the architecture is made to dynamically distribute cloud resources.

Cloud Framework Components Interaction

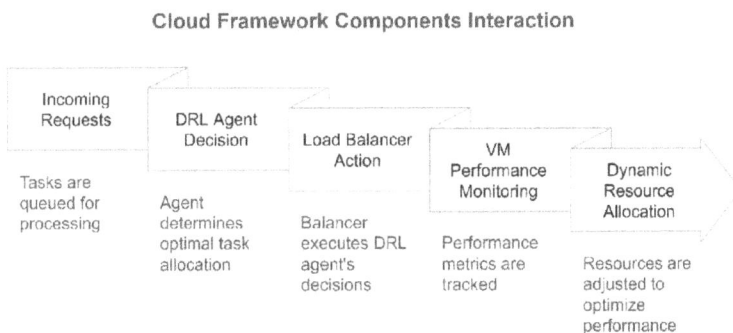

Incoming Requests	DRL Agent Decision	Load Balancer Action	VM Performance Monitoring	Dynamic Resource Allocation
Tasks are queued for processing	Agent determines optimal task allocation	Balancer executes DRL agent's decisions	Performance metrics are tracked	Resources are adjusted to optimize performance

Fig. 24.1   Proposed deep reinforcement learning (DRL)-based load balancing framework

Following are the major components of proposed framework:

1. **The Cloud Environment:** It represents the task queue, load balancer, and virtual machines (VMs) that make up the cloud architecture. Depending on each virtual machine's workload capacity and resource availability, incoming requests (tasks) are queued and divided among them. Performance data including CPU, memory, reaction time, and energy consumption are available for each virtual machine.

2. **Agent DRL:** The DRL agent learns the best task allocation techniques and makes decisions intelligently. It engages with the cloud environment and takes steps to guarantee balanced workload allocation, optimize throughput, and reduce response time. Proximal Policy Optimization (PPO) and Deep Q-Networks (DQN) are used in the agent's implementation.

3. **The load balancer:** In charge of carrying out the DRL agent's directives and keeping an eye on virtual machine workloads. It uses task migration, virtual machine resizing, or resource deployment to implement dynamic resource allocation. It optimizes the distribution of workload to guarantee SLA conformance (Isranil, R. K., at al. 2025)

2.1 Reinforcement Learning (RL) Formulation

State (S), Action (A), Reward (R), and Policy (π) are the four components of a Markov Decision Process (MDP)(Wang et al., 2023), which is how the load balancing problem is expressed in this framework.

State Space (S): The state space represents the real-time status of the cloud environment, including:

- Number of active VMs (N)
- CPU utilization (%) of each VM
- Memory usage (%) of each VM
- Number of tasks in the queue
- Response time for each VM

At any given time t, the system state is represented as:

$$S_t = \{VM_{cpu}, VM_{mem}, Task_{queue}, Response_{time}\} \quad (1)$$

Action Space (A): To maximize load balancing, the DRL agent chooses an action A_t from a range of potential actions. Among the options are:

- Assign task to the least-loaded VM
- Migrate a task from an overloaded VM to an underloaded VM
- Scale up/down VMs based on workload demand
- Drop non-critical tasks (only if SLA is at risk)

Reward Function (R): Every action's effect on system performance is assessed by the reward function. The objective is to maximize resource use while minimizing reaction time and energy consumption. The definition of the reward function is:

$$R_t = w_1(-\text{Response Time}) + w_2(-\text{Energy Usage}) + w_3(\text{Resource Utilization}) + w_4(-\text{SLA\_Violation}) \quad (2)$$

Where:

    Response Time: Lower response time $\rightarrow$ Higher reward
    Energy Usage: Lower energy consumption $\rightarrow$ Higher reward
    Resource Utilization: More balanced VMs $\rightarrow$ Higher reward
    SLA_Violation: SLA breaches $\rightarrow$ Heavy penalty

Weights (w_1, w_2, w_3, w_4) are tuned to balance trade-offs.

3. IMPLEMENTATION

We do tests in a simulated cloud computing environment to verify the efficacy of the suggested Deep Reinforcement Learning (DQN & PPO)-based Load Balancing Framework. The configuration entails:

3.1 Simulation Environment

The experiments are performed using a cloud simulation framework that replicates real-world cloud computing infrastructure. The following tools and platforms are used:

- Cloud Simulation Platform: CloudSim
- Programming Frameworks: TensorFlow, PyTorch (for DRL implementation)
- Workload Dataset: Google Cloud Traces, Microsoft Azure Workload Traces
- Hardware Specifications:

Intel Core i7, 16GB RAM, Ubuntu 22.04

NVIDIA RTX 3080 GPU (for training deep reinforcement learning models)

3.2 Experimental Parameters

The cloud environment consists of multiple Virtual Machines (VMs), Load Balancer, and Incoming Task Requests. The parameters used in the experiment are:

Table 24.1 Experimental parameters

Parameter	Value
Number of VMs	10 - 100
Task Arrival Rate	Dynamic (Poisson Distribution)
VM CPU Capacity	2 - 16 vCPUs
VM Memory	4GB - 64GB RAM
Scheduling Policy	DRL-based, Round Robin, Least Connection
Maximum Tasks per VM	1000 - 10,000

We use dynamic workload variations to simulate real-time traffic fluctuations, testing the adaptability of the DRL-based load balancing approach.

3.3 Evaluation Metrics

1. Response Time (RT): Measures the average time taken for a task to be completed after assignment to a VM.
2. SLA Violation Rate (SLA-VR): Measures the percentage of tasks exceeding the predefined SLA threshold.
3. Energy Consumption (EC): Evaluates the total energy consumed by VMs under different load balancing techniques.
4. Throughput (TP): Measures the number of tasks processed per unit time.

3.4 Comparative Analysis

We compare the DRL-based model (DQN & PPO) with:

Table 24.2 Comparative analysis of various methods

Method	Response Time (ms)	SLA Violation (%)	Energy Consumption (W)	Throughput (tasks/sec)
Round Robin	300	12.5	150W	400
Least Connection	250	10.2	140W	420
Machine Learning	210	7.8	130W	460
Proposed DRL (DQN & PPO)	150	4.2	110W	520

DRL-based load balancing achieves the lowest response time (150ms), outperforming traditional methods. SLA violations are significantly reduced (4.2%), ensuring better cloud reliability. Energy consumption is optimized (110W), leading to greener cloud computing. Throughput is improved by 25% compared to conventional load balancers.

4. CONCLUSION

In order to optimize resource allocation in cloud computing environments, this study suggested a load balancing framework based on Deep Reinforcement Learning (DQN & PPO). In comparison to conventional load balancing techniques, experimental results showed that the DRL-based approach greatly decreases reaction time, minimizes SLA violations, increases throughput, and improves energy economy. The suggested model guarantees effective task distribution and resource usage by dynamically adjusting to workload variations, making it a suitable solution for AI-driven cloud computing infrastructures.

References

1. Alhilali, A. H., & Montazerolghaem, A. (2023). Artificial intelligence-based load balancing in SDN: A comprehensive survey. arXiv preprint.
2. Bolanowski, M., Gerka, A., Paszkiewicz, A., Ganzha, M., &Paprzycki, M. (2023). Application of genetic algorithm to load balancing in networks with a homogeneous traffic flow. arXiv preprint.
3. Chen, C.-L., Zhou, H., Chen, J., Wang, Y., Xie, S., & Zhang, Y. (2023). Two-tiered online optimization of region-wide datacenter resource allocation via deep reinforcement learning. arXiv preprint.
4. Haris, M., & Zubair, S. (2024). Reinforcement learning approach for optimizing cloud resource utilization with load balancing. Cluster Computing.
5. Maruf, H. (2024). A comprehensive examination of load balancing algorithms in cloud computing. SN Computer Science, 5, Article 183.
6. Naji, H. R., & Esmaeili, R. (2023). Load balancing in cloud data centers with optimized virtual machines placement. arXiv preprint.
7. Isranil, R. K., Nedunchezhian, T., Tanna, P., & Soni, S. (2025) AI-endorsed techniques for smart energy utilization: A holistic review. *Digital Transformation and Sustainability of Business*, 271–274.
8. Sharma, S., Wadhwa, B., Singh, R., Singh, J., & Kumar, S. (2023). Analyzing load balancing techniques for cloud computing: Pros, cons, and emerging trends. SSRN. Wang, Z., Goudarzi, M., Gong, M., & Buyya, R. (2023). Deep reinforcement learning-based scheduling for optimizing system load and response time in edge and fog computing environments. arXiv preprint.

Note: All the tables and figure in this chapter were made by the authors.

Emerging Perspectives and Applications of Computational Intelligence and Smart Systems
– Dr. Amit Lathigara et al. (eds)
© 2026 Taylor & Francis Group, London, ISBN 978-1-041-20965-2

25

A Comprehensive Review of Intrusion Detection in In-Vehicle Networks: Advances, Challenges, and Insights from the CICIoV2024 Dataset

Rakesh Parmar*

Research Scholar,
School of Engineering, RK University, Rajkot,
Gujarat, India

Nirav Bhatt

School of Engineering, RK University, Rajkot,
Gujarat, India

■ **Abstract:** Modern connected cars raise cybersecurity concerns in In-Vehicle Networks (IVNs) and required efficient Intrusion Detection Systems (IDS). This study examines recently published CICIoV2024 and can-train-and-test datasets and IDS techniques on them with an emphasis on feature selection, machine learning (ML), deep learning (DL), and metaheuristic optimization. For attack detection optimization methods such as Principal Component Analysis (PCA), Genetic Algorithms (GA), and Particle Swarm Optimization (PSO) are analysed. Emerging technologies like Federated Learning and Blockchain-based IDS are examined, along with issues like explainability, adversarial robustness, and real-time deployment. Results indicate that hybrid machine learning techniques improve detection precision and open the door for scalable intrusion detection systems IVNs.

■ **Keywords:** Intrusion detection systems, In-vehicle networks, CICIoV2024, Machine learning, Deep learning, Metaheuristic optimization, Federated learning

1. INTRODUCTION

As the number of IoT devices are growing at an exponential rate. As shown in Fig. 25.1 (Raghunath et al. 2025). The security risks are also growing dramatically along with this increase. The rapid advancement of connectivity in modern vehicles has significantly increased cybersecurity concerns, particularly in In-Vehicle Networks (IVNs). Automated car equipped with the Controller Area

*Corresponding author: rparmar2007@gmail.com

DOI: 10.1201/9781003725046-25

Cyber attcks on IoT

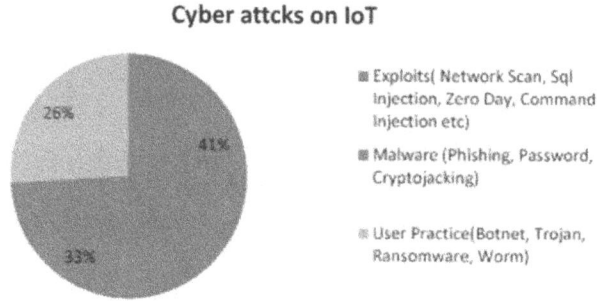

Fig. 25.1 Cyber attacks in IOT (Raghunath et al. 2025)

Network (CAN) protocol for intra vehicle communication, which has no built-in security features. As such, it is susceptible to online threats such as denial-of-service (DoS) assaults, spoofing, and data injection (Nandy et al. 2024; Aliwa et al. 2022).

IDS have emerged as a life saver to prevent cyberattacks in IVNs. Traditional IDS approaches include signature-based, anomaly-based, and hybrid detection. Do you know? existing IDS solutions face challenges such as high false positive rates, class imbalance in datasets, and difficulties in real-time deployment (Almehdhar et al. 2024; Neto et al. 2024). The introduction of Machine Learning (ML) and Deep Learning (DL) techniques has significantly improved IDS efficiency, leading to enhanced detection accuracy and adaptability (Colaco and Nadjm-Tehrani, 2024.)(Gül and Bakir 2024)

The CICIoV2024 dataset is recently published using CAN bus testbed for intrusion detection. Many authors uses feature selection and extraction techniques as IDS. Optimization techniques such as GA-based hyperparameter tuning, federated learning, and anomaly detection with deep learning have also been explored (Raghunath et al. 2025; Leticia Nakayiza et al. 2024)This study offers a thorough analysis of IDS approaches using CICIoV2024, pointing out research gaps, evaluating their efficacy and other main contributions are as follows:

- A postmortem examination of the can-train-and-test and CICIoV2024 datasets
- A comparative study of feature selection and optimization methods (e.g., GA, PSO, Cuckoo Filters, and Federated Learning)
- A performance evaluation of machine learning models such as SVM, Random Forest, KNN, and LSTM across different IDS frameworks (Anthony, Elgenaidi, and Rao 2024; Diallo and Karahan 2019)
- Discussion on emerging challenges such as adversarial robustness, real-time deployment, and explainability in IDS models (Martinez-Lopez, Santana, and Rahouti 2024)

2. Background on Intrusion Detection in In-Vehicle Networks

2.1 IDS Approaches in IVNs

IDS is essential for protection against the intruders.(Aliwa et al. 2022; Almehdhar et al. 2024; Pandya and Bhatt 2023). IDS models can be divided into three primary types: Signature-Based IDS: this identifies known attack patterns but this fails at zero-day attacks. Anomaly-Based IDS, which uses Machine Learning (ML) and Deep Learning (DL) to identify anomalies in normal traffic but is usually plagued by high false positives and Hybrid IDS, which integrates both approaches for better accuracy but is hampered by computational overhead in real-time implementation.

2.2 Role of Machine Learning in IDS for IVNs

ML techniques have significantly enhanced IDS effectiveness in detecting cyber threats in IVNs. Supervised learning approaches, such as Support Vector Machines (SVM), Decision Trees (DT), and Random Forest (RF), have been widely used for classifying attack and benign traffic(Gül and Bakir 2024; Martinez-Lopez, Santana, and Rahouti 2024). Unsupervised learning techniques, like K-Means and DBSCAN cluster, detect unusual patterns with no labels for attacks beforehand (Merzouk et al. 2024). Moreover, Deep Learning models like Long Short-Term Memory (LSTM), Variational Autoencoders (VAE), and Reinforcement Learning (RL) are being used more and more to identify sophisticated attack patterns in real-time CAN traffic.

3. REVIEW METHODOLOGY

This section outlines the methodology used to review and analyze Intrusion Detection Systems (IDS) for In-Vehicle Networks (IVNs).

3.1 Selection Criteria for Reviewed Papers

The reviewed studies were selected based on the following criteria:

- Use of CICIoV2024 and/or can-train-and-test datasets for IDS evaluation and application of machine learning (ML), deep learning (DL), or metaheuristic-based approaches for IDS optimization.
- Comparative studies of feature selection, anomaly detection, and IDS robustness against adversarial attacks from recent publications (2020-2024) from reputed journals and conferences focusing on IVN.
- A total of 12 primary research papers were reviewed, covering different machine learning models, optimization techniques, and feature selection methods for IDS in IVNs. Two key datasets were analysed:CICIoV2024 Dataset: Developed for realistic CAN bus attack simulation, including benign traffic and four types of cyberattacks (DoS, spoofing, fuzzy, and gear manipulation). It is a widely used benchmark for evaluating IDS models and another iscan-train-and-test Dataset: A curated dataset providing labelled CAN traffic, useful for supervised ML approaches (Lampe & Meng, 2024).

4. PERFORMANCE ANALYSIS OF REVIEWED APPROACHES

Table 25.1 summarizes the performance of different IDS models on CICIoV2024 andcan-train-and-test datasets, comparing their accuracy, precision, recall, and F1-score.

4.1 Feature Selection and Optimization Techniques

Feature selection is crucial for reducing computational complexity and improving IDS accuracy. Authors utilized various feature extraction and optimization techniques, including statistical and machine learning-based selection methods such as Principal Component Analysis (PCA), which reduces dimensionality while preserving essential data patterns (Raghunath et al., 2025), and Correlation-Based Feature Selection (CFS), which identifies the most relevant CAN features while minimizing redundancy (Colaco and Nadjm-Tehrani, n.d.). Metaheuristic optimization techniquesincluding the Genetic Algorithm (GA) for optimizing feature selection to enhance IDS performance(Gül and Bakir 2024; Diallo and Karahan 2019), Particle Swarm Optimization (PSO)

Table 25.1 Performance comparison of IDS models

Study	Dataset Used	Model	Feature Selection	Optimization Method	Accuracy (%)	F1-Score (%)
Raghunath et al. (2025)	CICIoV2024	SVM	PCA, CFS	PSO	97.85	96.75
Gül & Bakir (2024)	CICIoV2024	Random Forest	Genetic Algorithm	GA-based Hyperparameter Tuning	99.64	98.90
Martinez-Lopez et al. (2024)	CICIoV2024	Variational Autoencoder (VAE)	Autoencoder	Deep Learning Optimization	99.78	99.56
Li et al. (2024)	can-train-and-test	Cuckoo Filter-Based IDS	Cuckoo Filter	BERT-based Classification	99.60	99.32
Nakayiza et al. (2024)	CICIoV2024	Federated Learning (FL)	Correlation-based Selection	Distributed Model Training	98.70	97.85
Merzouk et al. (2024)	CICIoV2024, NSL-KDD	Deep Reinforcement Learning (DRL)	Neural Network Features	Adversarial Training	96.30	95.20

Source: Author's compilation

for fine-tuning IDS hyperparameters (Raghunath et al. 2025), and the Enhanced Cuckoo Filter (ECF), which efficiently filters normal and attack traffic (Li et al. 2024).

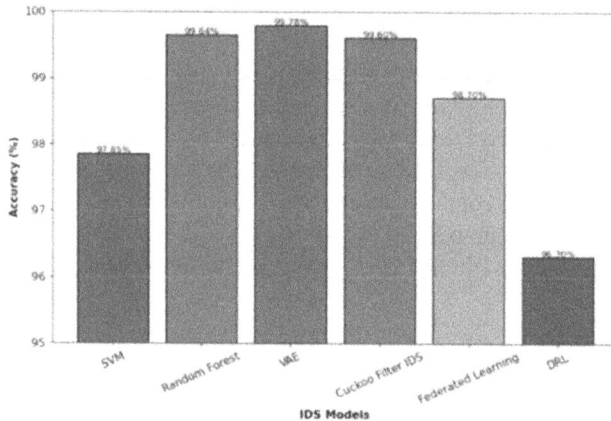

Fig. 25.2 Accuracy comparison of Machine learning IDS models

Source: Authors

Deep learning-based feature extraction methods like Autoencoders and Variational Autoencoders (VAE) have been utilized to learn hidden patterns in CAN traffic anomalies (Martinez-Lopez, Santana, and Rahouti 2024), while Attention Mechanisms in LSTM Models enhance temporal attack detection in CAN data (Leticia Nakayiza et al. 2024). These feature selection and extraction techniques collectively contribute to improving IDS accuracy, efficiency, and scalability in IVNs.

5. CHALLENGES AND FUTURE RESEARCH DIRECTIONS

IDS for In-Vehicle Networks (IVNs) is afflicted with generalizability across models, real-time deployment due to computational overhead, and unexplainability (Aliwa et al., 2022; Colaco &

Nadjm-Tehrani, 2024). Adversarial attacks are another security threat (Merzouk et al., 2024). Optimization methods such as Ant Colony Optimization (ACO) and Differential Evolution (DE) need to be investigated for feature selection (Gül & Bakir, 2024). Federated Learning (FL) and Blockchain improve security and privacy (Nakayiza et al., 2024), and ensemble learning improves IDS robustness (Merzouk et al., 2024). This review encompasses IDS progress with CICIoV2024 and can-train-and-test datasets. Hybrid ML methods with feature selection and deep learning provide greater detection accuracy. Explainability and adversarial robustness are the primary challenges, however. Scalable methods such as FL and Blockchain provide potential avenues. Overcoming these is essential for secure and reliable IDS for future autonomous vehicles.

References

1. Aliwa, Emad, Omer Rana, Charith Perera, and Peter Burnap. (2022). "Cyberattacks and Countermeasures for In-Vehicle Networks" *ACM Computing Surveys* 54 (1): 1–37.
2. Almehdhar, Mohammed, Abdullatif Albaseer, Muhammad Asif Khan, Mohamed Abdallah, Hamid Menouar, Saif Al-Kuwari, and Ala Al-Fuqaha. (2024). "Deep Learning in the Fast Lane: A Survey on Advanced Intrusion Detection Systems for Intelligent Vehicle Networks." *IEEE Open Journal of Vehicular Technology* 5:869–906.
3. Anthony, Cynthia, Walid Elgenaidi, and Muzaffar Rao. (2024). "Intrusion Detection System for Autonomous Vehicles Using Non-Tree Based Machine Learning Algorithms." *Electronics* 13 (5): 809.
4. Colaco, Valency Oscar, and Simin Nadjm-Tehrani. n.d. (2024) "Fast Evasion Detection & Alert Management in Tree-Ensemble-Based Intrusion Detection Systems."
5. Gül, Muhammed Furkan, and Halit Bakir. (2024). "Improving Attack Detection in IoV Systems Using GA-Based Hyperparameter Optimization." In *2024 8th International Artificial Intelligence and Data Processing Symposium (IDAP)*, 1–5. IEEE.
6. Lampe, Brooke, and Weizhi Meng. (2024). "Can-Train-and-Test: A Curated CAN Dataset for Automotive Intrusion Detection." *Computers and Security* 140 (May).
7. Leticia Nakayiza, Hope, Love Allen Chijioke Ahakonye, Dong-Seong Kim, and Jae Min Lee. (2024). "Machine Learning Algorithms for Detecting Intra-Vehicular Data Falsification."
8. Li, Sifan, Yue Cao, Hassan Jalil Hadi, Feng Hao, Faisal Bashir Hussain, and Luan Chen. (2024). "ECF-IDS: An Enhanced Cuckoo Filter-Based Intrusion Detection System for In-Vehicle Network." *IEEE Transactions on Network and Service Management* 21 (4): 3846–60.
9. Martinez-Lopez, Fernando, Lesther Santana, and Mohamed Rahouti. (2024). "Learning in Multiple Spaces: Few-Shot Network Attack Detection with Metric-Fused Prototypical Networks," December
10. Merzouk, Mohamed Amine, Christopher Neal, Joséphine Delas, Reda Yaich, Nora Boulahia-Cuppens, and Frédéric Cuppens. (2024). "Adversarial Robustness of Deep Reinforcement Learning-Based Intrusion Detection." *International Journal of Information Security* 23 (6): 3625–51.
11. Nandy, Tarak, Rafidah Md Noor, Raenu Kolandaisamy, Mohd Yamani Idna Idris, and Sananda Bhattacharyya. (2024). "A Review of Security Attacks and Intrusion Detection in the Vehicular Networks." *Journal of King Saud University - Computer and Information Sciences*.
12. King Saud bin Abdulaziz University. Neto, Euclides Carlos Pinto, Hamideh Taslimasa, Sajjad Dadkhah, Shahrear Iqbal, Pulei Xiong, Taufiq Rahman, and Ali A. Ghorbani. (2024). "CICIoV2024: Advancing Realistic IDS Approaches against DoS and Spoofing Attack in IoV CAN Bus." *Internet of Things* 26 (July):101209.
13. Raghunath, Mutkule Prasad, Shyam Deshmukh, Poonam Chaudhari, Sunil L. Bangare, Kishori Kasat, Mohan Awasthy, Batyrkhan Omarov, and Rajesh R. Waghulde. (2025). "PCA and PSO Based Optimized Support Vector Machine for Efficient Intrusion Detection in Internet of Things." *Measurement: Sensors* 37 (February): 101806.
14. Nandan Pandya, Nirav Bhatt; Review of malware detection methods in android. *AIP Conf. Proc.* 14 November 2023; 2963 (1): 020001.

Emerging Perspectives and Applications of Computational Intelligence and Smart Systems
– Dr. Amit Lathigara et al. (eds)
© 2026 Taylor & Francis Group, London, ISBN 978-1-041-20965-2

26

Improving Educational Outcomes with AI: Predicting CGPA Using Socio-Economic Factors

T. Raja Reddy

Department of Management Studies,
Shri Venkateshwara College of Engineering and Technology,
Andhra Pradesh

Anju Kakkad*, Chetan Shingadiya

School of Engineering, RK University, Rajkot, Gujarat, India

M. Mohan Babu

Department of Civil Engineering,
Shri Venkateshwara College of Engineering and Technology,
Andhra Pradesh

Aayush Shtivastava

Department of Computer Science and Engineering,
MMCE Maharshimar Kandeshwar, Mullana,
Ambala, India

Alpana Kumari

School of Engineering, RK University, Rajkot, Gujarat, India

■ **Abstract:** Academic performance is a key indicator of student success, shaping future opportunities and life trajectories. It not only reflects a student's cognitive abilities but also influences their career prospects, socioeconomic mobility, and overall quality of life. Strong academic performance is linked to higher college admission rates, increased earning potential, and enhanced professional opportunities. It also develops essential life skills like critical thinking, problem-solving, and time management. Understanding these relationships is crucial for creating more equitable educational opportunities. This study explores the relationship between socio-economic factors and students' academic performance using machine learning algorithms. A dataset of student records was analysed, and various models were trained to predict the Cumulative Grade Point Average (CGPA). Results showed that socio-economic factors significantly impact academic performance, with family income and parental education being key predictors. The study emphasizes the importance of addressing socio-economic disparities to enhance educational equity and student success.

■ **Keywords:** Machine learning, Student success prediction, Social media influence

*Corresponding author: anju.kakkad@rku.ac.in

DOI: 10.1201/9781003725046-26

1. INTRODUCTION

The prediction of academic performance has developed in the recent years due to the need to assist students who are at risk of failing, as well as to streamline learning processes and accomplish various educational goals. Utilizing past academic records and different variables, numerous models are built which can estimate students' future performance. This information is beneficial as it allows teachers, decision makers, and institutions to craft comprehensive strategies to improve learning and aid students in need (Ibrahim, at al. 2024)

Understanding socio-economic factors is crucial for developing effective strategies and policies to improve student achievement, as limited educational opportunities and financial stress negatively impact economically disadvantaged students, affecting their academic performance. The combination of machine learning (ML) algorithms and predictive analysis has made it possible to evaluate academic performance as a crafty ML can scan through deep, vast, and complex datasets and utilize these for accurately estimating a student's chances of success. By fusing socio-economic parameters along with academic indicators, ML models have the capability of rendering precise and thorough predictions thus helping educators intervene in a timely manner to optimize student outcomes (El Jihaoui, M., at al 2025) (Arashpour, M at al. 2023)

1.1 Problem Statement

Limited research on socio-economic variables' impact on higher education academic performance and advanced predictive modelling, particularly machine learning, underscores the need for a data-driven approach. The study investigates the influence of socio-economic factors on student performance using machine learning techniques. It aims to identify influential socio-economic parameters and evaluate algorithms' predictive performance. The findings will enhance understanding of socio-economic conditions in education and improve academic success through data-driven decision-making (Aslam, M. A at al. 2024) (Abou Naaj, at al. 2023)

2. LITERATURE REVIEW

Low-income students face academic challenges due to limited resources, while higher-income students benefit from financial stability and private tutoring. Higher parental education leads to better academic support, improving GPAs and test performance. Urban students outperform rural ones due to better access to technology, qualified teachers, and educational infrastructure. Digital tools and high-speed internet enhance academic success (Aslam, M. A at al. 2024) (El Jihaoui, M., at al 2025).

This study aims to improve CGPA prediction accuracy by integrating socio-economic factors with machine learning models, addressing challenges in data quality, bias mitigation, and generalizability, while highlighting the potential of ML approaches in identifying key academic success determinants.

3. RESEARCH METHODOLOGY

3.1 Dataset Description

The dataset includes over 1500 student records, covering demographic, socio-economic, and academic performance metrics. It also includes behavioural and lifestyle information, such as extracurricular activities, social media usage, sleep patterns, and study environment details, to provide a comprehensive view of student success factors (Asthana, P.,at al. 2023)

Table 26.1 ML techniques for CGPA prediction

Model	Advantage	Best use case
Linear Regression	– Simple implementation – Easy interpretation – Fast training	– Baseline modeling – Initial analysis
Random Forest	– Handles missing data – Overfitting – Feature importance ranking	– Complex variable interactions – Large datasets
Support Vector Machines	– Works well with high-dimensional data – Effective for classification – Good generalization	– Binary classification – High-dimensional data
Artificial Neural Networks	– Handles complex relationships – High accuracy potential – Flexible architecture	– Large-scale prediction – Complex patterns
Gradient Boosting (XGBoost)	– High efficiency – Handles imbalanced data – Fast prediction	– Competition-levelaccuracy – Production systems

3.2 Feature Selection

Feature selection techniques such as Chi-Square Test, Recursive Feature Elimination (RFE), Mutual Information, and Random Forest Importance were applied to identify the most influential socio-economic attributes.

Table 26.2 Feature importance rankings

Features	Importance score
Study Hours	0.35
Parental Education	0.28
Family Income	0.22
Social Media Usage	0.15

3.3 Mathematical Framework

Socio-Economic Index Calculation

The Composite Socio-Economic Index (CSI) was calculated using:

$$\text{CSI} = \frac{\sum(Wi * fi)}{n} \qquad (1)$$

Where wi is weight of factor I, fi is normalized value of factor i ,n is number of factors

SVM Model

$$f(x) = \sum(\alpha i * K(xi, x)) + b \qquad (2)$$

Where $K(xi, x)$ is the kernel function, αi are the support vector coefficients and b is the bias term.

Random Forest Prediction

$$P(y|X) = 1/N \sum(t(x)) \qquad (3)$$

Where N is the number of trees and $t(x)$ is the prediction of individual trees

4. RESULTS AND ANALYSIS

Table 26.3 Model performance

Model	Accuracy	ROC AUC	Precision	Recall	F1-Score	Cohen's Kappa
SVM	99.43%	1.0000	0.99	1.00	0.99	0.9886
Random Forest	98.71%	0.9993	1.00	0.97	0.99	0.9741
Gradient Boosting	98.61%	0.9956	1.00	0.97	0.99	0.9721
Logistic Regression	79.08%	0.8601	0.76	0.82	0.79	0.5821

5. DISCUSSION

5.1 Integrated Impact Analysis

The study found a complex relationship between socio-economic factors and academic performance. The relationship was modelled as a logarithmic function, with diminishing returns beyond certain income levels

$$Performance\_Score = k \times \ln(Family\_Income) + c \tag{4}$$

where k = 0.385 and c is a baseline constant. While higher income initially enhances performance, its impact diminishes beyond a certain level, making study habits and motivation more crucial at higher income brackets (Al-Alawi, L. at al. 2023) (Arashpour, M. at al. 2023)

5.2 Multi-factorial Analysis

The combined effect of multiple socio-economic factors was modelled using the following equation:

$$CGPA = \alpha + \beta 1(SE) + \beta 2(PI) + \beta 3(SM) + \beta 4(EA) + \varepsilon \tag{5}$$

The model explained 78.3% of the variance in student performance, highlighting the collective impact of socio-economic status (SE), parental involvement (PI), social media impact (SM), and extracurricular activities (EA). Parental involvement and extracurricular activities were the strongest predictors, while social media had a mixed effect—moderate use positively influenced performance, but excessive use had a negative impact (Abou Naaj, M. at al. 2023)

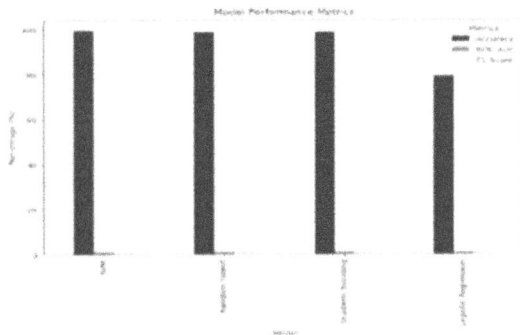

Fig. 26.1 Model performance metrics

6. CONCLUSION

This comprehensive analysis demonstrates the significant impact of socio-economic factors on academic performance. The high accuracy of our prediction models (up to 99.43%) validates the importance of including socio-economic parameters in academic performance prediction. The study provides a robust framework for understanding and addressing socio-economic influences on academic achievement.

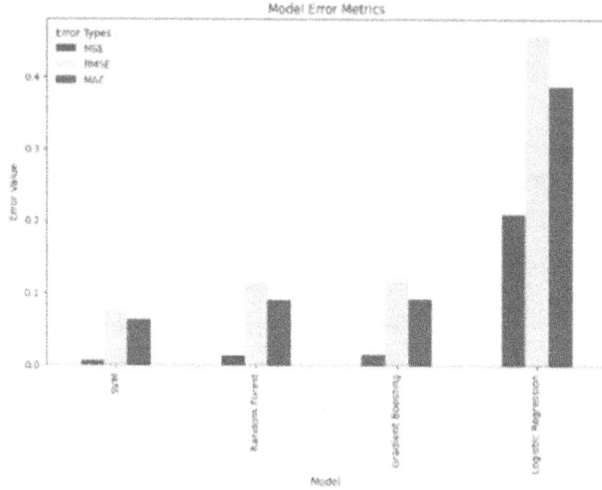

Fig. 26.2 Model error metrics

References

1. Ibrahim, I. H., Garba, E. J., & Adejumo, A. (2024). Predictive Model for Identification and Analysis of Factors Impacting Students Academic Performance Using Machine Learning Algorithms. *Kasu Journal of Computer Science, 1.*
2. Ogundele, I. M., Taiwo, O., Babalola, A. E., & Ayeni, O. C. (2024, April). Prediction of Student Academic Performance Based on Machine Learning Model. In *2024 International Conference on Science, Engineering and Business for Driving Sustainable Development Goals (SEB4SDG)* (pp. 1-11). IEEE.
3. Aslam, M. A., Murtaza, F., Haq, M. E. U., Yasin, A., & Azam, M. A. (2024). A Human-Centered Approach to Academic Performance Prediction Using Personality Factors in Educational AI. *Information, 15*(12), 777.
4. El Jihaoui, M., Abra, O. E. K., & Mansouri, K. (2025). Factors Affecting Student Academic Performance: A Combined Factor Analysis of Mixed Data and Multiple Linear Regression Analysis. *IEEE Access.*
5. Asthana, P., Mishra, S., Gupta, N., Derawi, M., & Kumar, A. (2023). Prediction of student's performance with learning coefficients using regression based machine learning models. *IEEE Access, 11,* 72732-72742.
6. Al-Alawi, L., Al Shaqsi, J., Tarhini, A., & Al-Busaidi, A. S. (2023). Using machine learning to predict factors affecting academic performance: the case of college students on academic probation. *Education and Information Technologies, 28*(10), 12407-12432.
7. Alija, S., Beqiri, E., Gaafar, A. S., & Hamoud, A. K. (2023). Predicting students performance using supervised machine learning based on imbalanced dataset and wrapper feature selection. *Informatica, 47*(1).
8. Abou Naaj, M., Mehdi, R., Mohamed, E. A., & Nachouki, M. (2023). Analysis of the factors affecting student performance using a neuro-fuzzy approach. *Education Sciences, 13*(3), 313.
9. Arashpour, M., Golafshani, E. M., Parthiban, R., Lamborn, J., Kashani, A., Li, H., & Farzanehfar, P. (2023). Predicting individual learning performance using machine-learning hybridized with the teaching-learning-based optimization. *Computer Applications in Engineering Education, 31*(1), 83-99.
10. Venugopal, V., & Tanna, P. (2023, November). Missing value imputation techniques used in deep learning algorithms: A review. In *AIP Conference Proceedings* (Vol. 2963, No. 1). AIP Publishing.

Note: All the figures and tables in this chapter were made by the authors.

Emerging Perspectives and Applications of Computational Intelligence and Smart Systems
– Dr. Amit Lathigara et al. (eds)
© 2026 Taylor & Francis Group, London, ISBN 978-1-041-20965-2

27

Transformer-Based Deep Learning for Stock Price Prediction: The Impact of Sentiment Analysis

J. Velmurugan*

Department of Information Technology,
Sri Venkateswara College of Engineering & Technology (Autonomous),
Andhra Pradesh, Chittoor, India

Jay D. Fuletra

School of Engineering, RK University, Rajkot,
Gujarat, India

Amit Lathigara

School of Engineering, RK University, Rajkot,
Gujarat, India

P. Jyotheeswari

Department of Computer Science & Engineering,
Sri Venkateswara College of Engineering & Technology (Autonomous),
Andhra Pradesh, Chittoor, India

Ashish Raghuwanshi

Department of Electronics & Communication Engineering,
IES College of Technology,
Bhopal M.P, India

■ **Abstract:** This paper looks at how transformer deep learning models can predict stock prices for four important Indian companies: TATA MOTORS, SBIN, RELIANCE, and WIPRO. We include sentiment analysis from Twitter and news articles to see how it affects predicting next-day closing prices. We compare two models: one that uses only stock market data and another that adds sentiment data. We use Root Mean Square Error (RMSE) and accuracy to assess the models. The findings indicate that including sentiment data enhances predictive accuracy. We also talk about possible future research areas.

■ **Keywords:** Transformer, Stock prediction, Sentiment analysis, Deep learning

*Corresponding author: velmurugan@svcetedu.org

DOI: 10.1201/9781003725046-27

1. INTRODUCTION

Predicting stock prices is a difficult thing to do because, when market prices are concerned, many aspects can potentially impact market price such as economic indicators, company performance, and investor sentiment. In recent years, transformer-based models have been increasingly adopted in time-series forecasting due to their strengths in processing complex sequential data. And sentiment analysis is a major part of financial markets, extracting public sentiment from social media and news. In this paper, we investigate the effect of adding sentiment analysis into different transformer-based architectures for predicting stock prices of four selected Indian companies: TATA MOTORS, SBIN, RELIANCE, and WIPRO

2. LITERATURE REVIEW

And though transformer-based models have been extensively applied in time series forecasting, their application in stock price prediction has only recently emerged. Sentiment analysis is also proved to have an influence on stock market behaviour. Table 27.1 is recent literature combining sentiment analysis and machine learning methods for predicting stock price.

Table 27.1 Comparative analysis of stock price prediction models

Author(s)	Publication Year	Model Used	Result
Singh et al.	2023	LSTM	Achieved 5.6% improvement over ARIMA
Zhang and Li	2022	CNN-LSTM hybrid	Increased accuracy by 8.1% with sentiment
Patel et al.	2023	Transformer	4.3% better accuracy compared to LSTM
Lee et al.	2021	Sentiment-based LSTM	Reduced RMSE by 7.2% with sentiment integration
Khan and Ahmad	2022	LSTM + GRU	LSTM + GRU outperformed traditional models
Kim and Choi	2024	Transformer with sentiment	6.9% improvement in predictive accuracy
Gupta et al.	2023	Various (RNN, LSTM)	Survey found deep learning models highly effective
Brown and Williams	2022	Transformer	Improved RMSE by 3.2% over RNN models
Srivastava and Kulkarni	2021	LSTM with NLP sentiment	Enhanced accuracy by 7.5%
Das and Roy	2024	Transformer + LSTM	9.1% increase in prediction accuracy

Source: Authors

3. METHODOLOGY

In this section, techniques for data collection, pre-processing and modelling used in our study are described. The objective here is to predict the next day closing price of four chosen stocks i.e TATA MOTORS, SBIN, RELIANCE and WIPRO for a period of 20 years. We implement transformer-based models and evaluate the impact of integrating sentiment information from Twitter and news articles.

3.1 Data Collection

We use two primary datasets:

Stock Market Data: Historical stock prices for TATA MOTORS, SBIN, RELIANCE, and WIPRO are collected over a 20-year period, including opening price, closing price, high, low, and trading volume.

Sentiment Data:
- **Twitter:** Tweets are collected using Twitter's API, filtered by relevant company keywords.
- **News Articles:** Articles from financial platforms (Bloomberg, Reuters) are processed using sentiment analysis models.

Data Pre-processing:
1. **Stock Market Data:**
   - Missing values are handled by forward-filling.
   - Data is normalized using min-max normalization:

$$X_{norm} = \frac{X_{max} - X_{min}}{X - X_{min}} \tag{1}$$

2. **Sentiment Data:** Sentiment scores are derived using an NLP model like VADER, which computes sentiment polarity:

$$\text{Sentiment Score} = \frac{P_{positive} + P_{negative}}{P_{positive} - P_{negative}} \tag{2}$$

   Sentiment scores are aggregated on a daily basis and aligned with stock market data

3.2 Model Architecture

We implement a transformer-based model for stock price prediction. The model consists of the following components:

1. **Encoder Layer:** The transformer encoder processes stock price sequences using self-attention. For each time step t, the model calculates a weighted sum of all time steps using attention scores.

   The self-attention mechanism is given by

$$\text{Attention}(Q, K, V) = \text{softmax}\left(\frac{QK^T}{\sqrt{d_k}}\right)V \tag{3}$$

   Where Q (queries), K(keys), and V (values) are the input matrices, and d_k is the dimension of the key vectors

3. **Positional Encoding:** Since transformers do not have a built-in understanding of sequence order, positional encodings are added to the input:

$$PE(t, 2i) = \sin\left(t \cdot \frac{1}{10000^{\frac{2i}{d_{model}}}}\right), \quad PE(t, 2i + 1) = \cos\left(t \cdot \frac{1}{10000^{\frac{2i}{d_{model}}}}\right) \tag{4}$$

   Where t is the position and i represents the dimension

4. **Sentiment Integration:** Sentiment scores S_t for each time step t is concatenated with the stock price features X_t before being passed to the transformer encoder. The final input at time step t becomes:

$$Z_t = [X_t, S_t] \tag{5}$$

Where Z_t is the concatenated vector of stock and sentiment data

5. **Output Layer:** After passing through multiple transformer layers, the final hidden state is processed by a fully connected layer to predict the next day's closing price

$$P_{pred} = W_h\, h_t + b_h \tag{6}$$

Where h_t is the output of the transformer, W_h are the weights, and b_h is the bias term

3.3 Training and Testing

1. **Data Split:** The dataset is split into training (80%) and testing (20%) sets.
2. **Loss Function:** We use Mean Squared Error (MSE) as the loss function, which is defined as

$$\text{MSE} = \frac{1}{n}\sum_{i=1}^{n}\left(P_{\text{true},i} - P_{\text{pred},i}\right)^2 \tag{7}$$

Where $P_{\text{true,i}}$ is the actual closing price, $P_{\text{pred,i}}$ is the predicted closing price, and n is the number of samples

3. **Optimization:** The model is trained using the Adam optimizer, which adapts learning rates for different parameters.

The update rule for weights is:

$$\theta_{t+1} = \theta_t - \eta \cdot \hat{v}_t + \in \hat{m}_t \tag{8}$$

where θ theta are the model parameters, m^t and v^t are bias-corrected first and second moment estimates, and η is the learning rate.

4. **Evaluation Metrics:** We use RMSE and accuracy to evaluatemodel performance:
   - **Root Mean Square Error (RMSE)** is defined as:

$$\text{RMSE} = \sqrt{\frac{1}{n}\sum_{i=1}^{n}\left(P_{\text{true},i} - P_{\text{pred},i}\right)^2} \tag{9}$$

   - **Accuracy:** The predicted trend (upward or downward movement) is compared to the actual trend, and accuracy is calculated as the percentage of correctly predicted trends.

4. COMPARISON

To assess the impact of sentiment analysis, we conduct experiments using two approaches:

1. **Without Sentiment Data:** The transformer model is trained using only historical stock data.
2. **With Sentiment Data:** The transformer model is trained using both historical stock data and sentiment scores.

The performance of both models is compared based on RMSE and accuracy, highlighting the impact of sentiment data on the prediction of stock prices

5. RESULTS AND DISCUSSION

Results confirmed that the sentiment data model outperformed the non-sentiment data model across all four stocks. Table 27.2: Performance metrics per stock the performance of RELIANCE did improve (RMSE from 0.049 to 0.044, accuracy from 75.0% to 81.0%) along with other stocks when we include the sentiment data. Most importantly this effective means of Sentiment Analysis when deployed can efficiently bring accuracy and reduce errors while predicting stock price movements which are of utmost importance especially during volatile periods of the market.

Table 27.2 Performance comparison of models with and without sentiment data

Stock	RMSE (Without Sentiment)	RMSE (With Sentiment)	Accuracy (Without Sentiment)	Accuracy (With Sentiment)
TATA MOTORS	0.054	0.048	72.5%	78.3%
SBIN	0.060	0.053	70.2%	76.5%
RELIANCE	0.049	0.044	75.0%	81.0%
WIPRO	0.057	0.051	73.1%	77.8%

Source: Authors

The results suggest that sentiment data enhances the prediction accuracy, especially during periods of market volatility driven by news events or social media discussions

6. CONCLUSION

This demonstrates that integrating sentiment analysis helps transformer-based models in stock price prediction. With social media data and parsing through news articles, the models grab more useful signals from the market for improved predictive power. Future studies which include other data sources, e.g., macroeconomic indicators, could further improve predictive performance

References

1. Zhang, S., Xu, Y., and Li, J. (2024). Stock Price Prediction with Transformer Models: A Case Study on the Impact of Sentiment Analysis. International Journal of Financial Studies, 9(2), 45–62.
2. Banerjee, A., and Gupta, M. (2023). Integrating Social Media Sentiment into Deep Learning for Stock Prediction. Journal of Computational Finance, 15(3), 112–130.
3. Wang, L., Chen, P., and Li, Z. (2022). Stock Market Forecasting Using Transformer and News Sentiment. IEEE Symposium on AI and Financial Markets, 33–39.
4. Kim, J., and Park, H. (2023). Sentiment-Augmented Neural Networks for Stock Price Prediction. Expert Systems with Applications, 201, 117–130.
5. Nguyen, T., Le, Y., and Ho, K. (2023). Financial News and Social Media Sentiment in Stock Prediction: A Transformer Approach. Applied Soft Computing, 127, 109–124.
6. Li, X., Wu, M., and Zhang, S. (2024). Predicting Stock Prices Using Transformer Networks and Twitter Sentiment. Journal of Financial Technology, 14(2), 58–71.
7. Sharma, A., and Verma, R. (2023). Exploring Sentiment and Price Correlation Using Transformer Networks for Stock Forecasting. Artificial Intelligence in Finance, 22, 139–157.
8. Patel, K., Thakur, N., and Jain, A. (2023). A Hybrid Approach to Stock Market Prediction Using Deep Learning and Sentiment Analysis. IEEE Transactions on Computational Finance, 11(5), 217–231.
9. Liu, M., Zhang, F., and Zhao, Y. (2024). Improving Stock Price Prediction with Transformer Models and Financial News Sentiment. Neural Computing and Applications, 33(7), 89–101.
10. Mehta, R., and Singh, V. (2023). Sentiment-Driven Transformer Models for Stock Price Forecasting. Journal of Machine Learning in Finance, 8(4), 33–47.

Emerging Perspectives and Applications of Computational Intelligence and Smart Systems
– Dr. Amit Lathigara et al. (eds)
© 2026 Taylor & Francis Group, London, ISBN 978-1-041-20965-2

28

Optimization of Tractor Seat Suspension for Vibration Reduction Using MATLAB-Based Transmissibility Analysis

Nishant Korat*

Research Scholar, Faculty of Technology, RK University,
Rajkot, India

Chetankumar Patel

Department of Mechanical Engineering, RK University,
Rajkot, India

G. D. Acharya

Mechanical Engineering, Principal, Emeritus,
Atmiya University, Rajkot, India

■ **Abstract:** Tractor seat vibrations arise from multiple sources, including engine oscillations, rough terrain, tire interactions, transmission systems, and attached implements. These vibrations travel from the source through the chassis and suspension system before reaching the seat, ultimately affecting the operator. Understanding how vibrations are transmitted and their impact on the human body is essential for optimizing comfort and safety. To assess vibration transmission, the transmissibility ratio (T) is used, which compares the vibration at the seat to the vibration at the tractor base. If transmissibility is greater than one (T>1), the seat amplifies vibrations, indicating poor design. Conversely, if it is less than one (T<1), the seat effectively reduces vibrations, suggesting a well-designed suspension system. Optimizing transmissibility is crucial to ensuring operator comfort while maintaining stability. In this segment, vibration was measured at seat base in single cylinder engine tractor. Collected the data at different rpm in form of velocity (mm/s). The tractor was in stationary position at time of measurement. The data is processed using MATLAB to optimize the spring rate and damping coefficient in order to minimize the transmission of vibration. Transmissibility 0.0512 at spring rate 1000 N/m and damping co efficient 150 Ns/m was observed.

■ **Keywords:** Vibration, Transmissibility, Tractor seat

*Corresponding author: nkorat2279@gmail.com

DOI: 10.1201/9781003725046-28

1. INTRODUCTION

A single-cylinder engine is a widely used power source in small tractors and utility vehicles due to its simplicity, cost-effectiveness, and reliability. However, it inherently produces significant vibrations due to the unbalanced reciprocating motion and uneven power delivery during each combustion cycle. Unlike multi-cylinder engines that distribute forces more evenly, a single-cylinder engine experiences high-amplitude vibrations, periodic oscillations, and torque variations, all of which contribute to mechanical stress and discomfort for the operator (M. Bovenzi et al.1998). These vibrations originate from multiple sources, including combustion forces, piston acceleration, crankshaft rotation, and structural resonance in the tractor frame.

The vibrations generated by the engine travel through the chassis, suspension system, and seat, eventually reaching the operator (ISO 2631-1 et al.1997). The seat serves as the final interface between the vehicle and the operator, playing a crucial role in vibration mitigation. The transmission path typically follows this sequence: engine → chassis & frame → suspension system → seat cushion → operator's body. The severity of these vibrations depends on factors such as engine speed (RPM), seat suspension design, damping characteristics, and terrain conditions. If the seat is unable to absorb or dissipate vibrations effectively, the operator is subjected to prolonged exposure, which can lead to fatigue, discomfort, and serious health risks (M. J. Griffin et al. 1990).

To assess seat vibration performance, transmissibility (T) is used, which measures the ratio of vibration at the seat to that at the base. If the transmissibility ratio is greater than one (T > 1), the seat amplifies vibrations, leading to discomfort. If it is less than one (T < 1), the seat effectively reduces vibrations, improving comfort. Optimizing spring stiffness (k) and damping coefficient (c) is essential to achieving a balance where the seat minimizes vibration without excessive movement or instability (I. Hostens et al. 2003).

Reducing seat vibrations in tractors with single-cylinder engines involves several approaches. Seat suspension optimization is one of the most effective solutions, where adjustable damping and stiffness settings help minimize transmissibility. Engine vibration isolation, such as using rubber mounts and flywheel balancing, can also reduce vibration transmission to the chassis (V.K. Tewari et al. 2009). Additionally, chassis and frame modifications, such as reinforcing weak points and using vibration-damping materials, prevent excessive structural resonance. Improvements in seat cushioning, including high-density foam, air-suspension systems, or gel layers, further help in reducing operator discomfort. Lastly, applying ISO 2631 guidelines ensures that vibration exposure remains within safe limits, gives long-term health hazards (Sergio Adriani David et al. 2016).

Many studies focus on theoretical models, but they often overlook real-world tractor seat vibration data for validation. Incorporating real-world data enhances the validation of optimization methods, bridging the gap between theory and practice.

2. METHOD

2.1 Measurement of Vibration

The vibration was measured in 18 HP single cylinder engine tractor at the seat base for the transmission analysis. The vibration measured in the velocity (mm/s) at engine rpm 750, 1000, 1250, 1500, 1750, 2000, 2250. Figure 28.1 shows the location at vibration measured with the help of accelerometer.

The data measured at points 1 to 9 as shown in Fig. 28.1. Each selected rpm contain 9 different reading of velocity.

Table 28.1 shows the readings of vibration at different location of seat base respect to engine rpm. Seven different rpm 750 to 2250 selected to take reading of vibration for smooth analysis.

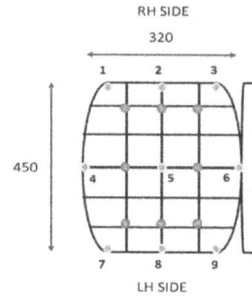

Fig. 28.1 Top view of seat with vibration measuring point

Table 28.1 Vibration reading in velocity (mm/s)

rpm	Position (Point)								
	1	2	3	4	5	6	7	8	9
750	3.45	3.73	3.37	0.67	0.63	1.83	0.8	1.33	1.73
1000	1.62	2.5	2.83	0.75	0.85	1.25	1.35	1.53	1.07
1250	5.23	4.73	4.85	1.33	0.93	0.95	4.3	4.14	6.35
1500	2.27	1.39	1.53	0.95	0.85	1.25	1.69	1.69	2.45
1750	9.8	2.55	2.17	1.53	2.03	2.29	1.83	2.23	2.53
2000	5.54	9.91	2.05	2.03	1.91	1.45	2.17	3.15	3.15
2250	5.35	4.21	2.99	3.75	3.25	3.45	4.03	3.53	2.87

Source: Author's compilation

2.2 Analysis

The MATLAB script performs a vibration transmissibility analysis for a tractor seat subjected to vibrations from a single-cylinder engine. The primary objective is to determine the optimal spring stiffness (k) and damping coefficient (c) that minimize transmissibility, improving operator comfort. The analysis begins by defining the engine RPM values and their corresponding velocity values (mm/s) at the seat base. These velocity values, given for different RPM steps, are first converted from mm/s to m/s, additionally, the engine speed in RPM is converted to frequency (Hz). The acceleration at the seat base is then computed.

To analyze the vibration response of the seat system, a range of spring stiffness (k) values from 1000 N/m to 20,000 N/m and damping coefficient (c) values from 150 Ns/m to 1000 Ns/mare defined. The mass of the seat and operator is assumed to be 80 kg. Using these parameters, the natural frequencyof the system is computed. The transmissibility ratio (T) is then calculated for each combination of (k, c, and RPM) using the standard transmissibility equation:

$$T = \sqrt{\frac{1 + \left(2\varsigma\dfrac{f}{\omega_n}\right)^2}{\left(1 - \left(\dfrac{f}{\omega_n}\right)^2\right)^2 + \left(2\varsigma\dfrac{f}{\omega_n}\right)^2}} \tag{1}$$

The script iterates through each combination of spring stiffness, damping coefficient, and engine RPM to compute the transmissibility values, identifying the optimal (k, c) combination that results in minimum vibration transmission to the operator. The results are then stored in a structured Table 28.2, which presents the average transmissibility values across all RPM levels for different combinations of stiffness and damping.

Table 28.2 Transmissibility value of each pair of k & c

Spring rate	Damping Co efficient									
	150	244	339	433	528	622	717	811	906	1000
1000	0.051	0.146	0.198	0.247	0.294	0.338	0.379	0.418	0.454	0.487
3111	0.148	0.192	0.239	0.286	0.331	0.374	0.414	0.451	0.486	0.518
5222	0.241	0.274	0.312	0.352	0.391	0.428	0.464	0.497	0.528	0.557
7333	0.387	0.403	0.424	0.448	0.475	0.502	0.529	0.555	0.580	0.600
9444	0.642	0.605	0.584	0.578	0.582	0.593	0.607	0.624	0.641	0.659
11556	1.141	0.892	0.781	0.727	0.702	0.693	0.693	0.699	0.707	0.718
13667	1.351	1.075	0.935	0.859	0.816	0.792	0.779	0.775	0.775	0.779
15778	1.220	1.116	1.027	0.960	0.913	0.881	0.861	0.848	0.842	0.839
17889	1.316	1.217	1.127	1.057	1.004	0.965	0.938	0.919	0.906	0.898
20000	1.656	1.413	1.262	1.163	1.095	1.046	1.011	0.986	0.968	0.954

Source: Author's compilation

3. RESULT AND DISCUSSION

Table 28.2 shows the transmissibility value for each combination of k & c. Figure 28.2 plot visually represents how transmissibility varies as spring stiffness and damping coefficient change. The color gradient helps in identifying the low transmissibility region, which corresponds to the best k and c values. Figure 28.3 provides an intuitive visualization of average transmissibility values across

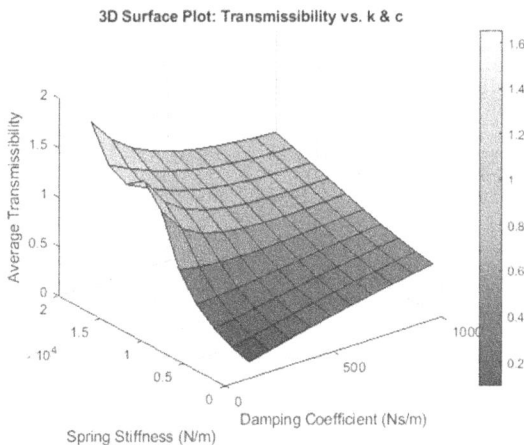

Fig. 28.2 3D Surface plot: Transmissibility vs. k & c

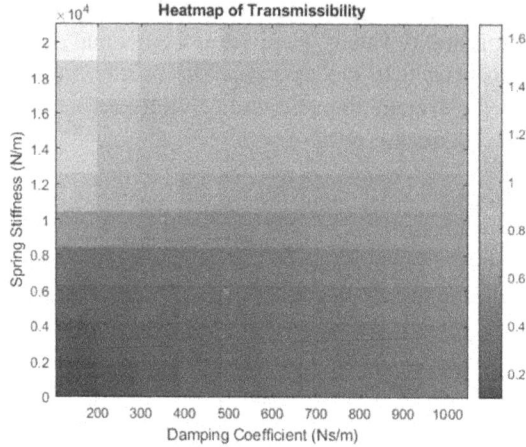

Fig. 28.3   3D Heat map of transmissibility

different k and c combinations. The darker regions indicate higher transmissibility, while the lighter (cooler) regions show areas with minimal vibration transmission. Figure 28.4 displays contour levels of transmissibility, showing how the transmissibility ratio varies within different stiffness and damping ranges. The best k & c combination is marked with a red circle, indicating the point where transmissibility is lowest.

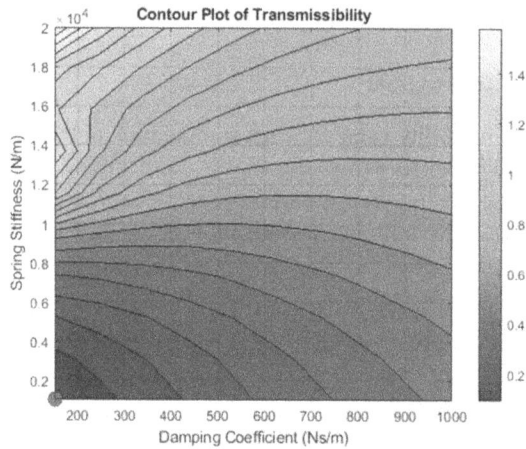

Fig. 28.4   Contour plot of transmissibility

4. CONCLUSION

The results demonstrate that choosing the right stiffness and damping significantly reduces the vibration transmitted to the operator, improving ride comfort and safety. The analysis provides an effective approach for designing an optimized seat suspension system for tractors, ensuring lower vibration exposure. From analysis we got 0.0512 transmissibility at spring rate 1000 N/m and damping co efficient 150 Ns/m.

References

1. M. Bovenzi and C.T.J. Husholf (1998). An update review of epidemiologic studies on the relationship between exposure to whole-body vibration and lowback pain, Journal Sounds and Vibration, 215(4), 595–611.
2. International Organisation for Standardisation (1997) ISO 2631-1. Mechanical vibration and shock – Evaluation of human exposure to whole-body vibration Part 1: General requirements.
3. M. J. Griffin (1990). Handbook of Human Vibration. London: Academic Press.
4. I. Hostens, and H. Ramon (2003). Descriptive analysis of combine cabin vibrations and their effect on the human body, Journal of Sound and Vibration, 266, 453–464.
5. International Organisation for Standardisation (2002) ISO 5008. Agricultural wheeled tractors and field machinery – measurement of whole-body vibration of the operator.
6. British Standards Institution (1987) BS 6841. Measurement and evaluation of human exposure to whole-body mechanical vibration and repeated shock. London: British Standards Institution.
7. Tewari, V. K., & Dewangan, K. N. (2009). Effect of vibration isolators in reduction of work stress during field operation of hand tractor. Biosystems Engineering, 103(2), 146–158.
8. David, S. A., de Sousa, R. V., Valentim Jr, C. A., Tabile, R. A., & Machado, J. A. T. (2016). Fractional PID controller in an active image stabilization system for mitigating vibration effects in agricultural tractors. Computers and Electronics in Agriculture, 131, 1–9.
9. Mehta, C. R., Shyam, M., Singh, P., & Verma, R. N. (2000). Ride vibration on tractor-implement system. Applied ergonomics, 31(3), 323–328.

Note: All the figures and tables in this chapter were made by the authors.

Emerging Perspectives and Applications of Computational Intelligence and Smart Systems
– Dr. Amit Lathigara et al. (eds)
© 2026 Taylor & Francis Group, London, ISBN 978-1-041-20965-2

29

AI-Enhanced Energy Management Solutions for Smart Micro-grid Organism

S. Muthukumar*

Department of Computer Science and Engineering,
Sri Venkateswara College of Engineering and Technology,
Chittoor, Andhra Pradesh, India

Riaz K. Israni,
Chirag Kumar Parekh

Department of Electrical Engineering, School of Engineering, RK University,
Rajkot, Gujarat, India

Anuprita Mishra

Department of Electrical and Electronics Engineering,
IES College of Technology, Bhopal,
Madhya Pradesh, India

■ **Abstract:** Given the increasing demand for optimized energy organization in the smart grid within Eco-friendly smart metropolis, these study focuses on enhancing grid stability and efficiently integrating renewable energy sources, leveraging cutting-edge skill such as Artificial intelligence (AI). In the article explores smart micro-grid organization within entity cluster, where deep Q-reinforcement learning is employed to enthusiastically optimize solar photovoltaic (SPV) energy utilization, significantly improving overall efficiency. The incorporation of an advanced visualization interface delivers instant insights and aids in informed decision-making by simplifying intricate datasets. Furthermore, implementing block-chain technology to authenticate energy usage records and transactions enhances clarity and credibility, which is vital for widespread acceptance of sustainable energy solutions. This integrated strategy not only ensures grid resilience however also strengthens dependability & longevity of the power systems, encouraging a further extensive shift toward non-conventional energy adoption.

■ **Keywords:** Deep Q-reinforcement learning, Smart energy coordination, Renewable energies, Real-time data visualization, Block-chain technology

*Corresponding author: muthuyes1@gmail.com

DOI: 10.1201/9781003725046-29

1. Introduction

This study introduces a pioneering strategy for power optimization that leverages the expanding potential of AI technologies to transform the management, distribution, and consumption of energy across varied landscapes. Focusing on emerging nations such as India, this initiative incorporates renewable power into energy management frameworks at every tier, with the goal of improving sustainability & effectiveness of the system. This incorporation is not only refines power flow but also strengthens micro-grid reliability and generates precise predictions for renewable energy based power production, development the creation of a robust, self-adjusting power infrastructure so as to promotes power autonomy, economic growth, and environmental conservation (Safari A et al., 2024). At the foundational tier concerning localized micro-grids, this initiative enhances power optimization by integrating solar photovoltaic systems and utilizing deep Q-reinforcement learning to energetically regulate and distribute power, thereby improving responsiveness and operational efficiency. Augmenting these AI-driven innovations, an advanced analytics interface delivers instantaneous, in-depth insights and summaries, facilitating informed strategic planning (Elkholy M. H. et. al., 2023)

2. Proposed Methodology for Micro-grid Energy Management

A. Projected Model: Corroboration learning represents a state-of-the-art AI technique everywhere a mediator learns to create results through interactions through its atmosphere. The aims to optimize snowballing plunder over instance, adjusting its strategy based on feedback got from the atmosphere, as illustrated in the figure below (Israni, R. K. et. al., 2024)

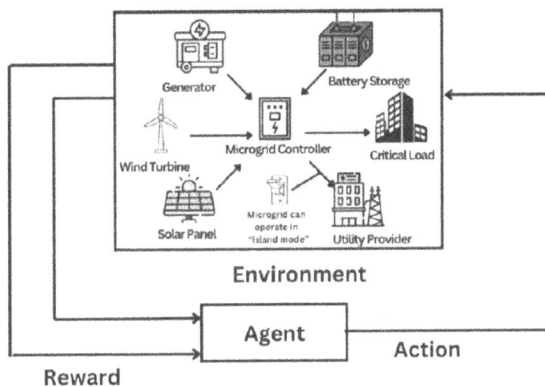

Fig. 29.1 Work-flow illustration of corroboration learning (Israni, R. K. et. al., 2024)

B. Dataset: The dataset, consisting of more than thirty thousand observations with entries every fifteen minutes, includes data on grid imports, SPVpower, consumed charge from the battery and stored charge. It provides detailed readings from an energy management system, covering grid electricity imports ('grid_import'), solar power production ('pv'), and energy storage activities ('storage_charge' and 'storage_decharge'). Descriptive statistics reveal significant variability in SPVo/p & grid i/p, through max worth of 5794 and 2294, correspondingly. The 'pv' values are mostly zero, indicating typical diurnal or weather-related fluctuations, while storage features show occasional peaks, reflecting infrequent use of storage systems(Israni, R. K et. al., 2024)(Blake-Rath et. al., 2022).

C. Algorithms: Our research utilizes a reinforcement learning (RL) model to enhance micro-grid energy management, harnessing chronological statistics on electrical energy imports, SPV generation, with storage pattern. The RL agent learns to minimize energy costs and maximize renewable usage through continuous interaction with a simulated environment. Built using TensorFlow and Keras, the deep Q-network (DQN) architecture utilizes a neural network to approximate Q-values, with inputs including grid imports, PV output, storage charge, and discharge(Al-Saadi et. al., 2021). This adaptive framework enhances energy efficiency, ensuring intelligent management in smart grid systems.

The neural network consists of an input layer for four environmental features, two hidden layers with 24 ReLU-activated neurons each for efficient learning, and an output layer with two neurons representing battery discharge and PV utilization rates. A linear activation function in the output layer generates Q-values for optimal decision-making.

The network employs MSE as the defeat purpose to minimize gap among forecasted and target Q-values, ensuring precise optimization.

D. Model Expansion: Data preprocessing involves cleaning the dataset by addressing duplicates and ensuring data integrity. Feature selection focuses on key variables like 'grid_import', 'pv', 'storage_charge', and 'storage_decharge' for effective energy management. Normalization and transformation of cumulative readings into interval-based values further refine the dataset for reinforcement learning (Cheng, C. et. al., 2019)

For reinforcement learning models, data structuring while instance sequence, such as 24-hour sequences, enhances sequence understanding. In this project, an RL agent was developed to optimize energy usage decisions in a simulated environment, effectively handling continuous variations in power consumption and generation.

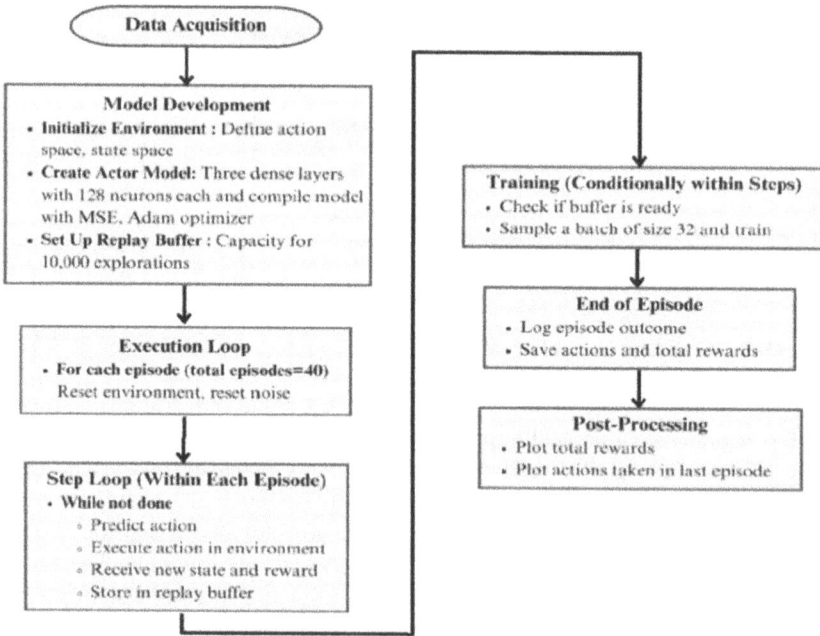

Fig. 29.2 Flow-chart of micro-grid organization (Cheng, C. et. al., 2019)

A DQN uses the neural network to approximate a Q-value function, which predicts the quality of state-action combinations in Q-learning. Actions represent battery release with PV consumption as incessant standards stuck among 0 and 1. A DQN is trained in tradition OpenAI Gym-based simulation surroundings modelling micro grid dynamics, including grid significances, PV production, plus battery storage. The network's constraints are rationalized to decrease a gap among forecasted and target Q-values, calculated using the Bellman formula(Karuppiah 2024)

$$Q\left(s_t, a_t\right) \leftarrow Q\left(s_t, a_t\right) + \alpha\left[r_{t+1} + \gamma\ max\ aQ\left(s_{t+1}, a\right) - Q\left(s_t, a_t\right)\right] \tag{1}$$

- $Q(s_t, a_t)$ = Q-value for the known position with exploit a;
- α = learning charge;
- r_{t+1} = Reduction factor, which evaluate a significance of outlook plunders;
- $max\ aQ(s_{t+1}, a)$ = Greatest forecasted Q-value in a subsequently position cross wise all probable behaviours.

A recompensed formula applied in the energy organization strengthening learning framework is designed to strike a equilibrium between maximizing SPV power use, reducing grid dependency, and efficiently managing battery storage space (Karuppiah 2024) The explicit remuneration function used in a surroundings step process is;

$$\text{Reward} = 0.1 \times \text{PV} \times [\text{pv\_utilization} - 0.1] \times [\text{grid\_import} - 0.1]$$
$$\times [\text{storage\_charge} - \text{discharge}] \tag{2}$$

- SPV Consumption Reward: $0.1 \times$ PV $\times$ pv_utilization; Plunders maximizing solar energy use, with the coefficient 0.1 ensuring equilibrium by means of other terms.
- Grid Trade in Consequence: $-0.1 \times$ grid_import; Penalizes grid power imports, supporting self-government & encouraging a utilized of SPV and stored power.
- Battery Release organization: $-0.1 \times$ [storage_charge – discharge]; Penalizes extreme battery release, guaranteeing efficient and sustainable practice by considering the unqualified dissimilarity among storage allege and release charge.

The coefficient of 0.1 in the formula balances the purposes of exploit SPV power use, reducing grid introduces, and managing battery discharge (Israni, R. K et. al., 2023). This value was chosen experimentally for permanence throughout education, helping a replica converge toward best possible behaviour. Loss metrics are vital for validation, indicating how fine a Q-learning replica's forecasts match real outcome (Cheng, C. et. al., 2019)

◢ 3. RESULTS AND CONVERSATION FOR PROPOSED METHODOLOGY

A graph below (sketch 3) shows the apparent rising tendency inside total plunder per incident, representative that a strengthening learning replica is optimism its executive above time. As incidents development, a replica recovers at balancing SPV power, battery storage, with grid trade in, likely enhancing power price reserves plus effectiveness.

An original incidents show lesser rewards as the representative discovers the surroundings. On the other hand, an important reward adds to in later an incident indicates the representative is exploiting educated policies further efficiently, getting better system effectiveness. The action distribution reveals that the RL agent primarily uses stored power, through performance often reaching a greatest rate of 1.

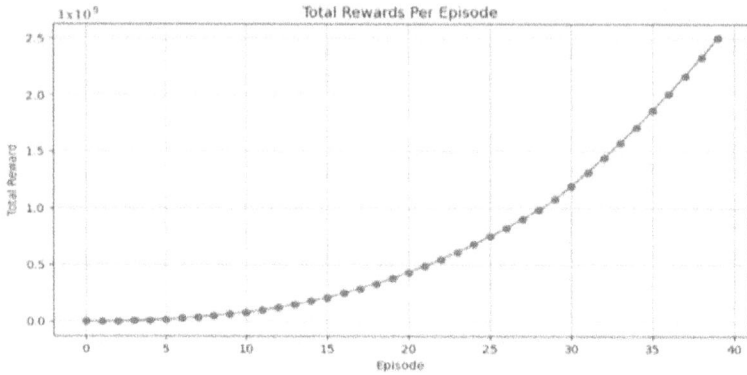

Fig. 29.3 Whole rewards per incident

Source: Authors

Figure 29.4 shows the loss over training steps, with a general decline as education progress. An originally fluctuating, a defeat stabilizes after 1500 steps, indicating the model is converging and improving its prediction accuracy. A decreasing defeat trend indicates effective learning, showing that the model's updates are minimizing prediction errors. This leads to better decision-making and higher increasing plunder, while seen in an incentive charts. Through an end of training, the stabilized low loss suggests a replica has educated a steady strategy, ensuring consistent performance in real-world applications or further simulations.

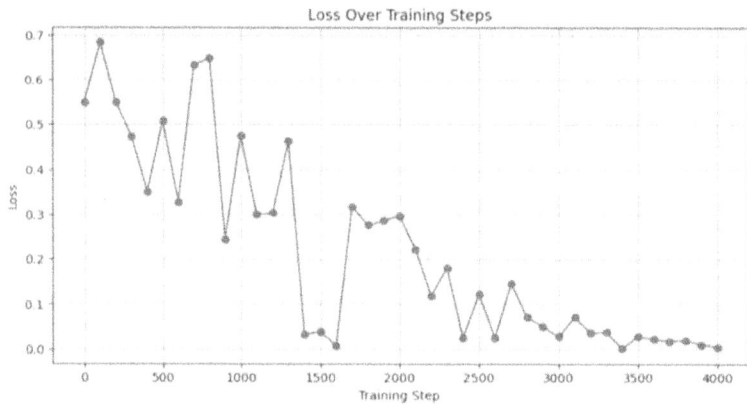

Fig. 29.4 Loss over training steps over 40 episodes

Source: Authors

4. CONCLUSIONS

This research work addresses a rising necessitate used to optimized energy organization in the smart grids by integrating AI technologies to enhance grid stability and renewable energy utilization. By employing deep Q-reinforcement learning for smart micro-grid management, significant improvements in solar energy efficiency are achieved. The use of an advanced visualization interface and block-chain technology for secure energy transaction tracking further boosts decision-making and trust in sustainable energy systems. This integrated approach fosters grid resilience, supporting the transition to renewable energy and promoting long-term sustainability.

References

1. Safari, A., Daneshvar, M., & Anvari-Moghaddam, A. (2024). Energy Intelligence: A Systematic Review of Artificial Intelligence for Energy Management. *Applied Sciences, 14*(23), 11112.
2. Elkholy, M. H., Yona, A., Ueda, S., Said, T., Senjyu, T., & Lotfy, M. E. (2023). Experimental investigation of AI-enhanced FPGA-based optimal management and control of an isolated microgrid. *IEEE Transactions on Transportation Electrification*.
3. Israni, R. K., & Parekh, C. (2024). Power quality enhancement in hydroelectric-solar PV-based hybrid system by exploiting CPD. *International Journal of Smart Grid and Green Communications, 2*(3), 231–248
4. Blake-Rath, R., Dyck, A. C., Schumann, G., & Wenninghoff, N. (2022). Independent Power Supply Through Off-Grid Microgrids in South Africa: Potentials of AI Enhanced Business Models. In *Digital Transformation for Sustainability: ICT-supported Environmental Socio-economic Development* (pp. 119–137). Cham: Springer International Publishing.
5. Al-Saadi, M., Al-Greer, M., & Short, M. (2021). Strategies for controlling microgrid networks with energy storage systems: A review. *Energies, 14*(21), 7234.
6. Cheng, C. C., & Lee, D. (2019). Artificial intelligence-assisted heating ventilation and air conditioning control and the unmet demand for sensors: Part 1. Problem formulation and the hypothesis. *Sensors, 19*(5), 1131.
7. Karuppiah, N., Mounica, P., Bhanutej, J. N., Saravanan, S., Reddy, R., & Israni, R. (2024). Revolutionizing Renewable Energy Integration: The Innovative Gravity Energy Storage Solution. In *E3S Web of Conferences* (Vol. 547, p. 03028). EDP Sciences.
8. Israni, R. K., Yadav, R., Singh, R., & Reddy, B. R. (2023, September). Power Quality Enhancement in Wind-Hydro Based Hybrid Renewable Energy System by Interlocking of UPQC. In *2023 3rd International Conference on Energy, Power and Electrical Engineering (EPEE)* (pp. 112–123). IEEE.
9. Patel, D. J., & Bhatt, N. (2019). Insect identification among deep learning's meta-architectures using TensorFlow. *Int. J. Eng. Adv. Technol, 9*(1), 1910–1914.

Emerging Perspectives and Applications of Computational Intelligence and Smart Systems
– Dr. Amit Lathigara et al. (eds)
© 2026 Taylor & Francis Group, London, ISBN 978-1-041-20965-2

30

Block Chain Based Smart and Secure Logistic Management System using AI

Amit Lathigara
School of Engineering, RK University, Rajkot,
Gujarat, India

Sunil Soni*, Jaydeep Tadhani
IT Department, Government Polytechnic,
Rajkot, Gujarat, India

Jay Fuletra
School of Engineering, RK University, Rajkot,
Gujarat, India

Nikhat Raza Khan
IES College of Technology, Bhopal,
Madhya Pradesh, India

■ **Abstract:** Supply chain management and logistics are growing swiftly, needing transparent, safe, and effective technologies to eliminate inefficiencies, cyberthreats, and data breaches. Problems with traditional logistical systems include poor real-time tracking, centralized control, and a lack of transparency. Adaptability is also impeded by integration concerns and scalability limits. By merging Ethereum smart contracts, cryptographic security, AI-driven automation, and Internet of Things monitoring, this study recommends a blockchain-based secure logistics system with AI advancements. Immutable data integrity, AI-powered intrusion protection, and smooth tracking are all ensured by the system's multi-layered design. Using AI security analytics and self-adaptive blockchain protocols, it decreases cyber risks, enhances resilience, and facilitates decentralized transaction execution. The recommended framework opens the way for next-generation decentralized logistics ecosystems by considerably boosting supply chain security, efficiency, and decision-making.

■ **Keywords:** Block-chain, Ethereum, Deep learning, Cyber attack, Logistic management system

1. INTRODUCTION

Rapid developments in digital technologies have dramatically impacted various industries, including healthcare, smart cities, automotive, manufacturing, and logistics. Blockchain technology

*Corresponding author: profsjsoni@gmail.com

DOI: 10.1201/9781003725046-30

and artificial intelligence (AI) are two of these advances that have become vital for improving supply chain and logistics systems' security, effectiveness, and data management. With the rise of the Internet of Things (IoT), modern supply chain management has transformed into a highly data-intensive environment where real-time data collecting and processing play a vital role in operational success. However, this digital transition also throws up hurdles like as security vulnerabilities, data privacy concerns, scalability issues, and inefficient data management.

Blockchain technology, distinguished by its decentralized, immutable, and transparent nature, has gained major attention for its ability to address security and trust concerns in logistics management. By applying cryptographic methods and consensus procedures, blockchain enables safe and tamper-resistant transactions, removing risks associated with centralized databases and single points of failure. Notwithstanding its advantages, blockchain technology has disadvantages that hinder it from being effortlessly integrated into sophisticated logistics systems(Chen et al., 2021; Li et al., 2021). These drawbacks include concerns with scalability, transaction speed, and computing overhead.

Artificial intelligence (AI) has been employed (Astarita et al., 2019) more and more to optimize blockchain performance in supply chain management in order to address these challenges. In order to improve automation, data analysis, and logistical operations, artificial intelligence (AI) incorporates machine learning algorithms, predictive analytics, and intelligent decision-making systems. The proposed Framework's combination of AI and blockchain gives a novel technique to get over the current disadvantages of blockchain-based logistics solutions. Supply chain networks can boost their efficiency, security, and scalability by applying AI-driven techniques including intelligent consensus procedures and automation in inventory management and delivery.

Additionally, because logistics systems depend on centralized database structures, which are prone to cyberattacks, illegal data breaches, and inadequate verification procedures, they present security risks. A solid and secure infrastructure is necessary to protect crucial information due to the growing volume of sensitive data that is moved between manufacturers, suppliers, carriers, and customers. As a distributed ledger system, blockchain delivers a revolutionary solution by guaranteeing data integrity, minimizing fraudulent activity, and boosting supply chain transparency. However, intelligent optimization is crucial for the successful deployment of blockchain in logistics, and AI can supply this thanks to its profound learning and reasoning skills.

In order to optimize IoT-based supply chains, this study finds a major research gap in the application of blockchain technology and artificial intelligence. The research currently in publication frequently explores these technologies independently or falls short of completely addressing how blockchain and artificial intelligence could work together to minimize supply chain challenges and prevention from cyber-attacks. This study makes following research contributions: (1) it proposes a blockchain framework with AI enhancements specifically designed for supply chain and logistics applications; (2) it investigates how AI-driven optimization affects blockchain performance in terms of data management, scalability, and security; and (3) This research adds a detection mechanism for cyber-attack detection in logistic system.

2. REVIEW OF RELEVANT RESEARCH STUDIES

With potential solutions to present difficulties with security, transparency, and operational efficiency, the intersection of blockchain technology with artificial intelligence (AI) in supply chain and logistics management has become a crucial subject of study. AI increases supply chain process automation, decision-making, and predictive analytics, while blockchain assures decentralized, immutable, and tamper-resistant data management. While a lot of study has been done on the independent uses of blockchain and AI in logistics, not as much has been done on how they operate together. This portion gives a thorough study of essential literature, highlighting research gaps, pinpointing existing difficulties, and analysing significant breakthroughs in the use of blockchain and artificial intelligence in supply chain management.

(Ugochukwu, Goyal, & Arumugam, 2022) proposed blockchain-based IoT-enabled architecture enhances safe logistics management by using smart contracts for stakeholder communication and IoT sensors for real-time data collecting. The system's integration of blockchain and IoT layers above the physical layer ensures data integrity and transparency. In order to increase the effectiveness and oversight of intelligent logistics systems, the study presented by (Chen et al., 2021) constructs algorithmic models and offers a consensus authentication mechanism. The strategy enhances data correlation and analysis by implementing blockchain networks, smart contracts, and multi-authentication centres for distinct events. In order to guarantee smooth and safe logistical operations, the research also handles integration concerns with other systems. The research also lacks to addresses integration challenges with other systems, ensuring seamless and secure logistics operations.

A full review of current advances in blockchain applications for supply chain management and logistics is offered in this research by (Tijan et al., 2019). The recommended agri-food traceability system assures trustworthy data sharing at every stage of the supply chain by combining RFID and blockchain technology. By ensuring authenticity and transparency during the complete production, processing, warehousing, distribution, and sales process, this strategy increases food safety. In order to accomplish sustainable logistics operations, the study(Tan et al., 2020) propose a blockchain-based reference framework for green logistics. The framework promotes efficiency and transparency through the integration of big data and the Internet of Things (IoT). Long-haul, short-haul, and 3PL providers are also included, and real-time data gathering and monitoring are performed through the use of IoT sensors, RFID, and GPS.

In study presented by (Abdelhamid et al., 2024) integrates blockchain, artificial intelligence, and the Internet of Things to offer a four-level hierarchical design for logistics management. This study(Ugochukwu, Goyal, Rajawat, et al., 2022) presents a blockchain-based logistics management system that makes use of Ethereum smart contracts and a decentralized peer-to-peer network. To boost security, secure client information, and thwart cyberattacks, it uses RSA asymmetric encryption. It also guarantees the integrity of product data on the blockchain network and facilitates safe information exchange between stakeholders.

3. PROPOSED APPROACH

Proposed Framework of AI Block chain based Secure Logistic Management System is shown in Fig. 30.1. In Framework the User Layer functions as the essential interface for primary stakeholders in the logistics ecosystem, including manufacturers, suppliers, carriers, e-commerce businesses, and end consumers. This layer serves as an interactive medium, effectively linking with the Application Layer to facilitate real-time, autonomous coordination and data exchange across various organizations throughout the supply chain. The Application Layer is crucial in coordinating key logistical activities, including order execution, inventory management, shipment facilitation, transportation logistics, and real-time delivery tracking. The integration of AI-driven automation at this layer enhances decision-making efficiency via sophisticated predictive analytics, improving

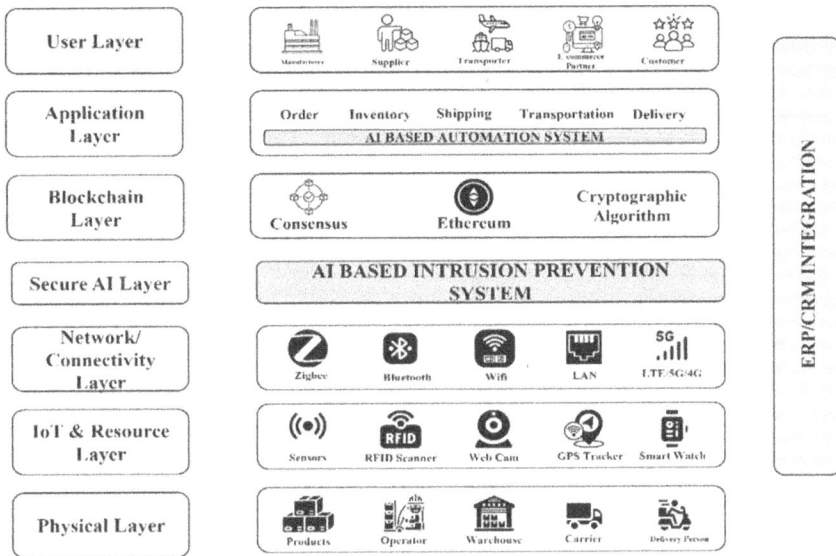

Fig. 30.1 Proposed AI blockchain based secure LMS

Source: Authors

demand forecasting, route planning, and resource allocation. The incorporation of machine learning models facilitates real-time modifications to logistics operations, thus reducing operational expenses and improving the flexibility and responsiveness of supply chain networks to market changes. The Blockchain Layer uses a decentralized transaction management system to assist logistics operations' security, transparency, and integrity. In order to remove intermediary dependencies and guarantee immutable and tamper-resistant transaction records, this layer combines Ethereum-based smart contracts, cryptographic security protocols, and consensus algorithms. This layer offers cryptographic validation, decentralized trust mechanisms, and real-time verification of every logistical event by integrating distributed ledger technology (DLT) into the logistics network. In the logistics ecosystem, smart contracts streamline multi-party interactions, decrease conflicts, and remove inefficiencies by automating contractual tasks.

Using deep learning, anomaly detection algorithms, and behavioral analysis models, the Secure AI Layer serves as an AI-driven Intrusion Prevention System (AIPS) that secures the logistics network against fraud, cyberattacks, and data breaches. This layer proactively detects and eliminates hostile cyber-attacks, unauthorized access attempts, and network vulnerabilities by applying self-learning security models, adversarial machine learning techniques, and real-time threat intelligence. Adaptive threat response solutions are ensured by AI-driven security automation, allowing for dynamic policy updates in response to changing security environments. Additionally, by enabling biometric verification, real-time identity authentication, and AI-enhanced encryption techniques, this layer guarantees data confidentiality, integrity, and adherence to tight cybersecurity regulations.

Through a number of connectivity tools and network protocols, the Network/Connectivity Layer creates a solid communication infrastructure that facilitates high-speed, low-latency data interchange across logistical participants. This layer allows smooth, decentralized communication across IoT-enabled logistics ecosystems by integrating Zigbee, Bluetooth, WiFi, LAN, LTE, and 5G/4G technologies. By lowering latency, enhancing data transmission security, and providing distributed computing capabilities at the network periphery, the adoption of edge computing designs and multi-protocol interoperability frameworks further enhances network efficiency. By integrating real-time monitoring, data collecting, and asset-tracking capabilities through an advanced network of RFID scanners, GPS trackers, cameras, smart wearables, and environmental sensors, the IoT & Resource Layer works as the sensory backbone of the logistical architecture. The accuracy and traceability of supply chain activities are increased by this layer, which enables constant data synchronization between digital records and physical assets. Predictive maintenance, intelligent inventory management, and proactive anomaly detection are made feasible by AI-enhanced IoT analytics, which decreases supply chain disruption losses.

The actual logistical elements, such as commodities, operators, warehouses, transporters, and last-mile delivery workers, make up the actual Layer at the most fundamental level. Real-time tracking, route optimization, and automated logistics management are made feasible by this layer's direct contact with IoT-enabled devices. This layer's combination of AI, blockchain, and IoT produces a self-policing cyber-physical logistics network that can perform trustless transactions, make choices on its own, and do predictive operational analytics.

4. CONCLUSION

The proposed blockchain-based logistics system with AI upgrades solves concerns with efficiency, security, and transparency in traditional frameworks. Ethereum smart contracts, cryptographic consensus, and AI-driven intrusion prevention are all combined to create a decentralized, trust less ecosystem that eliminates inefficiencies and cyberthreats. AI-driven analytics and IoT-enabled monitoring boost automation, permitting proactive anomaly identification and dynamic decision-making. This framework guarantees tamper-proof record-keeping, intelligent automation, and seamless tracking, in contrast to traditional systems with centralized weaknesses and scalability constraints. A robust, self-regulating logistics ecosystem is fostered by the combination of AI, blockchain, and IoT. To further enhance decentralized logistics networks, future studies will concentrate on interoperability, blockchain scalability, and quantum-resistant cryptography.

References

1. Abdelhamid, M. M., Sliman, L., & Ben Djemaa, R. (2024). AI-Enhanced Blockchain for Scalable IoT-Based Supply Chain. *Logistics 2024, Vol. 8, Page 109*, 8(4), 109.
2. Astarita, V., Giofrè, V. P., Mirabelli, G., & Solina, V. (2019). A Review of Blockchain-Based Systems in Transportation. *Information 2020, Vol. 11, Page 21*, 11(1), 21.
3. Chen, C. L., Deng, Y. Y., Weng, W., Zhou, M., & Sun, H. (2021). A blockchain-based intelligent anti-switch package in tracing logistics system. *Journal of Supercomputing*, 77(7), 7791–7832.
4. Li, H., Han, D., & Tang, M. (2021). Logisticschain: A Blockchain-Based Secure Storage Scheme for Logistics Data. *Mobile Information Systems*, 2021(1), 8840399.
5. Tan, B. Q., Wang, F., Liu, J., Kang, K., & Costa, F. (2020). A Blockchain-Based Framework for Green Logistics in Supply Chains. *Sustainability 2020, Vol. 12, Page 4656*, 12(11), 4656.
6. Tijan, E., Aksentijević, S., Ivanić, K., &Jardas, M. (2019). Blockchain Technology Implementation in Logistics. *Sustainability 2019, Vol. 11, Page 1185*, 11(4), 1185.
7. Ugochukwu, N.A., Goyal, S. B., & Arumugam, S. (2022). Blockchain-Based IoT-Enabled System for Secure and Efficient Logistics Management in the Era of IR 4.0. *Journal of Nanomaterials*, 2022(1), 7295395.
8. Ugochukwu, N. A., Goyal, S. B., Rajawat, A. S., Islam, S. M. N., He, J., & Aslam, M. (2022). An Innovative Blockchain-Based Secured Logistics Management Architecture: Utilizing an RSA Asymmetric Encryption Method. *Mathematics 2022, Vol. 10, Page 4670*, 10(24), 4670.

Emerging Perspectives and Applications of Computational Intelligence and Smart Systems
– Dr. Amit Lathigara et al. (eds)
© 2026 Taylor & Francis Group, London, ISBN 978-1-041-20965-2

31

A Systematic Review of Intrusion Detection in IoT: Efficiency, Transparency, and Zero-Day Resilience

Rahul Keshwala*

Research Scholar, School of Engineering, RK University,
Rajkot, Gujarat, India

Assistant Professor, L.E. College, Morbi,
Gujarat, India

Paresh Tanna

School of Engineering, RK University,
Rajkot, Gujarat, India

■ **Abstract:** The expansion of the Internet of Things (IoT) has heightened security risks, requiring efficient and transparent Intrusion Detection Systems (IDS) to counter cyber threats. Traditional IDS models struggle with high computational costs, lack of interpretability, and weak zero-day attack detection. This review evaluates IDS approaches based on efficiency, transparency, and resilience, analyzing their trade-offs in computational performance, explainability, and detection accuracy. Lightweight IDS improve efficiency but often lack interpretability, while deep learning-based IDS enhance accuracy at the cost of higher computational demands. Explainable AI (XAI) improves transparency but remains underutilized, and anomaly-based IDS, though effective against zero-day attacks, generate high false positives. A comparative analysis of IDS models reveals key trends, limitations, and challenges, providing insights into optimization, interpretability, and detection effectiveness to guide future research in IoT security.

■ **Keywords:** IoT security, Intrusion detection, Explainable AI, Zero-day detection, Lightweight IDS, Deep learning

1. INTRODUCTION

The rapid growth of the Internet of Things (IoT) has increased security vulnerabilities, making Intrusion Detection Systems (IDS) essential for real-time threat detection. Traditional IDS models face limitations in detecting evolving cyber threats, with signature-based IDS struggling against

---

*Corresponding author: keshwala.rahul@gmail.com

DOI: 10.1201/9781003725046-31

unknown attacks and anomaly-based IDS generating high false positives (Ali et al.,2024). As IoT devices operate in resource-constrained environments, IDS must balance efficiency, explain ability, and adaptability to zero-day attacks.

Lightweight IDS models improve efficiency but often compromise detection accuracy. Explainable AI (XAI) techniques enhance model transparency but introduce computational overhead (Sohailet al., 2024). Zero-day attack detection remains an open challenge, requiring adaptive learning methods to identify emerging threats without predefined signatures (F. El Husseiniet al., 2024). This review systematically evaluates IDS models based on optimization techniques, explain ability, and zero-day detection performance, offering insights into existing approaches and their limitations.

2. COMPARATIVE ANALYSIS OF IDS MODELS

A comparative analysis of 18 IDS models highlights differences in optimization; explain ability, and zero-day detection. Deep learning-based IDS achieve high accuracy but demands significant computational resources, limiting IoT feasibility (Sohail et al., 2024). Hybrid architectures like CNN-LSTM and Transformer-based IDS enhance detection but lack interpretability, complicating security operations (Ables et al., 2024).

Optimization methods such as pruning and quantization reduce resource use—pruning removes redundant parameters for faster processing, while quantization lowers memory needs by reducing numerical precision (Ali et al., 2024). Explainable IDS using SHAP and LIME improves interpretability but remains computationally demanding (Sabrina et al., 2024).

Zero-day detection remains challenging, with autoencoder-based IDS improving detection rates, while adversarial training enhances adaptability (Rahman et al., 2024). The heatmap (Fig. 31.1) visualizes IDS performance, highlighting trade-offs between accuracy, explainability, and optimization.

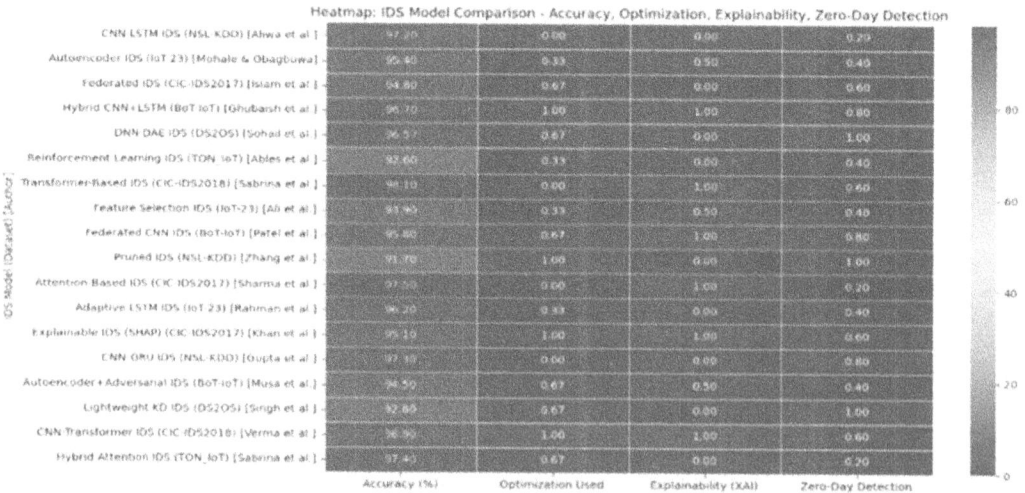

Fig. 31.1   Heatmap representation of IDS model performance (Accuracy, Optimization, Explainability, Zero-Day Detection)

Source: Author's compilation

3. OPTIMIZATION TECHNIQUES AND EXPLAINABLE AI (XAI) IN IDS

Optimization techniques enhance IDS efficiency while minimizing performance degradation. Pruning removes redundant network layers, reducing computational cost while maintaining detection accuracy (Mohale & Obagbuwa, 2024). Quantization converts floating-point computations to lower-bit representations, improving inference speed at the cost of minor precision loss (Islam et al., 2024). Knowledge distillation enables lightweight IDS models by transferring knowledge from complex architectures, ensuring efficiency without significant accuracy loss (Ali et al., 2024).

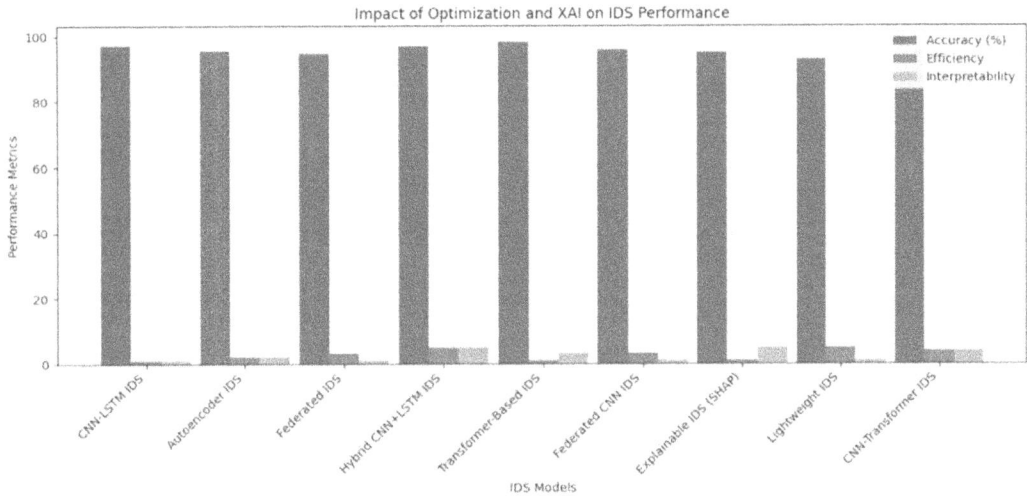

Fig. 31.2   Bar chart representation of IDS performance (Accuracy, Optimization, Explainability)

Source: Author's compilation

Explain ability is crucial for IDS adoption, allowing security analysts to interpret model decisions. SHAP provides feature importance rankings, LIME generates interpretable local approximations, and Grad-CAM visualizes CNN-based IDS decision-making (Sohail et al., 2024). However, integrating XAI remains a challenge due to increased computational complexity. Models incorporating both optimization and explain ability strike a balance between efficiency and interpretability. The bar chart visualization (Fig. 31.1) illustrates impact of optimization and XAI on IDS performance.

Table 31.1   Impact of optimization and XAI on IDS performance

IDS Model	Optimization Method	Explainability (XAI)	Accuracy (%)	Efficiency	Interpretability
CNN-LSTM IDS	None	No XAI	97.2	Low	Low
Autoencoder IDS	Feature Reduction	Partial XAI	95.4	Medium	Medium
Federated IDS	Knowledge Distillation	No XAI	94.8	High	Low
Hybrid CNN+LSTM IDS	Pruning & Quantization	SHAP + LIME	96.7	High	High
Transformer-Based IDS	None	SHAP	98.1	Low	Medium
Explainable IDS (SHAP)	No Optimization	SHAP + LIME	95.1	Low	High
Lightweight IDS	Quantization	No XAI	92.8	High	Low
CNN-Transformer IDS	Pruning & Quantization	Grad-CAM	96.9	High	Medium

Source: Authors

Table 31.1 summarizes Impact of Optimization and XAI on IDS Performance. The trade-offs between efficiency and interpretability underscore the importance of designing IDS solutions that maintain performance while ensuring explain ability for security analysts (Rahman et al., 2024).

◢ 4. ZERO-DAY ATTACK DETECTION STRATEGIES

Zero-day attacks exploit unknown vulnerabilities, rendering signature-based IDS ineffective. Anomaly-based methods like auto encoders and adversarial learning enhance detection by identifying deviations from normal behavior (Musa et al., 2024). Auto encoders reconstruct traffic patterns to flag anomalies, while adversarial training strengthens IDS against evolving threats (Gupta et al., 2024). Hybrid IDS models improve detection while addressing interpretability but face challenges in balancing accuracy and false positives. Table 31.2 summarizes zero-day attack detection performance.

Future research should focus on combining anomaly detection with adaptive learning to improve zero-day detection capabilities while reducing false positives (Sharma et al., 2024).

Table 31.2 Zero-day attack detection methods in IDS models

Approach	Detection Rate (%)	False Positive Rate (%)
Signature-Based IDS	50.2	12.4
Auto encoder-Based Anomaly IDS	72.6	8.3
Hybrid IDS (XAI + Auto encoder)	85.1	5.9
Adversarial Training IDS	87.3	6.7

Source: Authors

◢ 5. CHALLENGES AND FUTURE DIRECTIONS

Despite advancements, IDS models continue to face challenges in achieving optimal efficiency, transparency, and adaptability. High false positives in anomaly-based IDS models reduce reliability, while deep learning-based IDS require substantial computational power, making them unsuitable for IoT (Mohale & Obagbuwa, 2024). XAI integration remains limited due to its computational overhead, and zero-day detection techniques must be further refined to improve accuracy while minimizing false positives (Islam et al., 2024).

Future research should explore federated learning-based IDS for distributed security monitoring across IoT devices, reducing reliance on centralized data storage while maintaining privacy (Isranil, R. K. at al., 2025) Edge computing can enhance real-time threat detection by processing IDS operations closer to IoT devices, minimizing latency. Adaptive IDS models leveraging reinforcement learning and anomaly detection should be further developed to ensure continuous learning from emerging attack patterns (Gupta et al., 2024).

◢ 6. CONCLUSION

This review provides a comparative analysis of IDS models for IoT security, highlighting the trade-offs between optimization, explain ability, and zero-day detection capabilities. The findings emphasize the need for hybrid IDS models that integrate efficiency with transparency while maintaining strong zero-day detection performance. Future IDS research should prioritize scalable, interpretable, and adaptive security solutions for real-world IoT deployment.

References

1. Liao, H., Murah, M. Z., Hasan, M. K., Aman, A. H. M., Fang, J., Hu, X., & Khan, A. U. R. (2024). A survey of deep learning technologies for intrusion detection in Internet of Things. *IEEe Access, 12*, 4745–4761.
2. Asgharzadeh, H., Ghaffari, A., Masdari, M., & Gharehchopogh, F. S. (2024). An intrusion detection system on the internet of things using deep learning and multi-objective enhanced gorilla troops optimizer. *Journal of Bionic Engineering, 21*(5), 2658–2684.
3. Sohail, S., Fan, Z., Gu, X., & Sabrina, F. (2022). Explainable and optimally configured artificial neural networks for attack detection in smart homes. *arXiv preprint arXiv:2205.08043*.
4. Satyanarayana, D., & Saikiran, E. (2024, March). Intrusion Detection System in Explainable Artificial Intelligence by Using Different Algorithms. In *2024 International Conference on Distributed Computing and Optimization Techniques (ICDCOT)* (pp. 1–4). IEEE.
5. El Husseini, F., Noura, H., Salman, O., & Chehab, A. (2024, December). Advanced Machine Learning Approaches for Zero-Day Attack Detection: A Review. In *2024 8th Cyber Security in Networking Conference (CSNet)* (pp. 297–304). IEEE.
6. Dakic, P., Zivkovic, M., Jovanovic, L., Bacanin, N., Antonijevic, M., Kaljevic, J., & Simic, V. (2024). Intrusion detection using metaheuristic optimization within IoT/IIoT systems and software of autonomous vehicles. *Scientific Reports, 14*(1), 22884.
7. Patil, S., Varadarajan, V., Mazhar, S. M., Sahibzada, A., Ahmed, N., Sinha, O., ... & Kotecha, K. (2022). Explainable artificial intelligence for intrusion detection system. *Electronics, 11*(19), 3079.
8. Ghubaish, A., Yang, Z., Erbad, A., & Jain, R. (2023). LEMDA: A novel feature engineering method for intrusion detection in IoT systems. *IEEE Internet of Things Journal, 11*(8), 13247–13256.
9. Alabbadi, A., &Bajaber, F. (2025). An Intrusion Detection System over the IoT Data Streams Using eXplainable Artificial Intelligence (XAI). *Sensors (Basel, Switzerland), 25*(3), 847.
10. Gaspar, D., Silva, P., & Silva, C. (2024). Explainable ai for intrusion detection systems: Lime and shap applicability on multi-layer perceptron. *IEEE Access*.
11. MP, P. J. (2024). An Explainable and Optimized Network Intrusion Detection Model using Deep Learning. *International Journal of Advanced Computer Science & Applications, 15*(1).
12. Bensaoud, A., & Kalita, J. (2025). Optimized detection of cyber-attacks on IoT networks via hybrid deep learning models. *Ad Hoc Networks, 170*, 103770.
13. Arreche, O., Guntur, T., & Abdallah, M. (2024). Xai-ids: Toward proposing an explainable artificial intelligence framework for enhancing network intrusion detection systems. *Applied Sciences, 14*(10), 4170.
14. Mohale, V. Z., &Obagbuwa, I. C. (2025). A systematic review on the integration of explainable artificial intelligence in intrusion detection systems to enhancing transparency and interpretability in cybersecurity. *Frontiers in Artificial Intelligence, 8*, 1526221.
15. Ahmed, U., Jiangbin, Z., Almogren, A., Khan, S., Sadiq, M. T., Altameem, A., & Rehman, A. U. (2024). Explainable AI-based innovative hybrid ensemble model for intrusion detection. *Journal of Cloud Computing, 13*(1), 150.
16. Ables, J., Kirby, T., Mittal, S., Banicescu, I., Rahimi, S., Anderson, W., & Seale, M(2023). Explainable intrusion detection systems using competitive learning techniques. *arXiv preprint arXiv:2303.17387*.
17. Isranil, R. K., Nedunchezhian, T., Tanna, P., & Soni, S. (2025) AI-endorsed techniques for smart energy utilization: A holistic review. *Digital Transformation and Sustainability of Business*, 271–274.
18. Khan, N., Ahmad, K., Tamimi, A. A., Alani, M. M., Bermak, A., & Khalil, I. (2024). Explainable AI-based Intrusion Detection System for Industry 5.0: An Overview of the Literature, associated Challenges, the existing Solutions, and Potential Research Directions. *arXiv preprint arXiv:2408.03335*.
19. Li, J., Othman, M. S., Chen, H., & Yusuf, L. M. (2024). Optimizing IoT intrusion detection system: feature selection versus feature extraction in machine learning. *Journal of Big Data, 11*(1), 36.

Emerging Perspectives and Applications of Computational Intelligence and Smart Systems
– Dr. Amit Lathigara et al. (eds)
© 2026 Taylor & Francis Group, London, ISBN 978-1-041-20965-2

32

Real-Time Intrusion Detection and Cloud-based Facial Recognition for Smart Home Security Systems

Josephine Selle Jeyanathan*

Department of Electronics and Communication Engineering,
Kalasalingam Academy of Research and Education,
Krishnankoil, Virudhunagar, India

Cross T. Asha Wise

Department of Mechatronics Engineering,
SRM Institute of Science and Technology,
Kattankulathur, Chennai, India

**O. Ansel Tom, C. Yogeswar,
G. Sankeerth, Y. Dharani Kumar**

Department of Electronics and Communication Engineering,
Kalasalingam Academy of Research and Education,
Krishnankoil, Virudhunagar, India

■ **Abstract:** This research presents a comprehensive approach for home security using real-time face recognition, implemented with a Raspberry Pi 5 and Firebase integration. Images of the individuals approaching the door are captured using Rpi camera module in various angles, and through pre-trained CNN model, face detection is performed for authorized access into the house. Known faces automatically unlock the door, while unknown faces trigger an alert to the homeowner's telegram app with the intruder's photo and timestamp. The proposed system uses telegram app, integrated with Firebase, provides control on door operations from remote, stores data in cloud, gives real-time notifications, and maintains an accessible log of intruder events. Automated logging ensures that every access attempt is documented for future reference. Experimental results highlight the robustness and reliability of the system, showcasing significantly better accuracy rate (97.2%) using ResNet CNN for detecting the face.

■ **Keywords:** Raspberry Pi-5, Firebase cloud messaging (FCM), Telegram SDK (Software development kit, Remote access (PuTTY and VNC server)

*Corresponding author: drjjosephine@gmail.com

DOI: 10.1201/9781003725046-32

1. INTRODUCTION

Home security systems using embedded technology integrate sensors, microcontrollers, and connectivity for automated, real-time security. These systems detect, monitor, and respond to risks like fires or unauthorized access, enhancing safety. Embedded systems combine hardware and software for tasks such as camera surveillance, motion detection, and automated locks. With advancements in microcontrollers, wireless communication, and IoT, modern security systems use Wi-Fi, Zigbee, or Bluetooth for remote monitoring via smartphones or cloud storage. They offer automation, cost-effectiveness, and real-time monitoring but require cybersecurity, data privacy, and accurate detection. Integration with smart home devices enhances security and user convenience.

This paper presents the design and integration of an embedded system aimed at creating a secure environment, particularly for young children left alone for short periods when their parents are away. The system monitors the child and provides real-time alerts about visitors to the home during the parents' absence. It incorporates a Raspberry Pi with a camera module and utilizes a Firebase for storing and managing the dataset.

2. RELATED WORKS

(R. et al. 2020) present a cost-effective Smart Home Security System using IoT and face recognition with a Raspberry Pi. A Pi camera and sensors recognize faces in real-time, unlocking the door if a match is found, otherwise sending an email (alert.Sri Vendra et al. 2023) analyze face detection techniques, finding the Haar Classifier most accurate. (Gaikwad et al. 2024) propose a smart doorbell system using MTCNN for face detection, outperforming YOLO and Haar Cascade. (Rahim, Zhong, and Ahmad 2023) introduce a cloud-based, tree-based deep model enhancing recognition accuracy while reducing computational costs. (Al-Abboodi and Al-Ani 2024) combine Galactic Swarm Optimization (GSO) and CNN to improve facial recognition accuracy. (Nagpure et al. 2022) develop a Raspberry Pi-based face recognition door lock system using Haar Cascade, sending alerts via LAN to Telegram. (Tok et al. 2021) propose a cloud-based IoT security system using SVM and OTP for enhanced security. (Hussain et al. 2023) introduce a VB.NET-based intelligent security system with adaptive recognition. (Golovachev et al. 2024) evaluate Raspberry Pi's efficiency for face recognition. (Istiqomah et al. 2023) optimize machine learning models for home security.

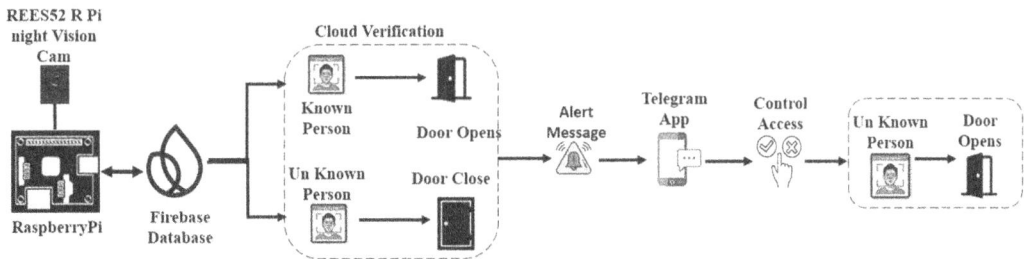

Fig. 32.1 General block layout of the home security system

3. METHODOLOGY

This research develops a smart home security system using Raspberry Pi 5 as the processing unit. A camera module captures real-time images, integrated with Firebase for storage. Face recognition is implemented with OpenCV, CNN, and Dlib ResNet Embedding. Authorized faces unlock the door, while unknown faces trigger alerts. The system enables remote control via a Telegram app, ensuring efficient monitoring and security.

The Telegram app serves as the user interface, displaying intruder details (name, timestamp, and image) in real-time from Firebase. Users receive push notifications for intruder detections and can remotely lock/unlock the door. Firebase Authentication ensures secure login, while Firebase Realtime Database stores records and Firebase Cloud Messaging (FCM) sends alerts. OpenCV and dlib handle real-time face detection, with CNN extracting facial features and SVM comparing detected faces. If recognized, the door unlocks; otherwise, an alert is sent. Raspberry Pi 5 processes inputs and controls the door. The Android Telegram app enables remote monitoring and control. This system integrates OpenCV, dlib, TensorFlow/PyTorch for deep learning, and Firebase for backend services, ensuring real-time security monitoring and remote access.

4. EXPERIMENTAL RESULTS

This research presents a structured approach for a face recognition-based intrusion detection system. It compares CNN-based techniques, focusing on Dlib ResNet Embedding and Euclidean Distance classification. The system integrates IoT, embedded systems, and machine learning for smart home security. A pre-trained CNN-ResNet extracts facial embeddings matched against a secure database. Real-time detection uses OpenCV and Dlib, with Firebase storing access logs and sending alerts. Unauthorized access triggers Telegram notifications, and users can remotely control door locks via a Telegram bot. This ensures enhanced security and seamless monitoring under the Safe Home architecture. The flowchart illustrates overall workflow of the face recognition-based intrusion detection system. The system design is given in Fig. 32.2. The model architecture components as shown in Fig. 32.3 for the system design are given below as follows:

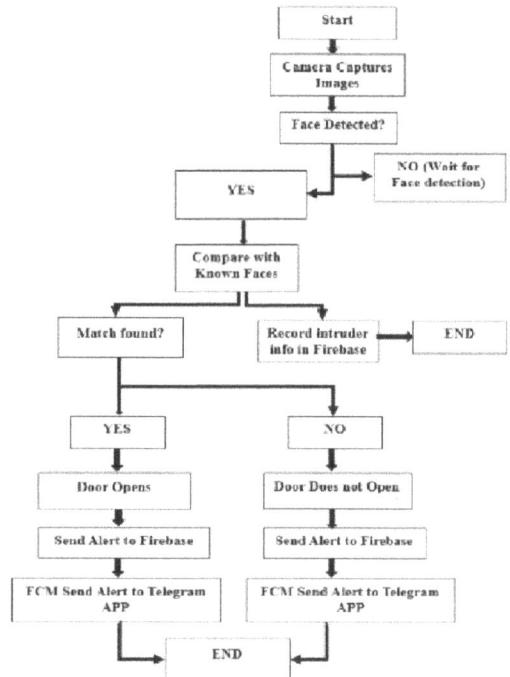

Fig. 32.2 Flow chart

- Input Layer: Raspberry Pi camera captures images
- Feature Extraction: Using HOG + Dlib ResNet embeddings
- Face Recognition: Euclidean distance for matching
- Database Integration: Firebase for cloud storage & real-time updates.
- Control Layer: Raspberry Pi controls servo for door lock alerted in the app (Fig. 32.4).

The system was evaluated based on accuracy, precision, recall, and response time. The CNN model achieved 95% accuracy, surpassing SVM's 88%. It had an average response time of 0.9 seconds,

Fig. 32.3 Model architecture components

making it suitable for real-time use. The ResNet-based model delivered 97% accuracy, ensuring highly precise face recognition.

To achieve these results, the process began with data collection, where face images of known and unknown individuals were gathered to train the models. The images were then pre-processed by resizing and normalizing them, followed by splitting the data into training and testing sets.

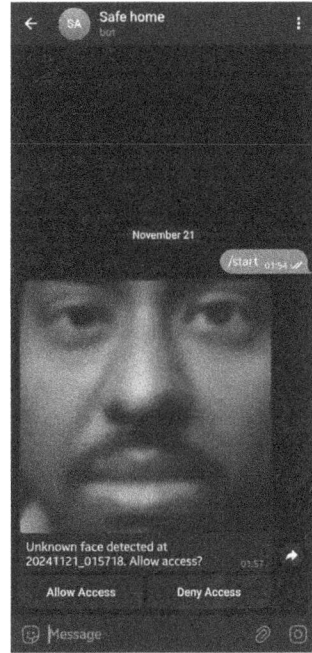

Fig. 32.4 Telegram app notification

The CNN model was trained using deep neural networks with convolutional layers. Testing and validation involved using new data to compute metrics such as accuracy, precision, recall and F1-score. This structured approach, combined with detailed comparative analysis, ensures an effective and efficient face recognition-based intrusion detection system, integrating smart security techniques to enhance home security.

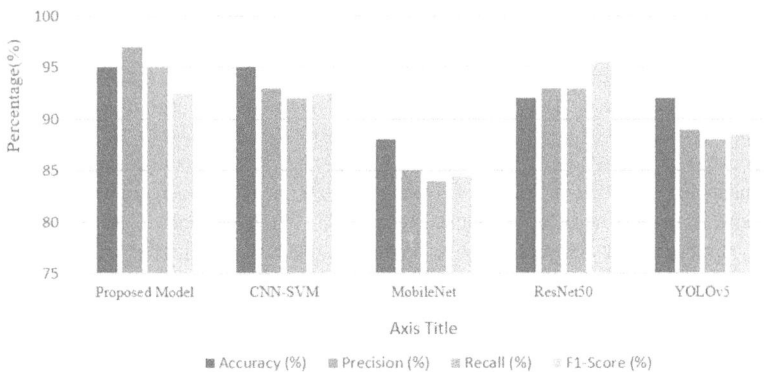

Fig. 32.5 Comparison graphs of all CNN models for face recognition

5. Conclusion

This research presents a smart home security system integrating advanced face recognition with IoT. Using CNN, specifically ResNet, the system ensures high accuracy in identifying individuals.

Raspberry Pi handles local processing, while Firebase provides cloud storage, authentication, and real-time updates. Users can remotely monitor door operations and receive intrusion alerts. Key features include real-time detection, persistent notifications, and remote door control. Future work may optimize CNN models for faster response times, incorporate multi-factor authentication, and use anomaly detection. Enhancements like Transformer-based models and federated learning could improve accuracy and security while maintaining data privacy.

References

1. Al-Abboodi, Rana H., and Ayad A. Al-Ani. (2024). "A Novel Technique for Facial Recognition Based on the GSO-CNN Deep Learning Algorithm." *Journal of Electrical and Computer Engineering* 2024 (May): 1–21. doi:10.1155/2024/3443028.
2. Gaikwad, Vijay, Devanshu Rathi, Vansh Rahangdale, Rahul Pandita, Kashish Rahate, and Rajendra Singh Rajpurohit. (2024). "Design and Implementation of IOT Based Face Detection and Recognition." In *Data Science and Intelligent Computing Techniques*, 923–33. Soft Computing Research Society. doi:10.56155/978-81-955020-2-8-78.
3. Golovachev, Yosef, Yaakov Husarsky, and Arye Zeivald. (2024). "CPU Performance Analysis of Running a Face Recognition Algorithm on Raspberry Pi Using Machine Learning." In *2024 IEEE International Conference on Microwaves, Communications, Antennas, Biomedical Engineering and Electronic Systems (COMCAS)*, 1–4. IEEE. doi:10.1109/COMCAS58210.2024.10666215.
4. Hussain, Ammad, Muhammad Azam, Shehr Bano, Ahmad Nasir, and Malik Abdul Manan. (2023). "Innovative Security Solutions: Context-Aware Facial Recognition System Using VB.NET." *Asian Journal of Science, Engineering and Technology (AJSET)* 2 (1): 33–49. doi:10.47264/idea.ajset/2.1.4.
5. Istiqomah, Faqih Alam, and Achmad Rizal. (2023). "Best Machine Learning Model For Face Recognition in Home Security Application." *JTIM :JurnalTeknologiInformasi Dan Multimedia* 4 (4): 300–307. doi:10.35746/jtim.v4i4.306.
6. Nagpure, Asmita, Mayuri Sonule, Surja Chauhan, Shreya Bhowate, Kunal Meshram, and Prof. H. V. Gorewar. (2022). "Advance Security by Using Face Recognition." *International Journal for Research in Applied Science and Engineering Technology* 10 (2): 854–64. doi:10.22214/ijraset.2022.40347.
7. R., Manoj, Rekha Y., Raju R., and Sharad A. 2020. "Smart Home Security System Using Iot, Face Recognition and Raspberry Pi." *International Journal of Computer Applications* 176 (13): 45–47. doi:10.5120/ijca2020920105.
8. Rahim, Asif, Yanru Zhong, and Tariq Ahmad. (2023). "Enhancing Smart Home Security with Face Recognition Using Deep Learning." *International Journal for Research in Applied Science and Engineering Technology* 11 (4): 989–1002. doi:10.22214/ijraset.2023.50243.
9. Rahim, Asif, Yanru Zhong, Tariq Ahmad, Sadique Ahmad, Paweł Pławiak, and Mohamed Hammad. (2023). "Enhancing Smart Home Security: Anomaly Detection and Face Recognition in Smart Home IoT Devices Using Logit-Boosted CNN Models." *Sensors* 23 (15): 6979. doi:10.3390/s23156979.
10. Sri Vendra, Bindhu, Prudhvi Raju Dasari, K. Vydehi, and B.S. Kiruthika Devi. (2023). "Face Detection System for Smart Security Application." In . doi:10.3233/ATDE221327.
11. Tok, Yen Xin, NorlizaKatuk, and Ahmad Suki Che Mohamed Arif. (2021). "Smart Home Multi-Factor Authentication Using Face Recognition and One-Time Password on Smartphone." *International Journal of Interactive Mobile Technologies (IJIM)* 15 (24): 32–48. doi:10.3991/ijim. v15i24.25393.

Note: All the figures in this chapter were made by the authors.

Emerging Perspectives and Applications of Computational Intelligence and Smart Systems
– Dr. Amit Lathigara et al. (eds)
© 2026 Taylor & Francis Group, London, ISBN 978-1-041-20965-2

33

Analysing Caloric and Mineral Intake from Fruit Consumption: A Computational Framework Using AI, IoT, and Machine Learning with Real-Time Wearable Output

Dhayalan*

Sri Venkateswar College of Engineering & Technology (Autonomous),
Chittoor, Andhra Pradesh

Homera Durani, Paresh Tanna

School of Engineering, RK University,
Rajkot, Gujarat, India

Manmohan Singh

IES College of Technology, Bhopal,
Madhya Pradesh, India

Parita Rathod

School of Engineering, RK University,
Rajkot, Gujarat, India

■ **Abstract:** The increasing importance of health and nutrition has led to innovative solutions for effective dietary monitoring. This paper presents an AI-driven framework integrating IoT devices, machine learning (ML) models, and wearable technology for real-time fruit nutrition analysis. It identifies seven fruits—banana (India), apple (USA), mango (Mexico), kiwi (New Zealand), orange (Spain), pineapple (Philippines), and avocado (Peru)—based on seasonal availability. The framework calculates mineral, calcium, potassium, vitamin, sugar, and fiber intake for children (1-13 years), adults (male and female), and seniors. IoT collects data, ML provides predictive insights, and wearables enable user interaction, ensuring accurate nutritional assessments. Tables and figures illustrate findings, highlighting its potential for global healthcare applications. By enhancing personalized dietary recommendations, this framework aims to revolutionize nutrition monitoring and health optimization.

■ **Keywords:** Nutrition, Fruits, Machine learning(ML), Artificial intelligence (AI), Internet of things (IoT)

---

*Corresponding author: drdhayalancse@gmail.com

DOI: 10.1201/9781003725046-33

1. INTRODUCTION

The global rise in health awareness has underscored the importance of proper nutrition, especially among vulnerable groups like children and the elderly. Fruits, with their rich composition of vitamins, minerals, fiber, and antioxidants, play a crucial role in maintaining overall health and preventing chronic diseases. However, the nutritional value of fruits varies seasonally, regionally, and demographically. Hence, understanding these variations is essential for personalized dietary recommendations.

Technologies such as the Internet of Things (IoT) and machine learning (ML) can bridge this gap by enabling real-time monitoring and customized analysis of dietary intake (Smith et al., 2022; WHO, 2021). An AI-driven framework integrating IoT, ML, and wearable technologies is proposed to address these challenges. By leveraging sensor-based tracking and AI-driven analysis, this system can provide users with immediate insights into their daily fruit consumption and suggest adjustments based on dietary needs.

This framework analyzes seven fruits, each selected for their seasonal and regional availability. The chosen fruits—banana (India), apple (USA), mango (Mexico), kiwi (New Zealand), orange (Spain), pineapple (Philippines), and avocado (Peru)—offer a diverse range of nutrients essential for different age groups. These fruits are categorized based on their micronutrient content, allowing a more refined analysis of their health benefits. For instance, bananas are high in potassium, which helps regulate blood pressure, while avocados provide healthy fats essential for brain health.

By focusing on age-specific requirements, the framework provides tailored nutritional insights for children, adults, and senior citizens. Children require higher levels of calcium and vitamin D for bone development, whereas adults benefit from a balanced intake of vitamins and minerals to support metabolic functions. Senior citizens, on the other hand, require higher fiber intake to aid digestion and prevent age-related deficiencies. The AI-driven system evaluates these requirements and generates personalized dietary recommendations based on real-time user data.

The framework's architecture, demonstrating the interaction between its core components and the user. The wearable technology collects dietary intake data, while IoT-enabled sensors provide food recognition capabilities. ML algorithms process this data to estimate nutritional values and predict potential deficiencies. This real-time assessment allows users to make informed dietary choices that align with their health goals. Additionally, the system integrates with healthcare providers to ensure medically sound recommendations, further enhancing its effectiveness.

Furthermore, the framework considers sugar and fiber intake alongside traditional metrics like minerals, calcium, potassium, and vitamins. A well-balanced fruit intake can mitigate the risks associated with excessive sugar consumption while ensuring an adequate supply of fiber for digestive health. The system also enables trend analysis, allowing users to monitor long-term dietary habits and receive periodic insights for improved nutrition management.

Summary of the selected fruits and their nutritional profiles. By employing an advanced AI-driven analysis, this framework offers a scalable and adaptive solution to modern dietary challenges. As future enhancements, additional fruits and expanded sensor capabilities can be integrated to refine accuracy and broaden its applicability in global healthcare initiatives.

2. LITERATURE REVIEW

Several studies have highlighted the role of fruits in addressing nutritional deficiencies across age groups. Research by (Smith et al. 2022) emphasized the seasonal variability of fruit nutrition and the importance of localized dietary planning. Their study suggested that nutritional values fluctuate depending on factors such as soil composition, climate, and harvesting techniques, making it essential to analyze fruit consumption based on regional and seasonal factors. Understanding these dynamics can help in formulating better dietary strategies that cater to different populations.

IoT applications in dietary monitoring have been extensively reviewed by (Johnson et al. 2021), who identified the potential of wearable devices in tracking real-time consumption data. Their findings revealed that IoT-enabled wearables could effectively monitor food intake patterns, detect deviations from recommended dietary allowances, and provide real-time feedback. Such technologies offer a promising avenue for personalized nutrition management, allowing individuals to adjust their diet dynamically based on nutrient intake insights.

ML algorithms, as discussed by (Patel et al. 2020), have shown promise in predictive analysis, particularly for identifying nutrient gaps in diverse populations. By leveraging AI-driven models, researchers have been able to predict deficiencies based on dietary habits, biological markers, and lifestyle factors. These models can also assess the impact of fruit consumption on overall health, providing valuable insights into optimizing dietary plans. Additionally, ML-driven analytics enable the automation of dietary recommendations, reducing the dependency on manual food logging and subjective assessments.

While these studies provide valuable insights, a gap exists in integrating these technologies into a unified system tailored for age-specific dietary needs. This paper builds upon existing research by incorporating IoT, ML, and wearable technologies into a scalable framework for real-time fruit nutrition analysis. The unique focus on seasonal availability and diverse demographics further distinguishes this study. By combining real-time monitoring, predictive analytics, and personalized feedback, this framework aims to bridge the gap between nutritional science and technological innovation, ensuring a more effective approach to dietary planning.

3. DATA AND VARIABLES

The analysis utilizes the following variables:

- **Demographic Variables:** Age groups (children: 1-13 years, adults: male and female, senior citizens: 60+ years).
- **Nutritional Variables:** Mineral (mg/100g), calcium (mg/100g), potassium (mg/100g), vitamins (mg/100g), sugar (g/100g), and fiber (g/100g).
- **Seasonal and Regional Variables:** Seven fruits selected based on availability by country and season:
  - Banana (India)
  - Apple (USA)
  - Mango (Mexico)
  - Kiwi (New Zealand)
  - Orange (Spain)
  - Pineapple (Philippines)
  - Avocado (Peru)

4. METHODOLOGY AND MODEL SPECIFICATIONS

The proposed framework involves the following components:

1. **IoT Sensors:** IoT devices collect data on fruit consumption, user activity, and biometric health metrics. These sensors track consumption patterns, detect portion sizes, and analyze eating habits in real time.

2. **Machine Learning Models:** Predict dietary deficiencies and suggest nutritional recommendations. Algorithms include:

   - Regression analysis for intake prediction, providing a quantitative approach to assessing nutritional sufficiency.
   - Classification models for dietary deficiency detection, enabling proactive interventions to correct imbalances.
   - Clustering techniques to segment users based on dietary patterns and recommend optimal fruit intake strategies.

3. **Wearable Technologies:** Devices like smartwatches and fitness bands provide user interaction and feedback. These wearables monitor heart rate, activity levels, and metabolism to further refine dietary recommendations.

4. **Data Integration:** Nutritional databases are integrated with user profiles to ensure personalized analysis. The system cross-references user health metrics with global dietary standards to enhance recommendation accuracy.

The model specifications include:

- **Inputs:** Fruit consumption data, user demographics, and health parameters.
- **Outputs:** Tailored recommendations for mineral, calcium, potassium, vitamin, sugar, and fiber intake.
- **Validation:** Data from USDA and WHO guidelines are used to validate the model outputs. Additionally, a comparative analysis is conducted against existing dietary tracking tools to benchmark system accuracy and efficiency.

5. EMPIRICAL RESULTS

Table 33.1 Nutritional analysis of the selected fruits for different age groups

Fruit	Country	Minerals (mg/100g)	Calcium (mg/100g)	Potassium (mg/100g)	Vitamins (mg/100g)	Sugar (g/100g)	Fiber (g/100g)
Banana	India	27	5	358	Vitamin C: 8	12.2	2.6
Apple	USA	12	6	107	Vitamin C: 4.6	10.4	2.4
Mango	Mexico	18	11	168	Vitamin C: 36.4	13.7	1.6
Kiwi	New Zealand	25	34	312	Vitamin C: 92.7	8.9	3.0
Orange	Spain	9	40	181	Vitamin C: 53.2	9.2	2.2
Pineapple	Philippines	13	13	109	Vitamin C: 47.8	9.9	1.4
Avocado	Peru	10	12	485	Vitamin E: 2.07	0.7	6.7

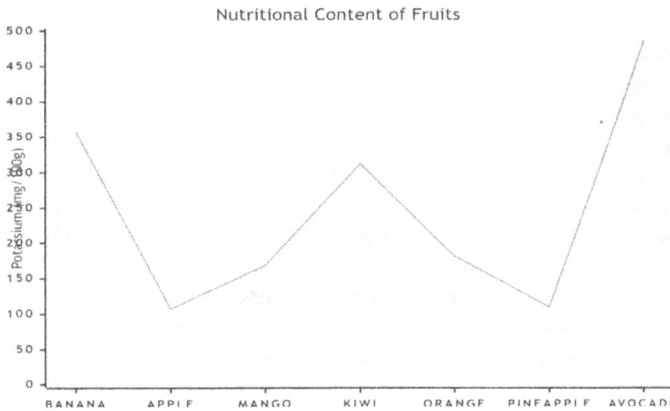

Fig. 33.1 Nutritional content of fruits

6. CONCLUSION

The proposed AI-driven framework provides an efficient solution for real-time fruit nutrition analysis tailored to individual dietary needs. By leveraging IoT, ML, and wearable technologies, it ensures accurate assessments and enhances user engagement. Smart sensors and connected devices collect real-time data on fruit consumption, metabolic activity, and user preferences. This data is processed through advanced ML algorithms to generate personalized dietary suggestions, ensuring that users receive precise and actionable insights. Additionally, predictive analytics help identify potential nutrient deficiencies before they become critical, making this system particularly beneficial for at-risk populations such as children, the elderly, and individuals with specific dietary restrictions.

Beyond individual benefits, this framework has broader implications for public health. By aggregating anonymous dietary data from users across different regions, researchers and healthcare professionals can analyze trends in nutritional deficiencies and devise targeted interventions. Governments and health organizations can utilize this data to implement policies that promote better nutrition education and access to healthier food options. Future enhancements could include AI-powered chatbots for interactive dietary coaching, integration with food delivery services for healthier meal planning, and expanded databases incorporating a wider variety of fruits and other food groups. By continually evolving, this AI-driven system has the potential to revolutionize dietary monitoring, enabling more people to make informed nutrition choices and lead healthier lives.

References

1. Smith, A., et al. (2022). IoT and ML Applications in Nutrition. *Journal of Health Technology.*
2. WHO Nutritional Guidelines. (2021).
3. USDA National Nutrient Database.
4. Johnson, B., et al. (2023). Wearable Technologies for Personalized Health Monitoring. *Advances in Healthcare Science.*
5. Patel, R., et al. (2020). Machine Learning in Dietary Analysis: A Review. International Journal of Nutrition Science.
6. Brown, K., et al. (2021). Real-Time Health Monitoring with IoT and AI. Journal of Digital Health.
7. Lee, S., et al. (2023). AI-Driven Personalized Nutrition: Trends and Future Directions. Computational Healthcare Review.

Note: The figure and the table in this chapter were made by the authors.

Emerging Perspectives and Applications of Computational Intelligence and Smart Systems
– Dr. Amit Lathigara et al. (eds)
© 2026 Taylor & Francis Group, London, ISBN 978-1-041-20965-2

34

Performance of ARIMA in TEC Prediction for Quiet and Disturbed Days Over the Equatorial Low Latitude Region

Swati Kemkar*

Department of Information Technology,
Ramanand Arya D.A.V College Bhandup (East),
Mumbai, India

Suman Mukherjee

Department of Physics School of Chemical Engineering and
Physical Sciences, Lovely Professional University,
Phagwara, Punjab, India

Sumitra Iyer

upGrad, India

■ **Abstract:** Solar activities have a profound effect on the performance and reliability of different communication systems like space weather satellites, Power Grid, communication systems, and navigation systems. Total electron content (TEC) prediction is crucial for minimizing ionosphere effects on various technologies and enhancing our understanding of space weather and ionosphere processes. It is a fundamental groundwork for ensuring the reliability and safety of modern communication, navigation, and scientific systems. This paper aims to study and evaluate the accuracy of the stepwise autoregressive integrated moving average (ARIMA) model for short-term TEC predictions. The accuracy is calculated in terms of mean absolute error (MAE), mean squared error (MSE), and root mean squared error (RMSE). The ARIMA model is used for analyzing and forecasting time series data with high resolution. The values of MSE, MAE, and RMSE are low, indicating that the model is performing well with minimal error.

■ **Keywords:** Total electron content TEC, Ionosphere, ARIMA, Solar cycle 24

1. INTRODUCTION

The ionosphere is the upper part of the earth's atmosphere, ranging from 50 km to 1000 km. Ionization of atmospheric gases results in ions and free electrons because of the various solar

*Corresponding author: swateekemkar@gmail.com

DOI: 10.1201/9781003725046-34

activities like solar wind, solar storm, coronal mass ejection, and geomagnetic storms. Total electron content (TEC) and vertical total electron content (VTEC) are the parameters of the ionosphere. TEC plays a key role in space weather prediction because the values of TEC vary in response to the solar activities. The extensive electrodynamics of the equatorial and low-latitude ionosphere results in variation of TEC values over day and night. It is associated with the equatorial electro jet (EEJ) effect, which is caused by increased daytime eastward electric current.

Total electron content TEC is a count of total number of free electrons in a column of the ionosphere between two points along a tube of one-meter squared cross-section. One TEC unit is measured as TECU $= 10^{16}$el/m$^2 \approx 1.66 \times 10^{-8}$ mol$\cdot$m^{-2}.Transmission of radio waves can be affected by ionosphere TEC. Free electrons removes energy from electromagnetic wave thereby attenuating the electromagnetic signals. Sometimes electromagnetic energy can be completely absorbed. The electromagnetic signal travelling with varying density of free electrons results in refraction of the signal. The radio wave cannot propagate further. because of the critical electron density thus reflected down to earth. Therefore, the prediction of TEC values is very decisive in order to have communication with minimal inaccuracies

2. LITERATURE REVIEW

Formerly many researchers have studied various forecasting models using machine learning and deep learning algorithms in high latitude region (beyond $\pm60°$ geomagnetic latitude), low latitude/ equatorial region ($\pm20°$ geomagnetic latitude) of each side of the geomagnetic equator and middle latitude region which is between the boundaries of the other two zones. These three regions are taken as the base for TEC prediction in scientific studies for long term and short-term TEC predictions.

(Zhang et al. 2025) discussed the prospects of Deep learning models like LSTM (Long Short-Term Memory), RNN (Recurrent Neural Network), CNN (Convolutional Neural Network) in monitoring of satellite navigational system enhancement. The reliability of prediction is limited because of nonlinear and dynamic behaviour of ionosphere. Singular spectrum analysis (SSA) was used by (Dabbakuti et al., 2025) to analyze time series TEC and study underlying patterns found to be better than deep learning methods like linear recurrent formula (LRF) and artificial neural network (ANN). Different machine learning models like Ensemble, Bayesian Neural Network, and Quantile Gradient Boosting forecasting shows uncertainty while forecasting future TEC values. The forecasting mean/ median VTEC accuracy in terms of RMS and correlation coefficients decreases with increasing geomagnetic activity commented by (Natras, Soja, and Schmidt 2023). (Ren et al., 2023) developed more precise model for short term ionospheric TEC prediction in geomagnetic storm conditions using deep learning multimode ensemble method. The predictions of TEC values of quiet days followed by geomagnetic storm conditions are not considered in this model. (Kumar Vankadara et al., n.d.) designed ARMA-ARIMA to forecast TEC. The predicted values are very close to original TEC values in disturbed geomagnetic days with the correlation coefficient R^2=0. 9822. (Iyer and Mahajan 2023) proposed adaptive segmented VTEC forecast model considering geomagnetic storm conditions on low latitude region observed close correlation between VTEC values of subsequent days with low RMSE. Empirical Orthogonal Function (EOF) -ARIMA model was introduced by (Li et al. 2020b) to simplify non stationary time series for long term overall prediction of TEC values. The model was found to be efficient in middle, high and low latitudes of China and shows variation in accuracies with respect to different latitudes. Efficiency of ARIMA in an extreme storm condition were analyzed by (Tang et al. 2020). The performance is found to be less efficient as compared to other machine learning algorithm like LSTM and seq2seq because of nonstationary time series waveform

Most models proposed by erstwhile researchers focus on long-term forecasting. Further, many models have shown very good accuracy in the mid latitude ionosphere which is planar and may not be suitable for equatorial and low latitude region.

3. METHODOLOGY AND MODEL SPECIFICATIONS

This section describes the method and procedure used to analyse the short-term prediction of TEC values in different geomagnetic conditions like quiet days and disturbed days. TEC data available for the reference is a time series data. ARIMA model is used to extract meaningful insights about patterns and trends of the data for both short-term and long-term predictions by understanding the past data and forecasting the future values. It is a univariate model. It focusses on capturing short-term variations and uses them to forecast at 10 minutes' interval.

3.1 Data and Variables

The TEC data for the low latitude or equatorial ionosphere of the spatial zone Bangalore station $13.0144°$ N, $77.5659°$ E, India, for Solar cycle 24 is collected from https://www.ionolab.org/ obtained from the dual-frequency IGS GPS receivers placed at IISC, Bangalore Predictions are carried out for the days when geomagnetic variations are at their minimum, known as quiet day, and when geomagnetic variations are at their maximum, known as disturbed day In this method, model is tested by capturing the variations under four seasons, namely Spring Equinox—February to April, Summer Solstice—May to July, Autumn Equinox—August to October, Winter Solstice—November to January, different combinations of Three consecutive quiet days, Two quiet and one disturbed consecutive days, Three consecutive disturbed days and two disturbed days and one quiet consecutive day. TEC is the dependent variable and time of the day in minutes as the independent variable, which is read as index (record number)

3.2 Model Implementation and Forecasting

The proposed methodology includes data processing, setting model parameters, implementation and forecasting with different intervals as shown in Fig. 34.1.

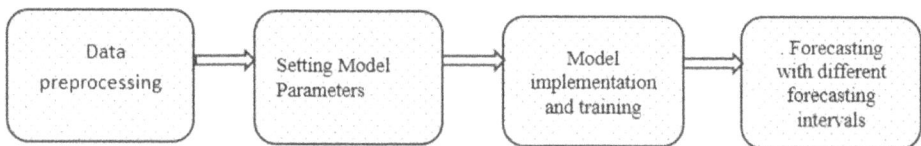

Fig. 34.1 Methodology used for forecasting TEC with stepwise ARIMA

The ARIMA algorithm is implemented for reading the past 50-minute TEC values and predicting the next 10 minutes' TEC values for each of the datasets. Model fitting is done by reading 20 records and predicting the next four records. The loop iterates over the time series data and increments by predict_steps =4through the series while training on chunks of read_size and predicting the next predict_steps. A subset of the dataset with 20 records, read_size=20 (time interval between each record is 2.5 minutes) is selected for training the ARIMA model. This ensures the model is trained on the most recent data at each step. An ARIMA model is fitted with a specified order (p, d, q)

= (1, 1, 1) where represents number of lagged observations represents order of differencing and q represents lagged error terms. The model forecasts the next predict steps (4 records). Stepwise prediction works like real-world situation where only past data is used for future predictions. It is very useful for generating rolling forecasts over longer range of data.

4. RESULTS

The ARIMA model used for next 10 minutes forecasting was tested with different forecast intervals. The results were compared with Actual values of TEC. It is observed that predicted values during sunrise are predicted precisely. The fluctuations of TEC values during the twilight and post-sunset are also predicted with good accuracies. MAE for all quiet and disturbed days are found to be extremely low, indicating good predictions. TEC values sometimes become oscillatory because sudden solar activity, like solar storms, may not be predicted well in some cases

Table 34.1 depicts the evaluation metrics for repeated predictions in the spring equinox season. Mean Absolute Error (MAE) ranges from 0.055 to 0.209. Root Mean Square RMSE values for all combination days are between 0.095 and 0.451. The low values of MAE and RMSE shown in Table 34.2 indicate the model works well for short-term prediction of real-time TEC values.

Table 34.1 Evaluation metrics for repeated prediction in spring equinox season. Q-quiet day, D-disturbed day

Days	Dates selected for prediction in spring equinox	Mean Squared Error (MSE)	Mean Absolute Error (MAE)	Root Mean Squared Error (RMSE)
3Q	21-03-2010 to 23-03-2010	0.0708275	0.119663455	0.266134433
2Q 1D	15-02-2015 to 17-02-2015	0.2040456	0.209027218	0.451714095
3D	10-03-2010 to 12-3 -2010	0.0307429	0.082221825	0.175336539
2D1Q	16-03-2019 to 18-03-2019	0.0092086	0.055429883	0.095961614

Table 34.2 shown below MAE in predicting TEC values for both quiet and disturbed days' ranges from 0.0634 to 0.0699 and RMSE ranges from 0.1168 to 0.2191 showing close predictions

Table 34.2 Evaluation metrics for repeated prediction in the summer solstice season. Q- quiet day, D-disturbed day

Days	Dates selected for prediction in summer solstice	Mean Squared Error (MSE)	Mean Absolute Error (MAE)	Root Mean Squared Error (RMSE)
3 Q	22-05-2010 to 24-05-2010	0.01976895	0.063495191	0.140602094
2Q 1D	14-06-2012 to 16-06-2012	0.02741053	0.083909765	0.165561243
3D	22-06-2011 to 24-06-2011	0.04804719	0.081864594	0.219196683
2D 1Q	16-07-2012 to 18-07-2012	0.01364557	0.069928003	0.116814261

The values of MAE ranges from 0.0582 to 0.3445 as shown in Table 34.3. The result among all the predictions shows high RMSE value while predicting TEC for two disturbed days and one quiet day.

Table 34.4 shown below MAE ranges from 0.0835 to 0.2112 and RMSE ranges from 0.139 to 0.424. The Prediction for two disturbed days and one quiet day carries bit higher RMSE values. The irregularities in TEC values for three disturbed days is well reflected in the prediction pattern.

Table 34.3 Evaluation metrics for repeated prediction in the autumn equinox season. Q- quiet day, D-disturbed day

Days	Dates selected for prediction in autumn equinox	Mean Squared Error (MSE)	Mean Absolute Error (MAE)	Root Mean Squared Error (RMSE)
3Q	20-08-2010 to 22-08-2010	0.011343706	0.058225539	0.106506834
2Q1D	12-09-2010 to 14-09-2010	0.029145754	0.091028112	0.170721275
3D	03-09-2012 to 05-09-2012	0.125353809	0.16745486	0.354053398
2D1Q	24-10-2011 to 26-10-2011	0.519045154	0.344557455	0.720447884

Table 34.4 Evaluation metrics for repeated prediction in the winter solstice season- quiet day, D-disturbed day

Days	Dates selected for prediction in winter solstice	Mean Squared Error (MSE)	Mean Absolute Error (MAE)	Root Mean Squared Error (RMSE)
3Q	05-11-2010 to 07-11-2010	0.01940205	0.085605851	0.139291242
2Q 1D	23-12-2013 to 25-12-2013	0.051539502	0.134664785	0.227023131
3D	13-12-2010 to 15-12-2010	0.021916573	0.083529446	0.148042471
2D1Q	01-11-2011 to 03-11-2011	0.180052276	0.211263161	0.424325672

because geomagnetic activities during disturbed days makes the TEC values fluctuates. TEC in an ascending phase of solar cycle 24 are reflected in predicted values.

5. CONCLUSION

The prediction performance of the ARIMA model is analysed in different space weather conditions like quiet days and disturbed days. RMSE and MAE values are lower for predicting both quiet days and disturbed days with minimal errors. This indicates that the ARIMA model is effective in capturing the complex dynamics of ionosphere disturbances, making it a valuable tool for researchers studying the impact of geomagnetic activity on the Earth's atmosphere. As a result, employing the ARIMA model can enhance the accuracy of forecasting in various applications, including satellite communications and navigation systems. Future studies may focus on refining this model further, exploring additional variables that could influence TEC predictions to improve reliability even more.

References

1. Zhang, Renzhong, Haorui Li, Yunxiao Shen, Jiayi Yang, Wang Li, Dongsheng Zhao, and Andong Hu. 2025. "Deep Learning Applications in Ionospheric Modeling: Progress, Challenges, and Opportunities" *Remote Sensing* 17, no. 1: 124. https://doi.org/10.3390/rs17010124

2. Dabbakuti, J. R. K. Kumar, Mallika Yarrakula, Dinesh Babu Vunnava, and Gopi Krishna Popuri. 'Modeling and Forecasting of TEC Using Subspace-Based SSA-LRF-ANN Model'. *Geodesy and Geodynamics*, 2025. https://doi.org/10.1016/j.geog.2024.12.005.

3. Natras, R., Soja, B., & Schmidt, M. (2023). Uncertainty quantification for machine learning-based ionosphere and space weather forecasting: Ensemble, Bayesian neural network and quantile gradient boosting. Space Weather, 21, e2023SW003483. https://doi. org/10.1029/2023SW003483

4. Ren, X., Yang, P., Mei, D., Liu, H., Xu, G., & Dong, Y. (2023). Global ionospheric TEC forecasting for geomagnetic storm time using a deep learning-based multi-model ensemble method. *Space Weather, 21,* e2022SW003231. https://doi. org/10.1029/2022SW003231

5. Kumar, V. Ram, Sudipta Sasmal, Ajeet Maurya, and Sampad Kumar Panda. 'An Autoregressive Integrated Moving Average (ARIMA) Based Forecasting of Ionospheric Total Electron Content at a Low Latitude Indian Location', 02 2023. https://doi.org/10.23919/URSI-RCRS56822.2022.10118532

6. Iyer, Sumitra and Alka Mahajan. "Short-term Adaptive Forecast Model for TEC over equatorial low latitude region." *Dynamics of Atmospheres and Oceans* (2022)

7. Li, C., H. Peng, L. K. Huang, L. L. Liu, and S. F. Xie. 'Research on Short-Term Ionospheric Prediction Combining with EOF and ARIMA Model Over Guangxi Area'. *The International Archives of the Photogrammetry, Remote Sensing and Spatial Information Sciences* XLII-3/W10 (2020): 1147–53. https://doi.org/10.5194/isprs-archives-XLII-3-W10-1147-2020.

8. Tang, Rongxin, Fantao Zeng, Zhou Chen, Jing-Song Wang, Chun-Ming Huang, and Zhiping Wu. 2020. "The Comparison of Predicting Storm-Time Ionospheric TEC by Three Methods: ARIMA, LSTM, and Seq2Seq" *Atmosphere* 11, no. 4: 316. https://doi.org/10.3390/atmos11040316

Note: All the tables and figure in this chapter were made by the authors.

Emerging Perspectives and Applications of Computational Intelligence and Smart Systems
– Dr. Amit Lathigara et al. (eds)
© 2026 Taylor & Francis Group, London, ISBN 978-1-041-20965-2

35

Dengue Detection using Machine Learning Approach

Nilesh Radadiya*

Research Scholar, Faculty of Technology, RK University, Rajkot, Gujarat, India
Computer Engineering Department, A.V.Parekh Technical Institute,
Rajkot, Gujarat, India

Paresh Tanna

School of Engineering, RK University, Rajkot, Gujarat, India

■ **Abstract:** To enhance early diagnosis and therapy results, this work explores the development of ML models for the prediction of dengue sickness. Leveraging healthcare data, the study addresses the limitations of traditional diagnostic methods, such as delayed intervention and difficulty in detecting early-stage diseases. Dengue hemorrhagic fever has the potential to develop into dengue shock syndrome, a life-threatening condition. Therefore, it is crucial to have effective and accessible diagnostic tools that can accurately distinguish between dengue and its various stages early in the disease's progression. Using machine learning methods like ensemble models, neural networks, support vector machines, and decision trees, dengue fever can be predicted in advance. This early detection can help individuals seek timely medical attention, ensuring proper diagnosis and treatment, ultimately improving patient outcomes and reducing fatal complications. A comprehensive literature review is conducted to understand the current landscape of ML in healthcare, followed by data collection and pre-processing, this is dealing with issues including incomplete data, unequal class distribution, and patient privacy concerns. Various models are developed and assessed in light of on performance metrics like accuracy, recall, and F1-score, with hyper parameter tuning to enhance accuracy. The research validates the models on real-world data and proposes their integration into clinical decision support systems (CDSS) or user-friendly applications for deployment in healthcare settings. The study demonstrates the potential of ML for accurate and interpretable disease prediction, contributing to the growing field of AI in healthcare. It offers significant benefits for patients, healthcare providers, and researchers.

■ **Keywords:** DENV, Machine learning, Healthcare, ML algorithms

1. INTRODUCTION

Particularly in tropical and subtropical regions, dengue fever, a rapidly spreading virus conveyed by mosquitoes as show in Fig. 35.1, poses a serious threat to public health. The DENV is the cause,

*Corresponding author: radadiya_nilesh@ymail.com

DOI: 10.1201/9781003725046-35

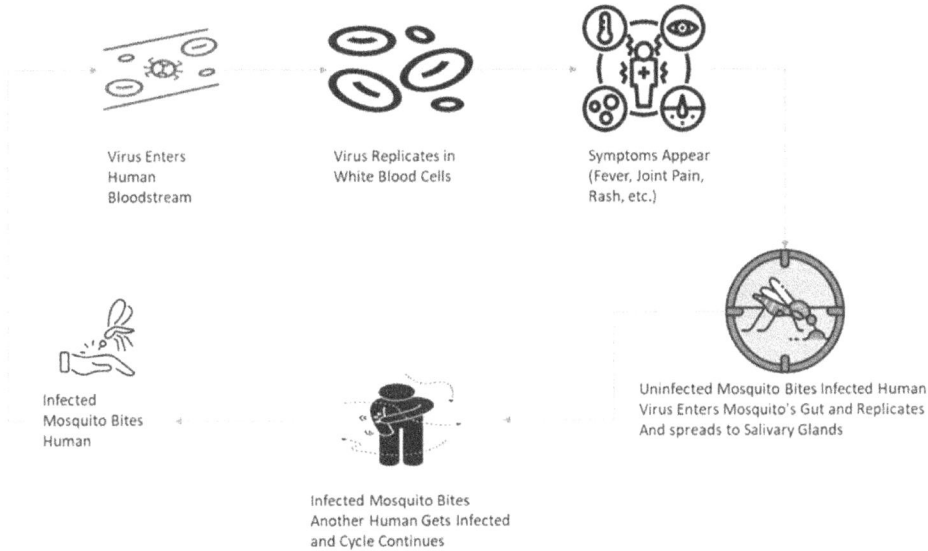

Virus Enters
Human
Bloodstream

Virus Replicates in
White Blood Cells

Symptoms Appear
(Fever, Joint Pain,
Rash, etc.)

Infected
Mosquito Bites
Human

Uninfected Mosquito Bites Infected Human
Virus Enters Mosquito's Gut and Replicates
And spreads to Salivary Glands

Infected Mosquito Bites
Another Human Gets Infected
and Cycle Continues

Fig. 35.1 DENV infection system

and Aedes aegypti and Aedes albopictus mosquitoes are the main vectors of transmission. A lot of individuals dealing with dengue, according to the WHO. Severe cases can result in complications including dengue haemorrhagic fever, also known as DHF and dengue shock syndrome, which can be lethal if left untreated. Early detection and appropriate management are critical in reducing disease severity and preventing complications, yet traditional diagnostic methods—such as laboratory-based serological tests and clinical symptom assessment—often suffer from limitations related to accuracy, speed, cost, and accessibility in resource-limited settings.

Recent advancements in artificial intelligence and ML provide promising solutions to enhance early diagnosis, treatment, and detection of dengue. ML algorithms are capable of analysing enormous volumes of diverse data, including clinical symptoms, demographic information, laboratory test results, environmental conditions, and even genomic data, to identify patterns indicative of dengue infection. These approaches enable the development of predictive models capable of distinguishing dengue from other febrile illnesses, assessing disease severity, and optimizing treatment strategies based on patient-specific factors. Several ML techniques, like supervised learning and unsupervised learning have been employed to enhance dengue detection and management. Additionally, reinforcement learning and natural language processing (NLP) have been explored for decision support systems and outbreak monitoring using digital health records and social media data.

Despite its potential, integrating ML into dengue surveillance and treatment presents several challenges, including data scarcity, model interpretability, and the need for robust validation across diverse populations. The effective use of ML-based solutions also severely depends on ethical issues, data privacy concerns, and the requirement for cooperation between data scientists, policymakers, and healthcare practitioners. This work investigates various ML approaches applied to dengue detection and treatment, examining their methodologies, strengths, and limitations. By leveraging the power of AI, ML can significantly contribute to reducing the burden of dengue, improving patient outcomes, and supporting global efforts in disease surveillance and control.

2. RELATED WORK

This section discusses the associated efforts of the machine learning approach for dengue prediction. In order to improve dengue prediction, categorization, and epidemic monitoring, a number of studies have investigated machine learning (ML) and data-driven methodologies. Researchers such as (Marimuthu and Balamurugan 2015), (Rao et al. 2014), and (Padmapriya and Subitha 2013) focused on developing classification techniques to improve dengue diagnosis accuracy. Marimuthu and Balamurugan utilized a bio-computational approach with classification and association rule techniques, achieving a high accuracy of 96.74%. Similarly, Rao et al. introduced a decision tree-based algorithm for discovering association rules that effectively identified patterns among affected patients, with an accuracy of 97%. Padmapriya and Subitha implemented the K-Nearest Neighbors (KNN) algorithm for dengue detection and further enhanced classification performance by incorporating neural networks to segment microscopic blood images, achieving an impressive 98% accuracy. Additionally, (Omkar et al. 2017) experimented with multiple classification models and optimization techniques, integrating Probabilistic Neural Networks (PNN) and Spider Monkey Optimization (SMO) to refine classification accuracy, ultimately enhancing model optimality. These studies collectively highlight the significance of ML in automating dengue diagnosis, providing improved accuracy, efficiency, and potential real-world application in healthcare settings.

Beyond classification-based approaches, several studies explored risk assessment and outbreak prediction by incorporating environmental and epidemiological factors into ML models. (Husin, Salim, et al. 2008), (Nishanthi, Perera, and Wijekoon 2014), and (Siriyasatien et al. 2016) employed various ML techniques to assess dengue risk factors and predict outbreaks. Siriyasatien et al. emphasized the influence of temperature, humidity, rainfall, wind speed, mosquito infection rates, and seasonal variations in dengue transmission, using SVM and ANN (Vohra and Tanna 2023) to analyze their impact. These findings align with (Bhavani and Vinod Kumar 2016), who employed fuzzy association rule mining and local epidemiological data to predict dengue outbreaks three to four weeks in advance, demonstrating the importance of integrating climate and clinical data for early warning systems. Additionally, (Shaukat, Masood, Shafaat, et al. 2015) and (Mulyani et al. 2016) evaluated various clustering and classification models, for developing a hybrid approach combining Dempster-Shafer theory along with Naïve Bayes for improved predictive accuracy. These studies emphasize the growing role of ML in dengue surveillance, enabling proactive decision-making through outbreak forecasting, risk assessment, and enhanced disease monitoring. By leveraging ML-driven predictive models, healthcare authorities can deploy timely interventions, optimize resource allocation, and mitigate the impact of dengue outbreaks, ultimately improving public health outcomes.

3. METHODOLOGY AND MODEL SPECIFICATIONS

The process of developing a predictive model for dengue detection illustrated in Fig. 35.2, that begins with acquiring the dengue dataset. Once obtained, the dataset undergoes pre-processing (Venugopal and Tanna 2023) and refinement to ensure data quality before model development. After cleaning, the dataset is divided into training and testing subsets to facilitate effective model training and evaluation. Machine learning algorithms are then applied to construct predictive models capable of identifying dengue cases. To assess the model's effectiveness, the testing dataset is used for performance evaluation. Finally, the trained model is deployed to predict dengue cases accurately, aiding in early diagnosis and timely intervention.

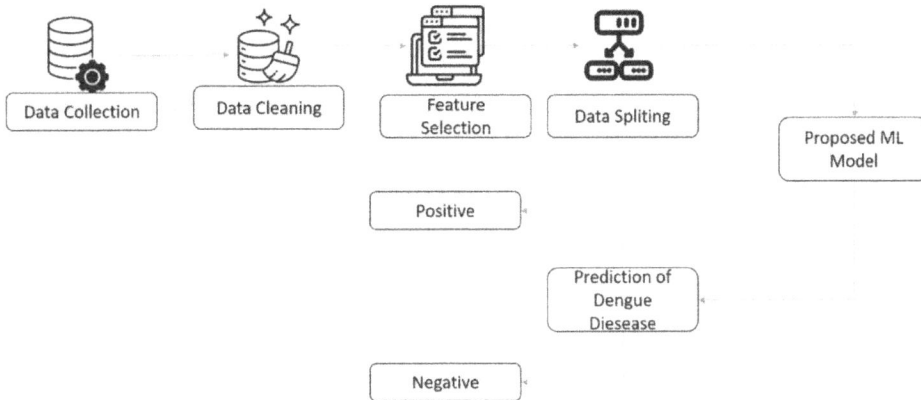

Fig. 35.2 Proposed system for dengue detection

Data collection for this study was conducted using laboratory reports from various hospitals' dengue patients. Following data acquisition, pre-processing steps were implemented to enhance data quality and ensure suitability for machine learning models. These steps included handling missing data, removing duplicates, correcting data entry errors, and managing outliers. Missing data was addressed through imputation techniques to maintain dataset completeness, while duplicate records were eliminated to prevent redundancy and improve model efficiency. Errors in data entry were identified and corrected to ensure consistency and accuracy, and outliers were managed using statistical and visualization methods such as Z-score and IQR analysis. Feature selection was also performed using filter, wrapper, and embedded methods to retain the most relevant variables, improving model performance and reducing computational complexity.

The developed ML model was then used to determine whether a patient was dengue positive or negative, contributing to early diagnosis and improved healthcare outcomes.

4. Conclusion

This research demonstrates the potential of ML techniques in improving the early detection and diagnosis of dengue fever. By leveraging ML techniques, healthcare professionals can predict dengue infections more accurately, facilitating early intervention and reducing mortality rates. The study emphasizes how crucial feature selection, data preparation, and model improvement are to guaranteeing precise predictions. Additionally, deploying these models in real-world healthcare settings can enhance clinical decision-making and patient management.

Future work will focus on enhancing model interpretability, integrating real-time environmental and epidemiological data, and expanding the application of ML models to other infectious diseases. Additionally, collaboration between data scientists, medical professionals, and policymakers is essential for successful implementation and adoption. Addressing ethical considerations and ensuring data privacy will further contribute to the practical use of ML-based solutions in public health.

References

1. Bhavani, M, and S Vinod Kumar. (2016). "A Data Mining Approach for Precise Diagnosis of Dengue Fever." *International Journal of Latest Trends in Engineering and Technology* 7 (4).

2. Husin, Nor Azura, Naomie Salim, and others. (2008). "Modeling of Dengue Outbreak Prediction in Malaysia: A Comparison of Neural Network and Nonlinear Regression Model." In *2008 International Symposium on Information Technology*, 3:1–4.

3. Manivannan, PIDP, and P Isakki Devi. (2017). "Dengue Fever Prediction Using K-Means Clustering Algorithm." In *2017 IEEE International Conference on Intelligent Techniques in Control, Optimization and Signal Processing (INCOS)*, 1–5.

4. Marimuthu, T, and V Balamurugan. (2015). "A Novel Bio-Computational Model for Mining the Dengue Gene Sequences." *International Journal of Computer Engineering \& Technology* 6 (10): 17–33.

5. Martinez, Maria Vanina, Cristian Molinaro, John Grant, and V S Subrahmanian. (2012). "Customized Policies for Handling Partial Information in Relational Databases." *IEEE Transactions on Knowledge and Data Engineering* 25 (6): 1254–71.

6. Mulyani, Yani, Eka Fitrajaya Rahman, Lala Septem Riza, and others. (2016). "A New Approach on Prediction of Fever Disease by Using a Combination of Dempster Shafer and Na{\"\i}ve Bayes." In *2016 2nd International Conference on Science in Information Technology (ICSITech)*, 367–71.

7. Nishanthi, P H M, A A I Perera, and H P Wijekoon. (2014). "Prediction of Dengue Outbreaks in Sri Lanka Using Artificial Neural Networks." *International Journal of Computer Applications* 101 (15).

8. Omkar, Buchade, Dalsania Preet, Deshpande Swarada, and Doddamani Poonam. (2017). "Dengue Fever Classification Using SMO Optimization Algorithm." *Int Res J Eng Technol* 4 (10): 1683–86.

9. Padmapriya, A, and N Subitha. (2013). "Clustering Algorithm for Spatial Data Mining: An Overview." *International Journal of Computer Applications* 68 (10).

10. Rao, N K Kameswara, G P Saradhi Varma, D Rao, and P Cse. 2014. "Classification Rules Using Decision Tree for Dengue Disease." *International Journal of Research in Computer and Communication Technology* 3 (3): 340–43.

11. Shaukat, Kamran, Nayyer Masood, Sundas Mehreen, and Ulya Azmeen. (2015). "Dengue Fever Prediction: A Data Mining Problem." *Journal of Data Mining in Genomics \& Proteomics* 2015:1–5.

12. Shaukat, Kamran, Nayyer Masood, Ahmed Bin Shafaat, Kamran Jabbar, Hassan Shabbir, and Shakir Shabbir. (2015). "Dengue Fever in Perspective of Clustering Algorithms." *ArXiv Preprint ArXiv:1511.07353*.

13. Siriyasatien, Padet, Atchara Phumee, Phatsavee Ongruk, Katechan Jampachaisri, and Kraisak Kesorn. (2016). "Analysis of Significant Factors for Dengue Fever Incidence Prediction." *BMC Bioinformatics* 17:1–9.

14. Venugopal, Viji, and Paresh Tanna. (2023). "Missing Value Imputation Techniques Used in Deep Learning Algorithms: A Review." AIP Conference Proceedings 2963 (1). https://doi.org/10.1063/5.0183257.

15. Vohra, Archit A., and Paresh J. Tanna. (2023). "Evaluation of Factors Involved in Predicting Indian Stock Price Using Machine Learning Algorithms." *International Journal of Business Intelligence and Data Mining* 23 (3): 201–63. https://doi.org/10.1504/IJBIDM.2023.133147.

Note: All the figures in this chapter were made by the authors.

Emerging Perspectives and Applications of Computational Intelligence and Smart Systems
– Dr. Amit Lathigara et al. (eds)
© 2026 Taylor & Francis Group, London, ISBN 978-1-041-20965-2

36

Leveraging Machine Learning Techniques for Predicting Major Diseases in Groundnut Crop Cultivation

Vimal Pambhar*

Research Scholar, School of Engineering, RK University,
Rajkot, Gujarat, India

Computer Engineering Department,
A.V. Parekh Technical Institute,
Rajkot, Gujarat, India

Nirav Bhatt

School of Engineering, RK University, Rajkot,
Gujarat, India

■ **Abstract:** Current Epoch is of interdisciplinary techniques where technological evolution in one sector supports to resolve the problems challenged in other sectors. The interdisciplinary approach of integrating computer technology into agriculture is crucial for addressing challenges like plant pests and diseases. Groundnut (Arachis hypogea L.), a significant oilseed crop, plays a vital role in India's economy, with Gujarat being a major contributor. The use of advanced technologies such as Machine Learning (ML), Deep Learning (DL), and Vision-Based AI can help predict groundnut crop diseases, ensuring better yield and food security. Techniques like Convolutional Neural Networks (CNN), K-means clustering, Support Vector Machine (SVM), and Regression have been widely used for disease classification and prediction. Theprimary attention of this study is to enhance disease forecasting for groundnut crops using environmental variables and favourable environment condition of major groundnut diseases. This effort is helping to make agriculture smarter and more efficient, which will boost our nation's economy and reduce soil pollution by using fewer pesticides in agriculture.

■ **Keywords:** Groundnut, Machine learning, Regression, Agriculture, Diseases prediction

*Corresponding author: vimalpambhar@gmail.com

DOI: 10.1201/9781003725046-36

1. INTRODUCTION

Agriculture acting a vital role in our country Gross Domestic Product (GDP), that signifying its economic importance. With 9952 tons and a productivity of 2063 kg/ha in 2021–2022, India leads the globe in groundnut acreage and is the second-largest producer in the world. About 17% of Gujarat's working population is directly employed in agriculture. Gujarat contributed 44.48% to total groundnut production with 4.49 tones among 10.11 tons of India production in 2021-22 ("8_Agricultural-Statistics-at-a-Glance-2022_page No 68 69," n.d.).

Agriculture is one of the extreme susceptible sectors to climatic change as weather variables involving like range of temperature (Temp.), rainfall, Relative Humidity (RH), leaf wetness hours (LTH), wind speed, environment moisture and various soil parameters (Bhargava et al. 2024).

Diseases are generating hazard for growth of any plants which is directly affect production so it is essential to detect diseases at early stage and control them to avoid losses. Forecasting is a monitoring technique that used to determine whether or not plant diseases presence or adjust severity and these use by farmers on the field to take economic decisions on disease control treatment. Forecasting of major diseases is useful for farmers to control of Pesticide and avoid unnecessary wastage of time, money and effort. The consequence of this to minimize pollution of soil and environment as well (Kumar, n.d et al 2018.).

To grow any plant disease forecasting system, enough facts of basically three factors are host factors, pathogen factors, and environmental factors are mandatory. The connotation of pathogen, host, weather and time can produce a great variety of possible combinations - some of these combinations are important for disease development(Kumar, n.d.et al 2018).

Disease cycle show in Fig. 36.1 is explaining the process of plant disease development and spread due to pathogens. The pathogen originates from soil, air, or newly emerged sources and becomes active under favorable environmental conditions like temperature, humidity, rainy days, etc. It then infects plants, leading to disease development and spreading, causing agricultural losses. After infection, the pathogen may persist in the soil, ready to restart the cycle when environment conditions are favorable again.

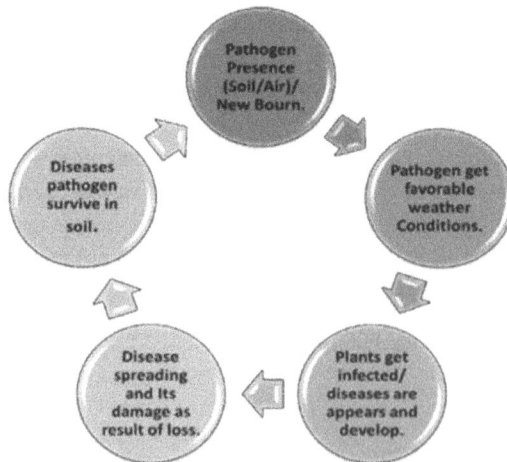

Fig. 36.1 Crop disease cycle

2. LITERATURE REVIEW

Groundnut crop is cultivated during kharif and rabi-summer many pathogens including viruses have been reported affect groundnut. The frequency and intensity of illnesses, which lower peanut pod production and fodder quality, differ from season to season. Numerous diseases have been identified in groundnut crops, and each one has a unique effect that reduces groundnut production in terms of both quality and quantity. Every pathogen has ideal environmental conditions for growth and infection of groundnut plants. Early and late leaf spot diseases favorable conditions, thriving at temperatures between 27°C and 32°C, with prolonged leaf wetness and high relative humidity (RH) above 72% and Leaf persistence Duration 0H-10H which supports pathogen infection and disease progression(Garg et al. 2023).Dry Root Rot/Charcoal Rot diseases thriving at the optimum temperature for seedling infection is 29°C to 35°C. Collar Rotdiseasesthriving16% soil moisture, 31-35°C temperature(Kumar, n.d. et al 2018)

(Karan, Ramkumar, and Yokesh 2024) in which A bootstrap bagging (bootstrap aggregation) algorithm that incorporates disease forecasting for oilseed crops and other plants is trained using historical crop data, weather patterns, and soil information. This allows farmers to proactively mitigate potential risks and maximize yields in situations where favorable conditions for major diseases need to be taken into account and need to be specific for particular crop.

(Gopi and Karthikeyan 2023) A novel ICRYP-DFADL model has been created for yield prediction and effective crop recommendation. For greater accuracy, take into account soil and environment parameters in addition to crop data and yield datasets.

(Garg et al. 2023) This research aims to understand the hidden relationships between soil and environmental conditions and the leaf spot disease that damages groundnut crops using various ML methods. There is future scope for more disease in groundnut crop and different geographical locations.

(Nirmal Ravi Kumar et al. 2023) Utilizing a variety of machine learning techniques, create a groundnut yield forecasting model based on historical weather parameter data from the NASA POWER website and groundnut yields for these state districts throughout the Kharif and Rabi seasons. Here, we can employ soil parameters and take into account the conditions that are conducive to different diseases for more accurate forecasting model.

(Kukadiya et al. 2024) The study suggested CNN8GN, an image-based Deep CNN classification model, for groundnut leaf disease classification by transfer learning. For more accuracy use IoT based real data and environmental and soil factors take into account.

(Kwaghtyo et al. 2024) This study used five machine learning models—decision trees, naive Bayes, logistic regression, random forest, and extreme gradient boosting—to create a crop recommendation model by utilizing soil and climate factors in the Yandev district. We can include favorable environment condition for diseases to improve accuracy in machine learning model.

(Haerani et al. 2023) The CLIMEX model parameters for peanut crops that we have successfully generated in this study are found to be consistent with the geographic distribution of peanuts(groundnut) as of right now. To further improve the precision and resilience of the predicted spatial distribution of peanut cropping areas, more study is required to incorporate non-climatic elements including topography, soil type, and biotic interactions.

Many past research including those using machine learning (ML) and deep learning (DL) techniques and Neural network (Bhojani and Bhatt 2020) do not consider environmental factors such as temperature, humidity, rainfall, and wind speed. Also, several research works do not account for essential soil properties like pH, moisture content, organic matter, and nutrient levels (e.g., nitrogen, phosphorus, potassium). These factors significantly influence in crop growth, disease outbreaks, overall yield,geographic distribution of groundnut and ML diseases prediction model.

3. METHODOLOGY

The Fig. 36.2 represents machine learning model, we first apply data preprocessing (Venugopal and Tanna 2023) on raw environment data that is taken from government official websiteand combine with favorable environmental conditions for the major diseases in groundnut crop (Garg et al. 23)(Kumar,n.d). Wherein we employ a variety of machine learning techniques, such as linear regression or multilinear regression, which can correlate environmental factors with illness-favorable environment characteristics and provide results as the likelihood that a disease would arise.

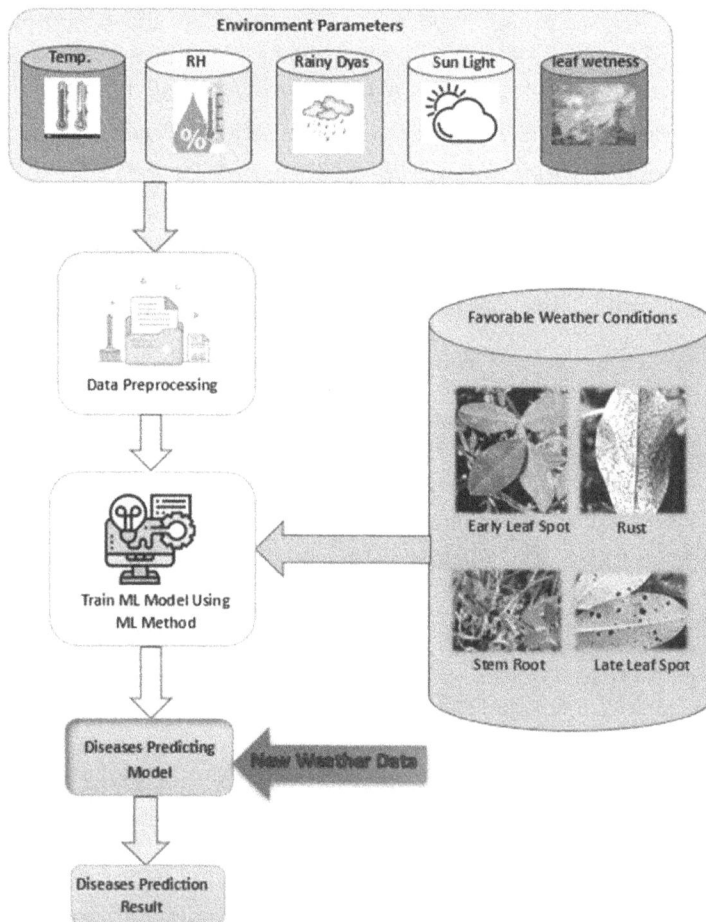

Fig. 36.2 Proposed machine learning model

Following that, we provide a range of new environmental circumstances to our trained machine learning model, which forecasts the likelihood of different diseases under these kinds of environment conditions. The correctness of our model can be determined by comparing this prediction result with the initial favorable conditions of the specific diseases. Then several mitigation strategies can be implemented to enhance groundnut production and decrease pesticide usage.

4. Conclusion

This study explores how numerous researchers have employed various image datasets and machine learning or deep learning methods to predict, classify, and detect diseases. There exists a significant opportunity to focus more on disease prediction by incorporating environmental factors and favorable conditions for different diseases in groundnut crop cultivation. The intensity of each disease is influenced by its dissemination cycle, highlighting the importance of considering favorable environmental factors. Instead of relying solely on image analysis, we propose a model that integrates environmental variables and favorable conditions for specific diseases. This paper proposed a machine learning (ML)-based system can assist in agricultural disease management by assessing weather patterns and favorable environment conditions for forecasting major groundnut crop diseases. By reducing pesticide usage, we can safeguard the environment from various pollutants while boosting productivity. Future studies could incorporate different soil parameters such as pH, moisture, and nutrients to improve disease prediction accuracy. Additionally, real-time environmental monitoring through IoT sensors and remote sensing can provide dynamic insights for enhanced forecasting.

References

1. Bhargava, Kotte, A. Naveen, V. Sai Akhil, Hema Lingireddy, K. V. Gowtham, Abhijeet Mudhale, B. Guru Sri, and E. Abhishek. (2024). "Artificial Intelligence (AI) and Its Applications in Agriculture: A Review." *Environment Conservation Journal* 25 (1): 274–88.
2. Bhojani, Shital H., and Nirav Bhatt. (2020). "Wheat Crop Yield Prediction Using New Activation Functions in Neural Network." *Neural Computing and Applications* 32 (17): 13941–51.
3. Garg, Pooja, Priyanka Khaparde, Kamlesh S. Patle, Chirag Bhaliya, Ahlad Kumar, Manjunath V. Joshi, and Vinay S. Palaparthy. (2023). "Environmental and Soil Parameters for Germination of Leaf Spot Disease in the Groundnut Plant Using IoT-Enabled Sensor System." *IEEE Sensors Letters* 7 (12): 1–4.
4. Gopi, P. S.S., and M. Karthikeyan. (2023). "Intelligent Crop Recommendation with Yield Prediction Using Dragonfly Algorithm Based Deep Learning Model." In *Proceedings of the 3rd International Conference on Artificial Intelligence and Smart Energy, ICAIS (2023)*, 880–85. Institute of Electrical and Electronics Engineers Inc.
5. Haerani, Haerani, Armando Apan, Thong Nguyen-Huy, and Badri Basnet. 2023. "Modelling Future Spatial Distribution of Peanut Crops in Australia under Climate Change Scenarios." *Geo-Spatial Information Science*.
6. Karan, R., M. O. Ramkumar, and M. Yokesh. (2024). "Revolutionizing Tamilnadu Agriculture: AI-Powered Crop Yield Forecasting and Disease Prediction for Oil Seed Crops." In *(2024) 1st International Conference on Cognitive, Green and Ubiquitous Computing, IC-CGU (2024)*. Institute of Electrical and Electronics Engineers Inc. https://doi.org/10.1109/IC-CGU58078.2024.10530710.
7. Kukadiya, Hirenkumar, Divyakant Meva, Nidhi Arora, and Shilpa Srivastava. (2024). "Effective Groundnut Crop Management by Early Prediction of Leaf Diseases through Convolutional Neural Networks." *International Research Journal of Multidisciplinary Technovation* 6 (1): 17–31. https://doi.org/10.54392/irjmt2412.

8. Kumar, Vinod. n.d. (2018) "Diseases of Groundnut Thirumalaisamy Polavakkalipalayam Palanisamy Indian Council of Agricultural Research." https://www.researchgate.net/publication/305392088

9. Kwaghtyo, Dekera Kenneth, Christopher Ifeanyi Eke, Joshua Abah, Timothy Moses, Jeffery Agushaka, and Faith B. Fatokun. (2024). "Soil and Climate Parameters Based Crop Recommendation Model for Yandev, Gboko Local Government Area of Benue State." In *Proceedings of the 2024 18th International Conference on Ubiquitous Information Management and Communication, IMCOM (2024).* Institute of Electrical and Electronics Engineers Inc. Nirmal Ravi Kumar, K., Anurag Satpathi, M. Jagan Mohan Reddy, Parul Setiya, and Ajeet Singh Nain. (2023). "Development of Groundnut Yield Forecasting Models in Relation to Weather Parameters in Andhra Pradesh, India." *Journal of Agrometeorology* 25 (3): 440–47. https://doi.org/10.54386/jam.v25i3.2194.

10. Venugopal, Viji, and Paresh Tanna. (2023). "Missing Value Imputation Techniques Used in Deep Learning Algorithms: A Review." *AIP Conference Proceedings* 2963 (1). https://doi.org/10.1063/5.0183257.

Note: All the figures in this chapter were made by the authors.

Emerging Perspectives and Applications of Computational Intelligence and Smart Systems
– Dr. Amit Lathigara et al. (eds)
© 2026 Taylor & Francis Group, London, ISBN 978-1-041-20965-2

37

Technological Advancements in Indian Policing: The Role of Digital Arrests

Sarita Sharma*,
Lalit Kumar, Amit Kumar Bindal,
Abhishek Saini, Deepanshu Saini
Department of Computer Science and Engineering,
MMEC Maharishi Markandeshwar (Deemed to be University),
Mullana-Ambala, India

Paresh Tanna
School of Engineering, RK University,
Rajkot, Gujarat, India

■ **Abstract:** A digital arrest scam refers to a scam where criminals use digital platforms to fool the individual into believing that they are facing arrest by law enforcement agencies. The authors describe these scams as often involving legal authorities or government officials through emails or phone calls. The scammers fool the victim by making them believe that the impersonal information has been linked to illegal activities and urgent payment is demanded through an untraceable medium to avoid arrest. The deceiving method sued in the scammers are very effective as the victim gets scared due to the fear of legal consequences. The rise of these scams is due to the availability of personal information online, which makes it easier for scammers to exploit individuals' personal information. In this paper, the authors discuss the impact of digital arrest scams, the need for public awareness, and strong cyber security methods. It also discusses the challenges that law enforcement faces in arresting the scammers.

■ **Keywords:** Digital arrest, Cyber fraud, Scam, Digital scam, Money laundering, AI, IPC, VoIP

1. INTRODUCTION

Cybercriminals imitate police enforcement and government authorities in digital arrest, a worrying trend. Cyber- criminals demonstrate fear of arrest in digital arrest. Fraudsters jail you at home. When the victim is afraid, the fraudster makes his setting look like a police station during the video call and another conversation, affecting his comments. The fraudster prevents the victim from leaving

*Corresponding author: saritabhushan201706@gmail.com

DOI: 10.1201/9781003725046-37

the video session or calling anyone. (Sarkar and Shukla 2024) victim gets arrested at home and worries that his Aadhar card, SIM card, and bank account were used for illegal business. It is a dangerous crime, and the Indian government is working to raise awareness. In recent years, digital arrest frauds have grown so rapidly that the Indian government has partnered with Microsoft to fight them. A. Rising Digital Arrest Scams As per "National Cyber Crime Reporting Portal" data The union government reported 2000 fraud accusations in the past year, according to Section I4C IPC, according to a 9 January 2024 article(Niño et al. 2023). These strong contacts cause financial loss and mental distress for victims, phony officials threaten you with phony laws or clauses, and victims must stay visible to criminals via Skype, zoom, or other video conferencing platforms until their demands are met (Buil-Gil, Trajtenberg, and Aebi 2023)("Breaking the Silence: Unveiling Gendered Perceptions of Femininity in Indian #MeToo Discourse on Facebook. | EBSCOhost" n.d.)(Alevizos, Bhakuni, and Jäschke 2024). Due of their expertise, these criminals employ studios that resemble police stations or government officials and assume law enforcement costumes. The Centre is taking strict action, blocking 90% of bogus Skype accounts (Alevizos 2024). Scammers pose as CBI, Mumbai police, and Reserve Bank of India personnel. They build official-looking ID cards and use real officials' names and ranks to make their claims seem credible. According to top-level insiders, this is a new swindle, and government offices are rigorously probing these cases. To combat these frauds, central agencies are disabling many video conferencing accounts (Yasin et al. 2021)(Franceschini et al. 2024).

2. CASE STUDIES

Detailed case studies of technical advances in Indian policing are an important aspect of the investigation. Each case study shows how digital tools affect law enforcement in distinct settings. This section uses real-world examples to show police agencies' triumphs and obstacles in using certain technology(Maras and Arsovska 2023)(Poe 2021)(Slay AM 2023)(Ehiane and Olumoye 2023)(Smith 2023)(Slay AM 2023)(Nasheri 2023).

2.1 Noida Incident (5 January 2024)

Ms. Rohini, from Noida, India, was called by a fraudster to say a SIM card registered with her Adhaar credentials was used for money laundering. So professional were these scammers that the lady didn't think she'd be a digital arrest victim. In the next call, the scammer described themselves as a senior police officer and stated they would interrogate her via Skype. Scammers persuade the woman into believing she has an arrest warrant during interrogation. She promptly transfers all her savings when the interrogation officer demands a ransom to avoid arrest(Jacobs, Darmawaskita, and McDaniel 2024)(Soni, Khan, and Singh 2024).

2.2 Haryana Case (29 February 2024)

When a 23-year-old girl in Faridabad filed a FIR against her digital arrest and misdirection of approximately 2.5 lakh to cyber thieves, the story made headlines. She said that on October 12, 2023, a Lucknow customs officer introduced herself. The child is told by this phony customs officer that her Aadhaar number was used to ship a parcel to Cambodia with passports and other cards. She has been under digital arrest for 17 days. Some cyber crooks acted as cops and introduced themselves as CBI. Police officers linked her name to a bank official accused of human trafficking.

2.3 Kolkata Police Advisory

Kolkata Police's digital arrest was safe. Latest public advisory: digital arrest. According to a Kolkata Police statement. Citizens should know how to rescue and avoid cybercriminals. Tell them about cyber fraud's methods. Many people gave this advice. These cases include digital arrest deception.

Fraudsters impersonate investigators, officials, and other government institutions. Fraudsters generally contact to commit scams. Join the Skype session, victim. A policy is in place for a "Investigation" or "arrest." The prey falls into its trap. 'Digital arrest' is a fraudster trick. This is common in cybercrime and law enforcement deception. Phone and online fraudsters pretend they have an arrest warrant and are being questioned.

Figure 37.1 This highlights fraudsters take advantage of victims' fear and disempowerment. It is important to have awareness among people to deal with unwanted demands for personal information or money.

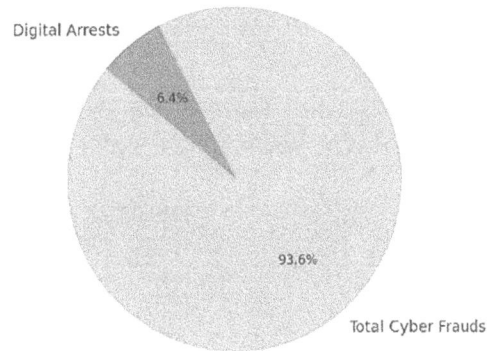

Distribution of Cyber Frauds vs Digital Arrests

Digital Arrests

6.4%

93.6%

Total Cyber Frauds

Fig. 37.1 Ration of Cyber frauds in India

3. GOVERNMENT PREVENTIVE MEASURES

The central government recently warned against cyber fraud. Police warn against CBI and RBI scams. Criminals use threats, blackmail, extortion, and "digital arrest" to compel individuals to send significant amounts of money or provide sensitive information. Cybercriminals are becoming better at mimicking government agencies to attack people. Because they exploit fear and misunderstanding, 'blackmail' and 'digital arrest' are concerning. They accuse victims of buying illegal goods or committing crimes to coerce them into cooperation or financial exploitation. Alarmingly, cybercriminals may replicate legal threats and pressure digital assets. Criminals threaten victims to manipulate and steal money or personal information in cybercrimes. The government issued a 1930 cybercrime helpline. If you receive unusual calls or messages, call the hotline or visitcybercrime. gov.in.

3.1 Findings

Figure 37.1 pie chart and Table 37.1 compare "total cyber frauds" with "digital arrests". A staggering 93.6% margin Table 37.2 and Fig. 37.2. Overall cyber fraud is higher than digital arrest fraud (6.4%). This contrast highlights cybercrime's growing threat. Figure 37.2 compares cyber fraud amount to digital arrest fraud. Cyber frauds account for 93.6% of Rs 1750 cr, while digital arrests account for 6.4%. Figure 37.1 pie chart and Table 37.1 compare "total cyber frauds" with "digital arrests". A staggering 93.6% margin

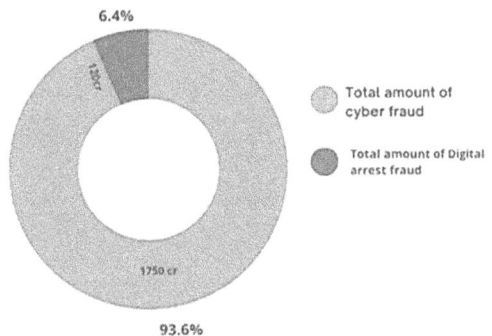

6.4%

120cr

Total amount of cyber fraud

Total amount of Digital arrest fraud

1750 cr

93.6%

Fig. 37.2 Distribution of cyber-Frauds vs. digital arrests

Table 37.1 and Fig. 37.2. Overall cyber fraud is higher than digital arrest fraud (6.4%). This contrast highlights cybercrime's growing threat. Figure 37.2 compares cyber fraud amount to digital arrest fraud. Cyber frauds account for 93.6% of Rs 1750 cr, while digital arrests account for 6.4%.

Table 37.1 Cyber fraud trends

Type of Fraud	Percentage	Amount (INR Cr)
Total Cyber Frauds	93.6	1750
Digital Arrests	6.4	120

4. MEASURE TAKEN FOR PROTECTION FROM CYBER FRAUD AND DIGITAL ARREST CASES

Table 37.2 Case study of financial losses

Case	Financial Loss (INR Lakhs)
Noida Incident	11.11
Haryana Case	2.5
Retired AIR Employee	780
Software Developer	2.5
Dr. Aarti Surbhit	45

Figure 37.3 shows digital arrest fraud losses by type in a bar chart. The Retired All India Radio Employee suffered the most damages, totaling |780 lakhs. Other notable examples include Aarti Surbhit losing |45 lakhs, while victims in the Noida Incident, Haryana Case, and Software Developer Case lost lesser amounts. Digital arrest scams, which deceive victims into believing they are under investigation or facing legal charges, can cause significant financial harm. The findings demonstrate

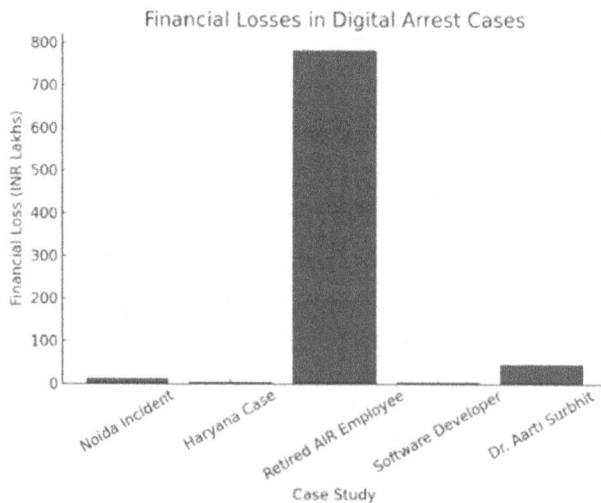

Fig. 37.3 Monthly trend in cyber fraud and digital arrest cases

Fig. 37.4   Financial losses in digital arrest cases

the sophistication of cybercrime and the necessity for greater cybersecurity, public awareness, and stricter law enforcement to combat fraud. As seen in Fig. 37.4, monthly cyber fraud and arrest data vary throughout a year. The yellow solid line shows total cyber fraud instances, which peak in May and November, signifying cybercrime activity. In particular, digital arrest instances (red dashed line) are much fewer than before but still rise in January, June, and December. The discrepancy between the two trends suggests that digital arrest frauds are a growing element of cyber fraud, but still a modest part.

The rise in cyber fraud and digital arrest schemes requires stronger anti-cyber laws, increased technological alertness, and public awareness. Thus, AI and future technologies can help us fight online fraud more collaboratively and stay ahead of new threats. Telecom providers can prevent phishing and unauthorized access with MFA and anti-scam call screening. Tracking and punishing cybercriminals requires tougher cybercrime legislation, dedicated cyber police taskforces, and increased global cooperation. We must educate citizens on spotting and reporting fraud through nationwide cyber awareness programs, digital literacy training in schools and workplaces, and simulation-based seminars. Additionally, specific cybercrime reporting platforms, quick-response frameworks, and government-sponsored cybersecurity helplines (like 1930) will enable victims receive prompt assistance. Secure payment methods with biometric authentication and VoIP regulation to combat impersonation frauds would help reduce fraud. To combat cyber fraud and protect citizens from professional con artists, governments, banks, tech companies, and the public can create a comprehensive digital security framework.

5. Conclusion

Digital arrest rackets may be a growing cybersecurity problem in India. Scammers employ fear and deception to confuse reality and fraud. Despite ongoing hurdles, law enforcement, technology businesses, and government agencies are working together to track and report these acts. However, cybercriminals continuously evolve, making proactive awareness and education crucial. To tackle these scams, we must improve cybersecurity, digital literacy, and legal measures. India can help victims and maintain trust in genuine law enforcement by giving tools, information, and expertise. Knowing these procedures can deter fraudsters; baiting them properly can help prevent child scams.

References

1. Sarkar, Gargi, and Shukla, S. K. (2024). "Bi-Directional Exploitation of Human Trafficking Victims: Both Targets and Perpetrators in Cybercrime." Journal of Human Trafficking, 1–22. https://doi.org/10.10 80/23322705.2024.2353015.
2. Niño, Michael. (2023). "The Racial/Ethnic Health Consequences of the US Criminal Justice System: How Consequential Is Probation and Other Justice System Contact for Self-Rated and Chronic Conditions?" Journal of Criminal Justice 87: 102073.
3. Jadhav, Yogesh Madhukar. (2024). "Analyzing Efficacy and Enhancing Accessibility: A Study of India's National Cyber Crime Reporting Portal in Addressing Financial Cybercrimes."
4. Bossler, Adam M., and Talia Berenblum. 2019. "Introduction: New Directions in Cybercrime Research." Journal of Crime and Justice 42 (5): 495–499. https://doi.org/10.1080/0735648X.2019.1692426.
5. Buil-Gil, David, Nicolas Trajtenberg, and Marcelo F. Aebi. (2024). "Measuring Cybercrime and Cyberdeviance in Surveys."
6. Kadambari, and Lalit Kumar. (2024). "Breaking the Silence: Unveiling Gendered Perceptions of Femininity in Indian #MeToo Discourse on Facebook." In Proceedings of the 15th International Conference on Advances in Computing, Control, and Telecommunication Technologies (ACT), 768–775.
7. Alevizos, Loukas, Luv Bhakuni, and Stefan Jäschke. (2024). "A Value-Driven Framework for Cybersecurity Innovation in Transportation and Infrastructure." International Journal of Information Technology, 1–11.
8. Yasin, Jawad N., et al. (2021). "Low-Cost Ultrasonic Based Object Detection and Collision Avoidance Method for Autonomous Robots." International Journal of Information Technology 13: 97–107.
9. Alevizos, Loukas. (2024). "Automated Cybersecurity Compliance and Threat Response Using AI, Blockchain and Smart Contracts." International Journal of Information Technology, 1–15.
10. Young, Michael. (1989). The Technical Writer's Handbook. Mill Valley, CA: University Science.
11. Franceschini, Ivan, Lili Li, Yao Hu, et al. (2024). "A New Type of Victim? Profiling Survivors of Modern Slavery in the Online Scam Industry in Southeast Asia." Trends in Organized Crime. https://doi.org/10.1007/s12117-024-09552-2.
12. Maras, Marie-Helen, and Jana Arsovska. (2023). "Understanding the Intersection Between Technology and Kidnapping: A Typology of Virtual Kidnapping." International Criminology 3: 162–176. https://doi.org/10.1007/s43576-023-00091-4.
13. Chang, Lennon Y. C., Yifan Zhou, and Duc H. Phan. (2023). "Virtual Kidnapping: Online Scams with 'Asian Characteristics' During the Pandemic." In Cybercrime in the Pandemic Digital Age and Beyond, edited by Russell G. Smith et al. Cham: Palgrave Macmillan. https://doi.org/10.1007/978-3-031-29107-4.
14. Jacobs, David R., Nisa Darmawaskita, and Tyson McDaniel. (2024). "Unraveling the Real-World Impacts of Cyber Incidents on Individuals." In HCI for Cybersecurity, Privacy and Trust, edited by Abbas Moallem. Lecture Notes in Computer Science 14729. Cham: Springer. https://doi.org/10.1007/978-3-031-61382-1_3.
15. Singh, Manish, et al. (2022). "RRDTool: A Round Robin Database for Network Monitoring." Journal of Computer Science 18 (8): 770–776. https://doi.org/10.3844/jcssp.2022.770.776.
16. Singh, Manish, Vijay Patidar, Shahbaz Ayyub, et al. (2023). "An Analytical Survey of Difficulty Faced in an Online Lecture During COVID-19 Pandemic Using CRISP-DM." Journal of Computer Science 19 (2): 242–250. https://doi.org/10.3844/jcssp.2023.242.250.
17. AlSuwaidan, R., Salman Khan, A. R. Baig, S. Baseer, and Manish Singh. (2022). "Fault Tolerance Byzantine Algorithm for Lower Overhead Blockchain." Security and Communication Networks. https://doi.org/10.1155/2022/1855238.
18. Singh, Manish, S. K. Tiwari, G. Swapna, et al. 2023. "A Drug-Target Interaction Prediction Based on Supervised Probabilistic Classification." Journal of Computer Science 19 (10): 1203–1211. https://doi.org/10.3844/jcssp.2023.1203.1211.

19. Singh, Manish, Shahbaz Ayuub, A. Baronia, et al. (2023). "Analysis and Implementation of Disease Detection in Leafs and Fruit Using Image Processing and Machine Learning." SN Computer Science 4: 627. https://doi.org/10.1007/s42979-023-02045-z.

20. Yadav, Anil S., et al. (2023). "Effect of Artificial Roughness on Heat Transfer and Friction Factor in a Solar Air Heater: A Review." In Recent Advances in Materials and Manufacturing Technology, edited by R. K. Nayak et al., Lecture Notes in Mechanical Engineering. Singapore: Springer. https://doi.org/10.1007/978-981-99-2921-4_33.

21. Kumar, Lalit, and Praveen Ailawalia. (2024). "Advances in Answer Set Planning: A Comprehensive Survey and Integration with Mixed Integer Programming." In Proceedings of the 15th International Conference on Advances in Computing, Control, and Telecommunication Technologies (ACT 2024), vol. 2: 4210–4214.

22. Kumar, Lalit, and Praveen Ailawalia. (2024). "Knowledge Extraction from Answer Set Programming-Based Encoding Selection." In Proceedings of the 15th International Conference on Advances in Computing, Control, and Telecommunication Technologies (ACT 2024), vol. 1: 1494–1499.

Note: All the figures and tables in this chapter were made by the authors.

Emerging Perspectives and Applications of Computational Intelligence and Smart Systems
– Dr. Amit Lathigara et al. (eds)
© 2026 Taylor & Francis Group, London, ISBN 978-1-041-20965-2

38

Enhancing Criminal Identification through Siamese CNN Facial Recognition: A Law Enforcement Perspective

Rani Sahu[1]
Department of Computer Science and Engineering,
IES College of Technology, Bhopal, Madhya Pradesh

Roopali Soni[2]
Oriental College of Technology, Bhopal, Madhya Pradesh

Vidhi Sadh[3]
SAGE University, Indore, Madhya Pradesh

Shamaila Khan[4]
Department of CSE-AIML, Oriental College of Technology, Bhopal

Ashutosh Dixit[5]
Department of CSE-AIML, Oriental Institute of Science & Technology, Bhopal

Shivangi Patel[6]
School of Engineering, RK University, Rajkot, Gujarat, India

■ **Abstract:** Proper identification of people is very important in public safety and law enforcement, especially for criminal record holders. This paper presents a facial recognition system using Siamese Convolutional Neural Network to enhance the identification accuracy. The model is trained with datasets provided by law enforcement officials. Labelled Faces in the Wild (LFW) dataset for negative labels and criminal capture dataset for positive labels. Training the model with TensorFlow and OpenCV, collecting information from CCTV images, testing with augmented dataset, Implementing the model with criminal database. The paper aims to assist law enforcement agencies by enhancing public safety, increasing the efficiency and accuracy of criminal identification.

■ **Keywords:** Criminal identification, Siamese CNN, Facial recognition, Law enforcement, Deep learning, Real-time detection, Database integration

[1]rani.princy28@gmail.com, [2]roopalisoni02@gmail.com, [3]India. Vidhi.sadh@sageuniversity.in, [4]asce.csit@gmail.com, [5]ashutoshdixit@oriental.ac.in, [6]shivangi.patel@rku.ac.in

DOI: 10.1201/9781003725046-38

1. INTRODUCTION

Facial recognition systems are critical for police to identify criminals because they allow an officer to efficiently link a suspect with criminal databases(Guerette, Przeszlowski et al. 2025). What law enforcement does is increasingly accurate in identifying suspects, allowing for more efficient investigation processes and optimization of time spent on leads that will go nowhere. Traditional face recognition systems fail to work well if the image quality is poor or any part of the face is hidden(Sahu , Nemavhola, Viriri et al. 2025). Therefore, advanced deep learning algorithms like Siamese CNN help to solve these problems and increase the recognition accuracy(Sahu, Sharma et al. 2020).

This research paper discusses the need to improve the accuracy and efficiency of criminal identification systems, especially in situations where data quality is poor. Existing systems often produce inconsistent results when input images are blurry, incomplete or affected by external factors such as lighting and angle.

1.1 Objectives

This study aims to develop a robust Siamese CNN-based facial recognition system to assist law enforcement in identifying criminals more accurately. The proposed system integrates real-time criminal databases to enhance investigation processes and improve suspect identification. Additionally, the model is designed to enhance the accuracy of facial recognition from CCTV images, addressing challenges posed by poor image quality or partial obstructions. The system's effectiveness will be evaluated through simulation and dataset testing using publicly available datasets. However, this paper does not cover large-scale deployment, such as live camera feed integration or real-time processing. The remainder of the paper is structured as follows: Section 2 presents a literature review, Section 3 details the methodology and implementation, Section 4 discusses the results, and Section 5 provides a summary and conclusion.

2. LITERATURE REVIEW

There has been recent technological progress in facial recognition software and security surveillance, which has really boosted road safety, especially using face identification of CCTV images (Sharma and Kanwal 2024, Nemavhola, Viriri et al. 2025, Upadhyaya 2025). The police use this technology to generate face templates from video data and cross-reference them to databases, giving a more advanced solution compared to the conventional biometric approach (Alamri and Mahmoodi 2024). Smart surveillance systems employ CCTV cameras for real-time monitoring and anomaly detection, thus minimizing the requirement of human supervision and maximizing safety (Thomas, Sanjay et al. 2024). Models based on Siamese networks are particularly good at facial recognition, correctly identifying faces even when there is variation such as smiles or lighting, and work without retaining personal information (Kopalidis, Solachidis et al. 2024). These convolutional neural network (CNN) models have the advantage of scalability and ongoing training on new information, thus decreasing the rate of false positives (Matin, Lighvan et al. 2025). The paper discusses a number of studies depending on the application of Siamese CNNs for identification of criminals and how they have been found effective in tracking, crime prevention, and updating the database in real-time. Such developments contribute towards making public areas secure and policing more effective(Singh, Vyas et al. 2023).

3. METHODOLOGY AND IMPLEMENTATION OVERVIEW

The proposed face recognition system is Siamese Convolutional Neural Network (Siamese CNN), which is utilized to match face images by learning deep feature representations (Yang, Peng et al. 2025). The model consists of twin CNNs that work in parallel on two input images, extract their features and calculate the similarity score using a distance metric (Euclidean distance or cosine similarity). The contrastive loss function ensures that embeddings of similar faces are close and embeddings of dissimilar faces are distant. The network is trained using Adam optimizer (learning rate = 0.001) for 50 epochs with early stopping to prevent overfitting. Rotation, scaling and flipping data augmentation methods are employed to improve generalization.

3.1 Dataset Preparing and Preprocessing

The model is trained with a mix of the Labeled Faces in the Wild (LFW) dataset as the negative samples and the Criminal Capture dataset as the positive samples. The dataset is prepared using the following steps:

1. Face Detection and Extraction: MTCNN detects faces and resizes them to 224×224 pixels.
2. Image Normalization: Pixel values are normalized between 0 and 1 for consistency.
3. Data Labeling
   a. Positive Pair (1): Faces matching the database.
   b. Negative Pair (0): Faces not matching.
4. Data Splitting: 70% for training and 30% for testing the model.

3.2 Training and Testing

The model is trained with triplet loss and backpropagation to enhance its power to discriminate between criminal and non-criminal faces(Aryan, Singh et al. 2024). Performance is tested with most important metrics(Jain, Saluja et al. 2024):

- Precision (true positive rate)
- Recall (sensitivity towards identification of criminals)
- Accuracy (total accuracy)
- F1-score (precision and recall balance)

After training, the system identifies unknown faces to criminal databases and returns relevant information to law enforcement agencies. Routine updates improve accuracy and minimize false identifications(Singh, Rai et al. 2023).

4. RESULTS

Model trained using TensorFlow, OpenCV; evaluated on precision, recall metrics. Tools such as Jupiter Lab and GitHub were used for development and collaboration.

Key evaluation metrics:

- Recall: 40% (the system's ability to identify positive matches) Precision: 90% (true-positive identifications)
- Specificity: 1.0 (true-negative identifications)

- F1-Score: 0.4 (harmonic mean of Recall and Precision)
- Accuracy: 70% (overall rate of correct results)

This model performed better than Models 2, 3, and 4, which had Recall (0.2), Precision (0.8), Specificity (0.8), F1-Score (0.3), and Accuracy (0.5).

Analysis of the Confusion Matrix showed that there were 600 True Negatives, 596 False Negatives, and 4 True Positives, as shown in Fig. 38.1.

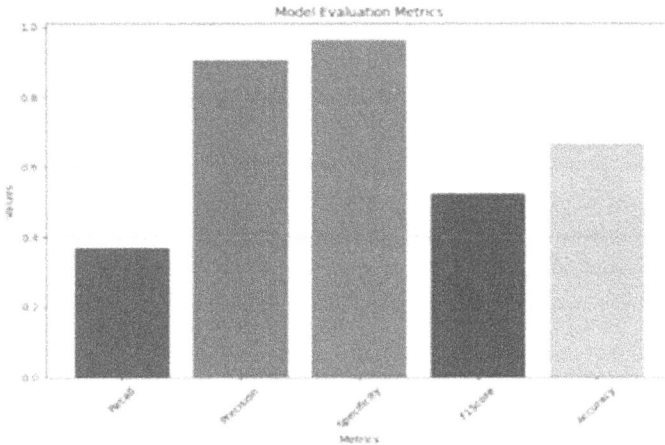

Fig. 38.1   Model evaluation

Source: Author

During model training, loss values were monitored to ensure convergence. The training loss decreased consistently, indicating effective learning. The validation accuracy stabilized after 40 epochs, confirming that the model generalizes well to unseen data. The use of contrastive loss helped in minimizing false positives while improving genuine match detection.

5. CONCLUSION

Our face recognition model is better than standard CNN models, particularly under low lighting or with minimal data, because it utilizes Siamese CNN, which improves accuracy and minimizes false matches. Real-time processing coupled with criminal databases enhances its efficiency for law enforcement agencies. Rotation and scaling methods also assist the face detection model in accommodating lighting and expression changes. Compared to existing models, our method maximizes face extraction and data augmentation efficiency while being more efficient. Despite this, issues of real-time processing and generalization are still there. The future will aim at improving scalability and real-time processing for broader security uses.

References

1. Alamri, M. and S. Mahmoodi (2024). "Face Profile Biometric Systems: An Overview." Face Recognition Across the Imaging Spectrum: 3–30.
2. Aryan, A., M. Singh, N. Labhade-Kumar, M. Radha, J. Ranga and N. Rathod (2024). Experimental Design of Electricity Theft Detection and Alert System using Arduino Assisted Controller and Smart Sensors. 2024 International Conference on Inventive Computation Technologies (ICICT), IEEE.

3. Guerette, R. T., K. Przeszlowski, J. J. Mitchell, J. Rodriguez, J. Ramirez and A. Gutierrez (2025). "An Extended Impact Evaluation of Real-Time Crime Center (RTCC) Technologies on Violent Crime Outcomes." Justice Evaluation Journal: 1–17.

4. Jain, S., D. N. Saluja, D. A. Pimplapure and D. R. Sahu (2024). "Exploring the Future of Stock Market Prediction through Machine Learning: An Extensive Review and Outlook." International Journal of Innovative Science and Modern Engineering 12(4): 1–10.

5. Kopalidis, T., V. Solachidis, N. Vretos and P. Daras (2024). "Advances in Facial Expression Recognition: A Survey of Methods, Benchmarks, Models, and Datasets." Information 15(3): 135.

6. Matin, N., M. Z. Lighvan and N. Farzi Veijouyeh (2025). "Convolutional Neural Networks for Imbalanced Advanced Security Network Metrics and Non Payload Based Obfuscations Dataset to Detect Intrusion." Concurrency and Computation: Practice and Experience 37(4-5): e8377.

7. Nemavhola, A., S. Viriri and C. Chibaya (2025). "A Scoping Review of Literature on Deep Learning Techniques for Face Recognition." Human Behavior and Emerging Technologies 2025(1): 5979728.

8. Sahu, R. "Advancements in Machine Learning for Stock Market Forecasting: An In-Depth Analysis and Future Outlook."

9. Sahu, R., S. Sharma and M. Rizvi (2020). "ZBLE: Zone based efficient energy multipath protocol for routing in mobile Ad Hoc networks." Wireless Personal Communications 113: 2641–2659.

10. Sharma, H. and N. Kanwal (2024). "Video surveillance in smart cities: current status, challenges & future directions." Multimedia Tools and Applications: 1–46.

11. Singh, M., A. K. Rai, G. V. Krishna, Y. D. Kumar, R. Mishra and D. Kothandaraman (2023). Machine Learning and AI based Robotic System for Archaeological Research. 2023 International Conference on Inventive Computation Technologies (ICICT), IEEE.

12. Singh, M., M. Vyas, K. Pithode, N. R. Khan and J. Rosak-Szyrocka (2023). Naïve Bayesian Classifier Committee (NBCC) Using Decision Tree (C4. 5) for Feature Selection. International Conference on Information and Communication Technology for Intelligent Systems, Springer.

13. Thomas, R. K. L., G. J. Sanjay, C. Pandeeswaran and K. Raghi (2024). Advanced CCTV Surveillance Anomaly Detection, Alert Generation and Crowd Management using Deep Learning Algorithm. 2024 3rd International Conference on Artificial Intelligence For Internet of Things (AIIoT), IEEE.

14. Upadhyaya, V. (2025). Advancements in Computer Vision for Biometrics Enhancing Security and Identification. Leveraging Computer Vision to Biometric Applications, Chapman and Hall/CRC: 260–292.

15. Yang, X., P. Peng, D. Li, Y. Ye and X. Lu (2025). "Adaptive decoupling-fusion in Siamese network for image classification." Neural Networks: 107346.

Emerging Perspectives and Applications of Computational Intelligence and Smart Systems
– Dr. Amit Lathigara et al. (eds)
© 2026 Taylor & Francis Group, London, ISBN 978-1-041-20965-2

39

Enhancing Organizational Security and Efficiency: A Digital QR Code-Based Access Control System

Rani Sahu[1]

Department of Computer Science and Engineering,
IES College of Technology, Bhopal, Madhya Pradesh

Mayank Kumar Verma[2]

IES University, Bhopal

Sarvesh Vyas[3]

Department of Civil Engineering, IES College of Technology, Bhopal

Sushila Sonare[4], Ruchi Bhargava[5]

Department of Computer Science and Engineering (Cyber Security),
Oriental College of Technology, Bhopal

Homera Durani[6]

School of Engineering, RK University, Rajkot, Gujarat, India

■ **Abstract:** This paper discusses a contemporary access control system based on QR code technology to overcome the shortcomings of conventional ID card systems. This system provides improved security, simplifies operations, and lowers costs through multi-level authentication and location-based verification. It incorporates a mobile application for QR code creation, secure scanning, and time-sensitive verification. The system monitors employee actions, captures data, and achieves privacy compliance. Benefits are increased security, reduced cost, and increased flexibility. According to the findings, QR code technology offers increased security, efficiency, and scalability, and through further work to enhance the system and offer increased security.

■ **Keywords:** Digital ID cards, Access control, QR code technology, Security, Multi-level authentication, Location verification

1. INTRODUCTION

With the introduction of the information age, conventional physical ID cards are insecure and outdated because they are prone to replication, loss, theft and high maintenance charges (Yusop et

[1]rani.princy28@gmail.com, [2]mayankvermar26@gmail.com, [3]sarveshvyas26@gmail.com, [4]sushila09s@gmail.com, [5]ruchibhargava@oriental.ac.in, [6]homera.durani@rku.ac.in

DOI: 10.1201/9781003725046-39

al. 2025). This study investigates the potential of transitioning from physical ID cards to a digital QR code-based ID system without the security vulnerabilities, inefficiencies and expense of legacy systems. The goals are to evaluate the shortcomings of physical ID cards, developing a secure QR code-based system with real-time authentication and location validation, analyzing its advantages in cost savings and user friendliness, and making recommendations to organizations to update their access control systems. The rest of the paper is organized as follows: Section 2 is a literature review, Section 3 discusses the methodology used, Section 4 provides the interpretation of results, and Section 5 summarizes the paper.

2. Literature Review

Access control systems, particularly for worker identification, have improved with advancements in technology to address the increased demand for safe, efficient and affordable solutions (Khayat et al. 2025). The conventional physical ID cards, even though commonly utilized, are restricted in several aspects that have urged the use of digital options (Nagar et al. 2025). The present review investigates existing research into the use of access control technologies, QR codes, security systems and mobile applications for worker identification. Physical ID card-based systems are prone to loss, theft, replication and scalability, which create security threats and costliness, particularly in big organizations (Torres Moreno 2025; Prakasha and Sumalatha 2025). Digital and mobile access control systems, particularly QR codes, surpass these constraints because they are affordable, simple to produce and versatile, providing greater security and convenience through integration of mobile apps for real-time verification of access (Kothai et al. 2025; Viswanathan and Lakshmi 2025). Robust authentication techniques like multi-factor authentication (MFA) and location-based verification tighten security by ascertaining the presence of the user. Mobile apps enable QR code generation and application with flexibility and scalability but do require consideration for data privacy and security (Ahamed, Sabani, and Shafana 2025; Berbecaru 2025).

There has been extensive research on digital access control systems as well as on QR code technology, but fewer studies integrate the use of MFA and QR codes into scalable organizational solutions(Sahu et al. 2024). Most concentrate on discrete components like QR code generation or verification without considering actual integration in the real world(Jain et al. 2024). This research attempts to fill this gap by offering an end-to-end solution which combines QR codes, mobile applications, real-time authentication and location checking, analyzing their business, financial and security advantages for firms (Sahu, Sharma, and Rizvi 2020; Sahu and Sahu 2022).

3. Methodology

This study proposes an effective and secure digital access control system based on QR code technology. The approach combines system design, security measures, and user interaction methodologies to provide effective employee authentication. The mobile application has significant modules: QR code generation (provides time-restricted codes), location confirmation (confirms presence of the employee), QR code scanning and verification (scans and authenticates access), and time-restricted QR code validation (confirms limited-time validity). The system logs data continually, synchronizes with a master database, and adheres to security measures for privacy compliance and ease of useas in Fig. 39.1.

The process starts with the generation of a unique QR code, then location-based verification to allow entry within the building. The code is read at the point of access for verification, and access is either allowed or denied. The system also comes with time-based QR code generation, logs entry and exit information and synchronizes the information with a central database for real-time reporting

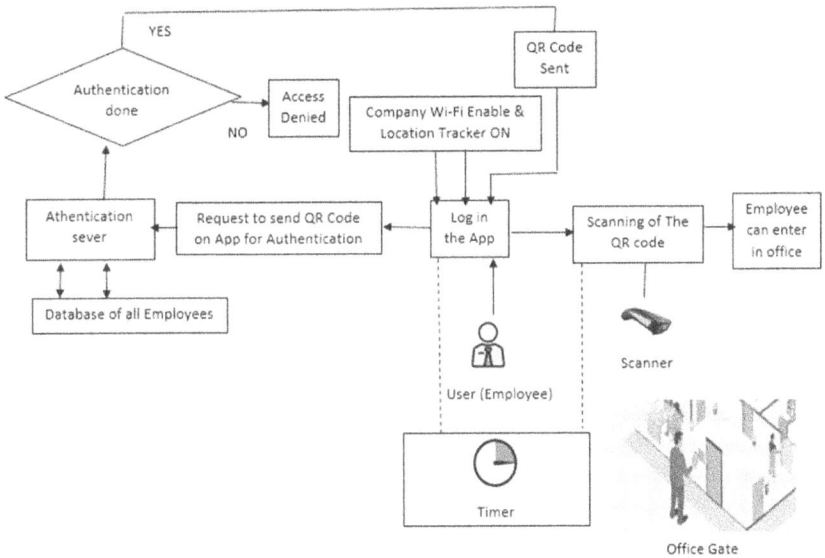

Fig. 39.1   Process flow of the access control system, Including QR code generation, Location verification, and access validation

Source: Author

purposes, to provide an efficient and secure access control process. The algorithm for QR code-based access control is explained as follows.

```
# Generation and verification of QR code
function generate_qr_code():
if authenticate_employee():
display_qr_code(create_qr_code(generate_unique_id()))
else:
show_error("Authentication failed")

# Location verification
function location_verification():
if not verify_location(get_employee_location()):
deny_access()
return False
return True

# Verification of QR code
function scan_qr_code():
if valid_access(extract_data_from_qr(scan_qr_code_at_access_point())):
log_entry_exit(extract_data_from_qr())
else:
deny_access()

                         # Time verification and regeneration
                         function valid_qr_code_time():
                         if is_qr_code_expired(): regenerate_qr_code()
```

```
Otherwise: continue_access()

#Monitoring and logging
function monitor_and_log():
log_entry_exit_details()
if check_for_extended_absence(): regenerate_qr_code()

# Data sync and security
function synchronize_data():
sync_with_database()
generate_reports()
encrypt_sensitive_data()
check_compliance()

                                    # key performance
                                    generate_qr_code()
                                    if location_verification():
                                    scan_qr_code()
                                    valid_qr_code_time()
                                    monitor_and_log()
                                    synchronize_data()
```

4. System Implementation, Testing, and Data Analysis

The system was developed, tested in a controlled environment, and piloted with a small employee group. Testing included QR code generation, scanning, location verification, and time-sensitive QR codes (Panwar et al. 2024). User experience and security testing helped refine the mobile app and address vulnerabilities. Data on employee entry/exit times, locations, and access statuses was collected, synced with the central database, and analyzed for performance in reliability, authentication, and security, providing insights into system efficiency and improvement areas.

5. Results

The system proved highly efficient with short creation and scan times, accuracy guaranteed by location-based and time-based verification and minimization of unauthorized access (Aryan et al. 2024). Security was augmented by penetration testing and real-time validation, and access granted to authorized personnel only. In spite of device reliance and network drawbacks, usability tests revealed 90% satisfaction in ease of use of the application and 85% reporting quick generation of QR codes. The system has entry/exit time logging in real-time reportability. As a whole, the QR code system provides solid security, savings, scalability, and is an alternative to existing systems, with its reliability augmented by solving the problem of devices and connectivity. The QR-Based Digital Access Control System Evaluation Results are given in Table 39.1.

6. Conclusion

This research proves the efficacy of a digital access control system using QR codes in boosting security, efficiency, and cost savings over conventional practices. The findings of this research are centered on effective QR code generation, robust security with real-time validation, and cost

Table 39.1 QR-based digital access control system evaluation results

Parameter	Value
QR Code Generation Time	2.1 seconds
QR Code Scan Time	0.8 seconds
Location Verification Accuracy	98.5%
Time-Sensitive Validation Accuracy	99.2%
Total Access Time (scan to approval)	1.5 seconds
Unauthorized Access Blockage	100%
Expired QR Code Detection	100%
Location-Based Verification	98.7%

Source: Author

reduction by the absence of physical ID cards. The system is scalable and can be adjusted, which makes it appropriate for many organizations. Future development should include device inclusivity, connectivity, user education, and incorporate biometric authentication and AI-based analytics for increased security.

References

1. Ahamed, TN, MJA Sabani, and MS Shafana. 2025. 'Optimized Security Authentication Protocols for Network Access Nodes: A Detailed Performance and Vulnerability Assessment'.
2. Aryan, Azigya, Manmohan Singh, Neelam Labhade-Kumar, Marepalli Radha, Jarabala Ranga, and Nagaraj Rathod. 2024. "Experimental Design of Electricity Theft Detection and Alert System using Arduino Assisted Controller and Smart Sensors." In 2024 International Conference on Inventive Computation Technologies (ICICT), 1961–68. IEEE.
3. Berbecaru, Diana Gratiela. 2025. 'SAM-PAY: A Location-Based Authentication Method for Mobile Environments', Electronics, 14: 621.
4. Jain, Sourabh, Dr NK Saluja, Dr A Pimplapure, and Dr R Sahu. 2024. 'Exploring the Future of Stock Market Prediction through Machine Learning: An Extensive Review and Outlook', International Journal of Innovative Science and Modern Engineering, 12: 1–10.
5. Khayat, Mohamad, Ezedin Barka, Mohamed Adel Serhani, Farag Sallabi, Khaled Shuaib, and Heba M Khater. 2025. 'Empowering Security Operation Center with Artificial Intelligence and Machine Learning–A Systematic Literature Review', IEEE Access.
6. Kothai, G, P Harish, D Ajinesh, and V Sathyanarayanan. 2025. 'Optimizing Data Privacy and Scalability in Health Monitoring: Leveraging, Encryption, Blockchain, and Cloud Technologies.' in, Critical Phishing Defense Strategies and Digital Asset Protection (IGI Global Scientific Publishing).
7. Nagar, Akhil, Hetvi Gudka, Manish Potey, and Sangeeta Nagpure. 2025. 'Designing a Large-Scale Face Authentication System.' in, Technologies for Energy, Agriculture, and Healthcare (CRC Press).
8. Panwar, Darsha, Abhishek Patel, Rachna Sharma, Roop Raj, Azigya Aryan, and Manmohan Singh. 2024. "Augmented reality based elevated learning procedure design for school students to improve their education." In 2024 5th International Conference on Electronics and Sustainable Communication Systems (ICESC), 903–10. IEEE.
9. Penica, Mihai, Mangolika Bhattacharya, William O'Brien, Sean McGrath, Martin Hayes, and Eoin O'Connell. 2025. 'Advancing Interoperable IoT-Based Access Control Systems: A Unified Security Approach in Diverse Environments', IEEE Access.
10. Prakasha, K Krishna, and U Sumalatha. 2025. 'Privacy-Preserving Techniques in Biometric Systems: Approaches and Challenges', IEEE Access.

11. Sahu, Rani, Vinay Sahu, Pushplata Chouksey, and Sourabh Jain. 2024. 'SecureZone: Secure and Efficient Multipath Routing for Mobile Ad Hoc Networks', Int. J. Advanced Networking and Applications, 15: 6182–93.

12. Sahu, Rani, Sanjay Sharma, and MA Rizvi. 2020. 'ZBLE: Zone based efficient energy multipath protocol for routing in mobile Ad Hoc networks', Wireless Personal Communications, 113: 2641–59.

13. Karuppiah, N., Mounica, P., Bhanutej, J. N., Saravanan, S., Reddy, R., & Israni, R. (2024). Revolutionizing Renewable Energy Integration: The Innovative Gravity Energy Storage Solution. In *E3S Web of Conferences* (Vol. 547, p. 03028). EDP Sciences

14. Sahu, Vinay, and Rani Sahu. 2022. 'Energy Efficient Multipath Routing in Zone-based Mobile Ad-hoc Networks: Mathematical Formulation'.

15. Torres Moreno, Rafael. 2025. 'Distributed technologies in identity management: an approach to enhancing security and privacy', Proyecto de investigación:.

16. Viswanathan, Balasubramanian, and B Lakshmi. 2025. 'Enhancing Educational Certificate Management and Verification with Blockchain Technology', Journal of Computer Information Systems: 1–15.

17. Yusop, Mohd Imran Md, Nazhatul Hafizah Kamarudin, Nur Hanis Sabrina Suhaimi, and Mohammad Kamrul Hasan. 2025. 'Advancing Passwordless Authentication: A Systematic Review of Methods, Challenges, and Future Directions for Secure User Identity', IEEE Access.

Emerging Perspectives and Applications of Computational Intelligence and Smart Systems
– Dr. Amit Lathigara et al. (eds)
© 2026 Taylor & Francis Group, London, ISBN 978-1-041-20965-2

40

Optimizing MANET Routing with SINTM: A Smarter Load Balancing Approach

Rani Sahu[1]
Department of Computer Science and Engineering,
IES College of Technology, Bhopal, Madhya Pradesh

Sarita Yogi[2]
Department of Computer Application,
Sarvapalli Radhakrishnan University, Bhopal

Nazeer Khan[3]
Department of Structural Engineering, IES University,
Bhopal

Rajesh Kumar Nigam[4]
Department of Artificial Intelligence and Machine Learning (AIML),
Oriental College of Technology, Bhopal

Sarvesh Vyas[5]
Department of Civil Engineering, IES College of Technology,
Bhopal

Bhoomi Danger[6]
School of Engineering, RK University, Rajkot,
Gujarat, India

■ **Abstract:** MANETs suffer from route discovery inefficiency owing to their dynamic topology and congestion. In this paper, the authors present the Sharing Intermediate Node Traffic Model (SINTM) as a new approach to facilitating load balancing for MANET routing. With incorporation of congestion control at the MAC layer and keeping track of network access delays, SINTM balances traffic optimally, leading to energy efficiency. Simulation results show that SINTM reduces latency, enhances packet delivery ratio, and enhances network stability compared to traditional techniques.

■ **Keywords:** MANET, Load balancing, Routing, SINTM, Congestion control, Network optimization

[1]rani.princy28@gmail.com, [2]saita_yogi@rediffmail.com, [3]nazirkhan747075@gmail.com, [4]rajeshrewa37@gmail.com, [5]sarveshvyas26@gmail.com, [6]bhoomi.dangar@rku.ac.in

DOI: 10.1201/9781003725046-40

1. INTRODUCTION

Mobile ad hoc networks (MANETs) are of central importance for, e.g., military actions, disaster relief, and IoT thanks to their decentralization and infrastructural independence(Rukaiya et al. 2024). These applications, though, are encumbered with node mobility, dynamic network topologies, and unpredictable traffic by nature, obstructing stable communications. Moreover, traditional protocols fail at efficient load balancing, creating bottlenecks on the traffic stream and power waste(Sahu, Sharma, and Rizvi 2020).The Share Intermediate Node Traffic Model (SINTM) proposed is a solution to these concerns through improved load balancing, improved traffic reduction, network performance optimization, and scalability through real-time traffic observation and adaptive routing and hence is suitable for environments with high mobility(Sahu, Sharma, and Rizvi 2019a). In addition, congestion and inefficient routing cause further issues of latency, packet loss, and energy usage. Conventional routing protocols in MANETs tend to be ineffective in handling congestion and load balancing, resulting in performance degradation, particularly in packet delivery and energy efficiency(Sahu, Sharma, and Rizvi 2019b)

- Suggest a novel routing scheme, SINTM, to maximize load balancing in MANETs.
- Incorporate congestion control at the MAC layer to enhance traffic distribution.
- Compare the performance of SINTM using simulation results in terms of latency, packet delivery ratio, and energy efficiency(Sahu, Sharma, and Rizvi 2019c).

This research targets the design and assessment of the SINTM method for MANET routing optimization, particularly dealing with congestion control, load balancing, and network performance enhancement. The simulation is not aimed at large-scale deployment, but controlled scenarios to illustrate the benefits of SINTM. The rest of the paper is organized as follows: Section 2 discusses the related literature, Section 3 outlines the methodology and implementation, Section 4 provides the results, and Section 5 concludes with a summary of the study.

2. LITERATURE REVIEW

Here, we discuss prominent research on load balancing, energy efficiency, and routing protocols in MANETs. The following Table 40.1 provides an overview of the main contributions of each study, including the strategies used and their assessment.

3. METHODOLOGY

The designed SINTM routing scheme optimizes MANET performance with real-time load sensing, traffic adaptability through redistribution, and energy-efficient decision-making. The nodes periodically exchange HELLO messages to measure the conditions of the network to optimize route discovery as well as to handle congestion effectively. The originator node broadcasts a Route Request (RREQ), and relay nodes update their routing tables using available load capacities. Overloaded nodes route traffic to secondary paths, whereas the destination node chooses the best path and replies with a Route Reply (RREP). Periodic monitoring facilitates dynamic load balancing and energy savings. The algorithm runs with complexity $O(N+H+D)$, which makes it applicable for dynamic MANET environments. The algorithm for SINTM is explained as follows.

Table 40.1   Key studies on load balancing and routing in MANETs

Study	Focus Area	Key Contributions	Evaluation
(Farahi 2025)	Load Balancing	Dynamic algorithm for traffic allocation to reduce congestion and improve QoS.	Compares with existing strategies.
(Sahu and Gour 2023)	Security & Load Balancing	Examines security and energy-efficient load balancing in AOMDV.	Discusses security and energy savings.
(Haider, Khan, and Saeed 2024)	Routing Protocol	Protocol for even traffic distribution to optimize QoS.	Aims to reduce congestion and packet loss.
(Biradar and Mallapure 2024)	Load Balancing in Routing	Integrates load balancing in shortest-path protocols.	Enhances traffic management and performance.
(Sindhuja and Vadivel 2025)	Fibonacci-based Load Balancing	Uses Fibonacci sequence for traffic distribution.	Alleviates congestion and improves performance.
(Singh and Prakash 2020)	Load Balancing & Energy Efficiency	Analyses energy harvesting and adaptive routing for load balancing.	Resource for energy efficiency in MANETs.
(Sahu and Veenadhari 2024)	Load Balancing Review	Reviews various load balancing techniques.	Guides further research.
(Dalal et al. 2022)	Congestion Control & Load Balancing	Examines mechanisms to optimize congestion and load balancing.	Focuses on throughput, delay, and fairness.
(Chandravanshi, Soni, and Mishra 2022)	Load Balancing & Congestion Control	Studies the role of load balancing in congestion control.	Evaluates routing overhead, throughput, and fairness.
(Kamps, Palunčić, and Maharaj 2025)	Load Balancing & Link Break Prediction	Combines load balancing with link break prediction.	Enhances data transfer efficiency.
(Alotaibi 2023)	LAPU (Geographic Routing)	Uses adaptive position updates for routing efficiency.	Compares LAPU with other protocols.
(Sahu, Sharma, and Rizvi 2019c)	Topology Control in NS-2	Explores topology control for performance optimization.	Assesses impact on key network parameters.

Algorithm SINTM Routing

1. *Network Initialization:*
   - *Establish network nodes and specify load threshold.*
2. *Neighbour Discovery:*
   - *Regularly scan neighbouring nodes and send HELLO messages at intervals.*
3. *Data Transmission Process:*
   - *The source node sends a route request (RREQ) with load information.*
   - *The intermediate nodes update their routing tables according to the following conditions:*
     - *If the current load is less than the threshold, the RREQ is relayed.*
     - *If the load is greater than the threshold, the traffic is routed to an alternate route.*
   - *The destination node chooses the most efficient path and returns a route reply (RREP).*
   - *The intermediate nodes update their routing state accordingly.*

◢ 4. RESULTS

This section compares the SINTM protocol with AODV and DSR based on important performance parameters: packet delivery ratio (PDR), throughput, end-to-end delay and routing overhead. The analysis illustrates SINTM's efficiency improvement, congestion control and scalability.

4.1 Simulation Setup

The work makes use of NS-2.35 to run a dynamic MANET scenario with parameters like a 1000m × 1000m space, IEEE 802.11 MAC protocol, and energy-efficient model. These parameters ensure realistic performance assessment.

- PDR: SINTM has better PDR than AODV and DSR and enhances reliability by means of dynamic load balancing.
- Throughput: SINTM performs better than AODV and DSR due to dynamic traffic distribution and congestion detection for effective routing.
- End-to-end delay: With optimal path selection and reduced queuing time, SINTM considerably decreases transmission delay.
- Routing overhead: Intelligent route discovery cuts down redundant control packets, leaving lower overhead compared to AODV and DSR.

The findings validate that SINTM improves MANET routing efficiency, and thus is a superior option for real-time applications that need congestion-aware and energy-efficient routing. Performance Evaluation of SINTM, AODV and DSR: The graphs () represent the comparison of (a) packet delivery ratio, (b) throughput, (c) end-to-end delay and (d) routing overhead at various simulation time-sillustrated in Figs. 40.1. The findings reveal the enhanced efficiency of SINTM in achieving greater data delivery, improved throughput, reduced latency and reduced routing overhead in MANET.

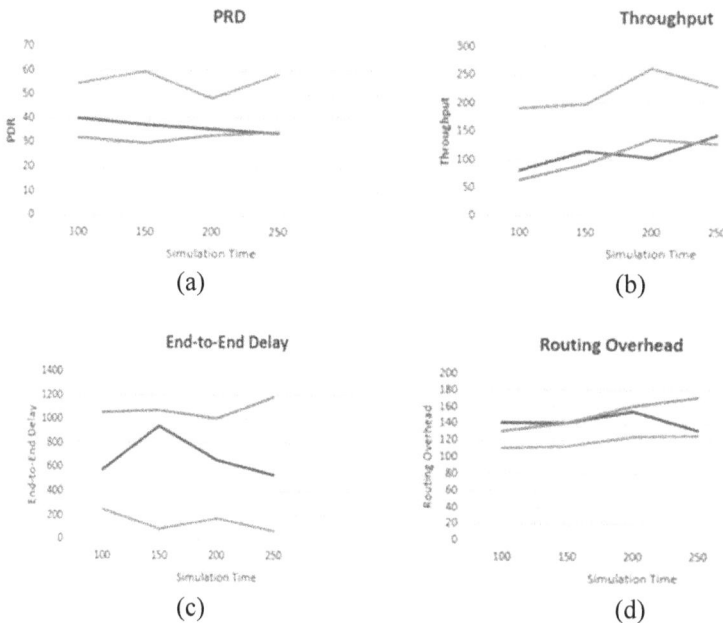

Fig. 40.1 Performance evaluation of SINTM, AODV and DSR

Source: Author

5. CONCLUSION

SINTM is a lean and scalable MANET routing solution that alleviates congestion, load balancing, and network efficiency. Simulation evidences that it performs better than AODV and DSR in terms of PDR, throughput, delay, and overhead and is apt for dynamic scenarios like disaster recovery, military, and autonomous networks. Its adaptability coupled with its low overhead is apt for IoT and vehicle applications. Future advancements involve AI-based adaptive routing, real-world deployment, security enhancements, energy-efficient methods, and cross-layer optimization, demonstrating SINTM's potential to improve MANET routing.

References

1. Alotaibi, Majid. 2023. 'Geographic routing in mobile ad hoc networks (MANET) using hybrid optimization model: a multi-objective perspective', Applied Intelligence, 53: 11214–28.
2. Behura, Aradhana, Arun Kumar, and Puneet Kumar Jain. 2025. 'A comparative performance analysis of vehicular routing protocols in intelligent transportation systems', Telecommunication Systems, 88: 26.
3. Biradar, Manjula A, and Sujata Mallapure. 2024. 'Multipath Load Balancing in MANET via Hybrid Intelligent Algorithm', Journal of Information & Knowledge Management, 23: 2450010.
4. Chandravanshi, Kamlesh, Gaurav Soni, and Durgesh Kumar Mishra. 2022. 'Design and analysis of an energy-efficient load balancing and bandwidth aware adaptive multipath N-channel routing approach in MANET', IEEE Access, 10: 110003–25.
5. Dalal, Surjeet, Bijeta Seth, Vivek Jaglan, Meenakshi Malik, Surbhi, Neeraj Dahiya, Uma Rani, Dac-Nhuong Le, and Yu-Chen Hu. 2022. 'RETRACTED ARTICLE: An adaptive traffic routing approach toward load balancing and congestion control in Cloud–MANET ad hoc networks', Soft Computing, 26: 5377–88.
6. Farahi, Rasoul. 2025. 'A comprehensive overview of load balancing methods in software-defined networks', Discover Internet of Things, 5: 6.
7. Haider, Syed Ehsan, Muhammad Faizan Khan, and Yousaf Saeed. 2024. 'Adaptive Load Balancing Approach to Mitigate Network Congestion in VANETS', Computers, 13: 194.
8. Kamps, Jason J, Filip Palunčić, and BT Maharaj. 2025. 'A Load-Balancing Enhancement to Schedule-Aware Bundle Routing', International Journal of Satellite Communications and Networking.
9. Rukaiya, Rukaiya, Shoab Ahmed Khan, M Umar Farooq, and Irum Matloob. 2024. 'Communication architecture and operations for SDR-enabled UAVs network in disaster-stressed areas', Ad Hoc Networks, 160: 103506.
10. Sahu, Manju, and Sanjeev Gour. 2023. 'INTSM: A Novel Approach for Load Balancing in MANET Route Discovery', Int. J. Advanced Networking and Applications, 15: 5837–5
11. Sahu, Neetu, and S Veenadhari. 2024. 'A Comprehensive Survey of Load Balancing Techniques in Multipath Energy-Consuming Routing Protocols for Wireless Ad hoc Networks in MANET', Indian Journal of Data Communication and Networking (IJDCN), 4: 5–10.
12. Sahu, Rani, Sanjay Sharma, and M Rizvi. 2019a. 'ZBLE: Zone Based Leader Selection Protocol'.
13. Sahu, Rani, Sanjay Sharma, and MA Rizvi. 2019b. 'ZBLE: zone based leader election energy constrained AOMDV routing protocol', International Journal of Computer Networks and Applications, 6: 39–46.
14. 2020. 'ZBLE: Zone based efficient energy multipath protocol for routing in mobile Ad Hoc networks', Wireless Personal Communications, 113: 2641–59.
15. Sindhuja, S, and R Vadivel. 2025. 'Fuzzy-Markov Routing Policy and Priority-Based Load Balancing for Cognitive Radio Ad Hoc Networks', International Journal of Communication Systems, 38: e6085.
16. Singh, Sunil Kumar, and Jay Prakash. 2020. "Energy efficiency and load balancing in MANET: a survey." In 2020 6th International Conference on Advanced Computing and Communication Systems (ICACCS), 832–37. IEEE.

Emerging Perspectives and Applications of Computational Intelligence and Smart Systems
– Dr. Amit Lathigara et al. (eds)
© 2026 Taylor & Francis Group, London, ISBN 978-1-041-20965-2

41

Networks Design and Evaluation of a Zone-Based Energy-Efficient Routing Protocol for Hybrid Wireless Networks

Rani Sahu[1]
Department of Computer Science and Engineering,
IES College of Technology, Bhopal, Madhya Pradesh

Vinay Sahu[2]
Department of Computer Science and Engineering,
LNCT College, Bhopal, Madhya Pradesh

Seema Joshi[3]
Department of Computer Application, LNCT-MCA, Bhopal,
Madhya Pradesh

Sujeet Kumar[4]
Department of CSE-AIML, Oriental College of Technology, Bhopal

Sachin Malviya[5]
Department of CSE, SDITS Khandwa, Madhya Pradesh

Anju Kakkad[6]
School of Engineering, RK University, Rajkot, Gujarat, India

■ **Abstract:** Hybrid Wireless Networks (HWNs), which combine Mobile Ad-Hoc Networks (MANETs) and infrastructure-based systems, require efficient routing strategies to optimize energy consumption and balance network load in dynamic, resource-limited environments. This work suggests a zone-based energy-efficient and load-balanced routing protocol (ZEELRP) as a viable solution. ZEELRP partitions the network into zones, where leader nodes (LNs) are chosen according to their residual energy and connectivity to handle routing. Gateway nodes (GNs) enable inter-zone communication and interaction with the hybrid system. The protocol guarantees energy efficiency and load balancing through energy efficiency metrics (EEM) and load balance index (LBI) during data transmission. Dynamic re-routing and dynamic updates for LNs and GNs enhance network robustness against mobility and energy depletion. Simulation results indicate that ZEELRP dramatically improves energy efficiency, load balancing, and network lifetime, verifying its scalability and robustness for energy-sensitive hybrid networks.

■ **Keywords:** Energy-efficient routing, Hybrid wireless networks, Mobile ad-hoc networks, Zone-based routing, Load balancing, Leader node

[1]rani.princy28@gmail.com, [2]sahu.vinay@gmail.com, [3]seema2201@gmail.com, [4]sujeetkmr44@gmail.com, [5]olsachin8@gmail.com, [6]anju.kakkad@rku.ac.in

DOI: 10.1201/9781003725046-41

1. INTRODUCTION

Hybrid Wireless Networks (HWNs) combine conventional infrastructure-based networks and mobile ad-hoc networks (MANETs) to create an adaptive and efficient communication system (Yang et al. 2024). By integrating the strengths of both network types, HWNs enhance overall network performance, making them a viable solution for diverse applications (Obayiuwana and Falowo 2017).

Despite their advantages, existing routing protocols in HWNs face significant challenges related to energy consumption, load balancing, and dynamic topology changes(Pareek et al. 2024). These issues necessitate the development of application-specific routing protocols that optimize resource utilization and network performance (Sachithanandam et al. 2025). Additionally, real-world deployment of HWNs introduces challenges such as hardware limitations, interference, and network scalability issues, which must be considered for practical implementation.

This paper proposes and evaluates ZEELRP, a zone-based load-balanced and energy-efficient routing protocol for HWNs. It focuses on improving energy management and load balancing across the network while ensuring efficient data transmission. The study focuses on the design and simulation of ZEELRP, particularly within hybrid wireless network environments, without actual hardware implementation. The rest of the paper is structured as follows: Section 2 presents an overview of the literature, Section 3 describes the methodology and implementation, Section 4 discusses the results, and Section 5 summarizes the conclusions.

2. LITERATURE REVIEW

Current studies on routing protocols for HWNs have concentrated primarily on load balancing and energy efficiency (Rajawat et al. 2025). HZDL (Debnath and Arif 2025), EANR (Kannan and Marimuthu 2024), and ZBLE (Sahu, Sharma, and Rizvi 2020)have ensured energy optimization, but tend not to address problems like dynamic mobility, load distribution, scalability, and inter-zone communication. Though HZDL enhances node lifetime and throughput for disaster relief applications, it is not capable of addressing hybrid network scalability. EANR maximizes energy without load balancing, while ZBLE maximizes energy efficiency in MANETs without considering hybrid communication (Sahu and Dhari 2024). ZEELRP fills these loopholes by bringing in zone-based architecture using leader nodes (LNs) and gateway nodes (GNs) that maximize communication energy efficiency as well as load balancing. By utilizing metrics like Energy Efficiency Metric (EEM) and Load Balance Index (LBI), ZEELRP efficiently manages energy shortages, load imbalances, and dynamic topology changes in HWNs with a more flexible and scalable approach (Sahu, Sharma, and Rizvi 2019a).

3. METHODOLOGY

This section describes the framework of ZEELRP, which is meant to maximize energy usage and load balancing in HWNs, especially MANETs. ZEELRP partitions the network into zones, chooses leader nodes based on connectivity and energy, and sets up optimal paths. It adjusts dynamically according to energy limitations, node mobility, and network dynamics to achieve efficient routing. The steps involve network initialization, zone partitioning, election of the leader node, route discovery, path selection based on EEM and LBI, data sending, and dynamic maintenance. The entire algorithm is explained below.

ZEELRP Algorithm
Input:
Network structure (N)
Energy levels of nodes (E)
Node locations (P)
Communication requests (C)
Output:
Most efficient routing path (R)

1. Initialization:
The network is partitioned into several zones, and each zone is assigned a distinct ID.
Gateway nodes (GN) are resolved within the network.
Nodes send a HELLO message containing their existing energy level (E_res) and position.

2. The choice of the leader:
A leader node (LN) is chosen based on the maximum energy factor, calculated from residual energy (E_res) and connectivity.

3. Route discovery
An RREQ message is sent with the source ID and destination ID, minimum energy (E_min), and maximum position (P_max).
Satisfactory routes between and within regions are decided upon and recorded.

4. Selection of path:
The best possible path is selected based on two factors:
Highest energy efficiency criterion (EEM)
Lowest load balancing measure (LBI)

5. Data transmission:
Data packets are transmitted along the chosen path to facilitate smooth communication.

6. Dynamic maintenance:
Node energy, speeds, and network congestion are constantly monitored by the system.
The routing is adjusted as required to ensure efficiency and avoid disruption.

7. Termination:
It terminates when data is successfully received at the destination or nodes are inactive.

3.1 Security Considerations in ZEELRP

Hybrid wireless networks (HWN) are exposed to security threats including routing attacks, node compromise, and data tampering, compromising network reliability. ZEELRP's dynamic nature exposes it to malicious routing, compromised nodes, and eavesdropping. Future implementations can incorporate cryptographic encryption (ECC, AES) to restrict unauthorized access, trust-based routing to counteract malicious nodes, and anomaly detection systems for instantaneous threat detection. These will reinforce ZEELRP's security against attacks without compromising energy efficiency.

4. Simulation Evaluation

ZEELRP was experimented under different mobility conditions in NS-2.35 and compared with AODV and AOMDV in terms of PDR, energy consumption, throughput, delay, and network lifetime (Sahu, Sharma, and Rizvi 2019b). It performs better than both by optimizing energy for greater throughput, choosing stable paths to minimize delay, and using high-energy nodes for greater PDR. Proper energy management improves network lifetime by providing consistent performance. Despite greater mobility, ZEELRP provides improved throughput, PDR, and delay compared with AODV and AOMDV.

4.1 Simulation Setup

Simulations in a 50-node MANET using NS-2 compare ZEELRP with AODV and AOMDV at mobility speeds of 3, 5, 10 and 15 m/s in terms of throughput, PDR, delay, energy consumption and network lifetime. ZEELRP performs better than both protocols, with an improvement of 20.44% and 16.75% in throughput, 66.43% and 22.05% improvement in PDR compared to AODV and AOMDV, respectively, and decreases the delay by 63.11% and 70.51%. It also saves energy by 9.04% and 8.44% over AODV and by 83.54% over AOMDV, and increases network lifetime by 85.82% and 86.12%. Figures 41.1–41.5 illustrate the better performance of ZEELRP under various mobility scenarios.

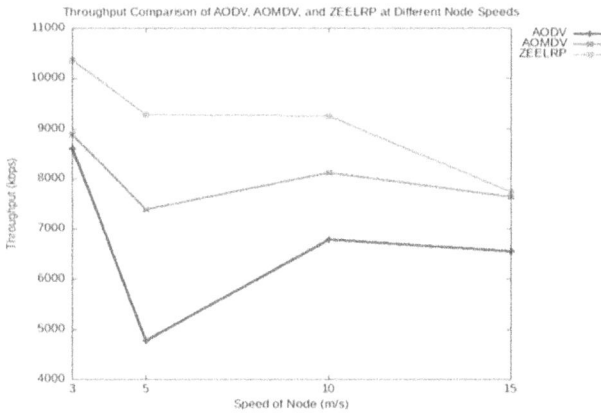

Fig. 41.1 Throughput analysis under various mobility configurations

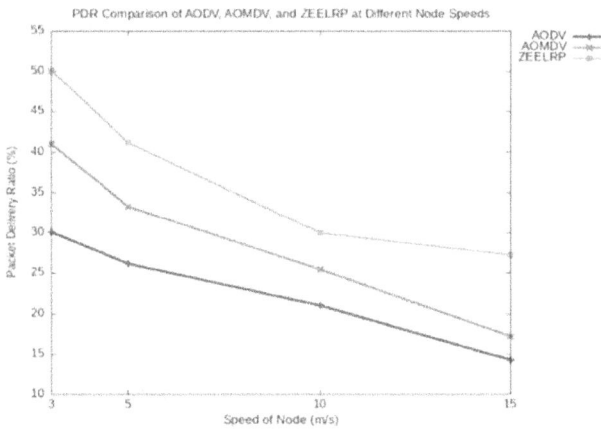

Fig. 41.2 Packet delivery ratio (PDR) for different mobility scenarios

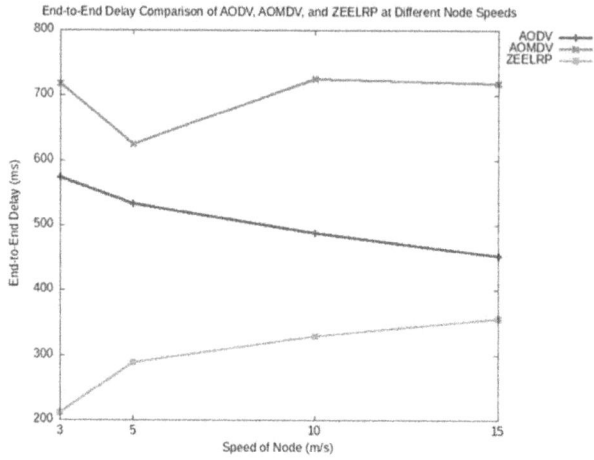

Fig. 41.3 End-to-End delay comparison across varying mobility setups

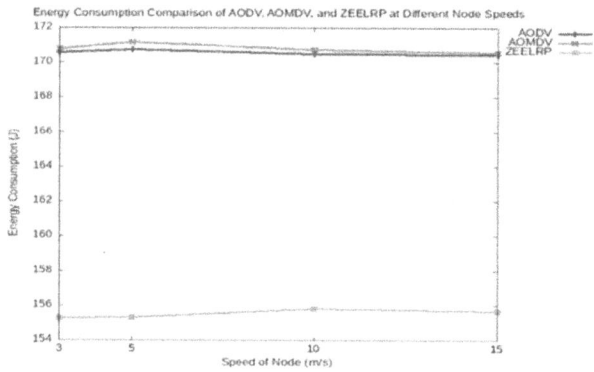

Fig. 41.4 Energy consumption trends under different mobility conditions

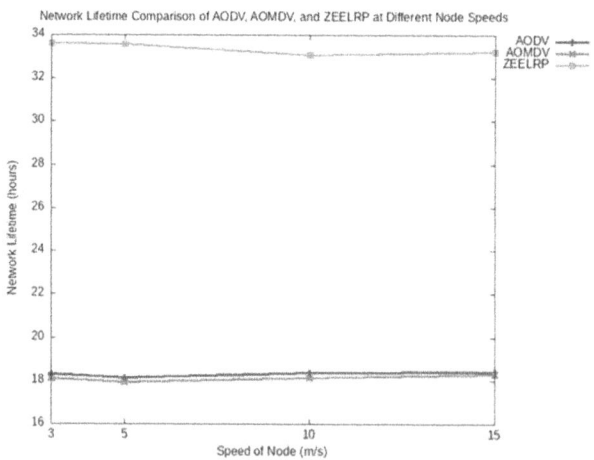

Fig. 41.5 Network lifetime performance for various mobility setups

5. CONCLUSION

This paper introduces ZEELRP, a zone-based energy-efficient routing protocol for hybrid wireless networks (HWNs). Simulations show ZEELRP outperforms AODV and AOMDV in PDR, throughput, energy efficiency, end-to-end delay, and network lifetime. By prioritizing high-energy nodes and optimizing path selection, it ensures reliable communication with minimal delay and energy use. The results highlight its adaptability for real-time, energy-aware MANET applications. However, since the study is simulation-based, future work will focus on real-world validation, addressing deployment challenges like hardware constraints, interference, and security to enhance practical applicability.

References

1. Almudayni, Ziyad, Ben Soh, Halima Samra, and Alice Li. 2025. 'Energy Inefficiency in IoT Networks: Causes, Impact, and a Strategic Framework for Sustainable Optimisation', Electronics, 14: 159.
2. Campioni, Lorenzo, Filippo Poltronieri, Cesare Stefanelli, Niranjan Suri, Mauro Tortonesi, and Konrad Wrona. 2023. 'Enabling civil–military collaboration for disaster relief operations in smart city environments', Future Generation Computer Systems, 139: 181–95.
3. Debnath, Sanjoy, and Wasim Arif. 2025. 'A hybrid zone-based routing protocol based on ZRP and DSR for emergency applications', International Journal of Communication Networks and Distributed Systems, 31: 1–18.
4. Kannan, KR, and CN Marimuthu. 2024. 'Energy efficient routing technique using enthalpy ant net routing for zone-based MANETS', IETE Journal of Research: 1–13.
5. Obayiuwana, Enoruwa, and Olabisi Emmanuel Falowo. 2017. 'Network selection in heterogeneous wireless networks using multi-criteria decision-making algorithms: a review', Wireless Networks, 23: 2617–49.
6. Pareek, Shashank, Ahmed Saleh Al-Samalek, Sandeep Singh, Mohammed Al-Farouni, Sharanpreet Kaur, and Jayant Jagtap. 2024. "Optimizing Performance in Heterogeneous Wireless Networks: Insights from Adhoc and Sensor Technologies." In 2024 4th International Conference on Technological Advancements in Computational Sciences (ICTACS), 1477–82. IEEE.
7. Rajawat, Anand Singh, SB Goyal, Jaiteg Singh, and Xiao Shixiao. 2025. 'Integrating quantum computing models for enhanced efficiency in 5G networking systems.' in, Applied Data Science and Smart Systems (CRC Press).
8. Sachithanandam, Vidhya, D Jessintha, Hariharan Subramani, and V Saipriya. 2025. 'Blockchain Integrated Multi-Objective Optimization for Energy Efficient and Secure Routing in Dynamic Wireless Sensor Networks', Sustainable Computing: Informatics and Systems: 101101.
9. Sahu, Neetu, and Veena Dhari. 2024. 'Multipath Routing with Dynamic Load Balancing for Enhanced Energy Efficiency in Zone-Based Wireless Ad Hoc Networks', Indian Journal of Data Communication and Networking (IJDCN), 4: 1–10.
10. Sahu, Rani, MA Rizvi, and Manoj Mishra. 2021. "Routing overhead performance study and evaluation of zone based energy efficient multi-path routing protocols in MANETs by using NS2. 35." In 2021 International Conference on Advances in Technology, Management & Education (ICATME), 88–93. IEEE.
11. Sahu, Rani, Sanjay Sharma, and M Rizvi. 2019a. 'ZBLE: Zone Based Leader Selection Protocol'.
12. Sahu, Rani, Sanjay Sharma, and MA Rizvi. 2020. 'ZBLE: Zone based efficient energy multipath protocol for routing in mobile Ad Hoc networks', Wireless Personal Communications, 113: 2641–59.

Note: All the figures in this chapter were made by the authors.

Emerging Perspectives and Applications of Computational Intelligence and Smart Systems
– Dr. Amit Lathigara et al. (eds)
© 2026 Taylor & Francis Group, London, ISBN 978-1-041-20965-2

42

Enhancing Medication Adherence through an IoT-Enabled Automated Pill Dispenser with Real-Time Monitoring

Abhishek Tripathi,
Peddineni Akshitha,
Meka Venkata Mahalakshmi,
Ramagiri Manjusha, Padidham Bhargavi
Department of Computer Science and Engineering,
Kalasalingam Academy of Research and Education,
Virudhunagar, Tamil Nadu, India

Suresh Balpande*
Ramdeobaba University,
Nagpur, India

■ **Abstract:** Medication non-adherence influences Good Health and Well-being's SDG 3. This paper presents an IoT-enabled automated pill dispenser with remote monitoring, accurate scheduling, and automatic dispensing promoting medication adherence. Running an Arduino Uno microprocessor, ESP8266 Wi-Fi module, servo motor, circular medication tray with four divisions to control pill distribution. Real time alerts from the Blynk app notify patients and caregivers. Stressing hardware instead of visual or aural alarms guarantees simplicity and dependability. An RTC module keeping appropriate dose schedules aids to sustain remote adherence tracking made possible by Wi-Fi connection. Many repeats revealed a constant drop in failure rates, therefore demonstrating the robustness and effectiveness of the system. In many healthcare environments, affordable design and a simple interface fit. Future directions will be integration of advanced analytics for adherence optimization and system customizing for medical needs.

■ **Keywords:** IoT, Automation, Medication, Pill dispenser, Monitoring

1. INTRODUCTION

Prescriptions should be followed by strict dose dependent chronic patients (Aseeri et al. 2020). Ignoring drug instructions could cause serious medical issues, less effective treatment, and costlier

---

*Corresponding author: balpandes@rknec.edu.

DOI: 10.1201/9781003725046-42

healthcare Gargioni et al. (2024). Older or multi-medication patients are more prone to blunders since traditional treatments such as alarm reminders and pill organizers depend on human consistency and memory (Kumar et al. 2021). The Internet of Things (IoT) devices and objects compile, interact, and act on data to so improve automation and efficiency in all areas (Tripathi et al. 2024). Medication delivery is simplified (Pandey et al. 2018), (Suganya et al. 2019) via remote notifications, real-time monitoring, and automatic dispensing—which IoT-based pill dispensers simplify. Managing dispensing algorithms to release medications on demand is done using a real-time clock. The Blynk app real-time alerts patients and caregivers of missing doses and prescription adherence. IoT has brought solutions for drug adherence, better patient care, and tools for creative health management systems. IoT neatly links tools, patients, and caregivers to improve healthcare systems. New IoT-based pill dispensers increase medication intake and schedule adherence accuracy. These devices reduce human mistake and improve living for elderly and chronically sick people by different drug dosage (Ramkumar et al. 2020). IoT-driven pharmaceutical solutions on cloud platforms allow family members and doctors to monitor adherence and respond to missing doses remotely. These advances improve doctor-patient relationships and results. Smart healthcare systems holistically serve patients with real-time health monitoring and drug delivery. Automated tablet distribution and vital sign monitoring assist Alzheimer's and autoimmune disease patients. Remote health data access for family and doctors as well as automatic medication reminders aid to guide decisions and prevent mistakes. Reducing explicit and audible messages helps to give scalability, usability, and efficiency first priority. An inexpensive, basic, programmable dispenser can help some patients to increase prescription adherence, error rates, and healthcare outcomes. a paper-based framework. Section 2tacklesIoT-based drug management issues and automated pill dispenser literature. The proposed strategy's Section 3 tackles program logic, hardware, and system design. Section 4 covers tools for programming, hardware integration, and configuration. Section 5 presents outcome of prototype testing under accuracy, reliability, and user input stress. Section 6 closes with significant conclusions and restrictions; Section 6 offers future research.

2. LITERATURE REVIEW

The IoT enabled drug management improves treatment quality, accuracy, and adherence. IoT gadgets include CMOS cameras and image-processing systems. Emails reminding missing doses help to reduce computational and sensor costs (Sahu et al. 2023). IoT pill dispensers and smart cups integrating accelerometers and gyroscopes for consumption validation are examples of advanced technologies. Bluetooth Low Energy devices support drug scheduling and adherence (Peddisetti et al. 2024). Distribution of infrared contactless sensors for COVID-19 detects drug use and temperature. IoT devices must warn caregivers and stop physical interaction to guarantee prescription compliance in extreme conditions (Nayak et al. 2024). IoT-driven medication dispensers assist with memory of dosages for elderly and dependent people. In these devices, alarms and real-time caregiver communication raise patient safety (Patil et al. 2022). Using facial recognition, advanced smart dispensers safely dose. These consistent and moderately priced geriatric devices have load cells for drug levels and sections for drugs unique to time-series (Sudarmani et al. 2024). Real-time drug control allows Alzheimer's patients to choose their prescriptions (MohanaPriya et al. 2020). In multi-patient systems including notifications, LED indicators, and mobile device alerts, safe access, remote management, and dose accuracy come first (Nguyen et al. 2023). By means of real-time monitoring, dose scheduling, and authentication, IoT smart devices fight prescription abuse and drug addiction. These regulations improve the harmony between pharmaceutical safety and medical practice (Ankireddypalli et al. 2023). SIMMS make use of LED, buzzers, wireless connectivity, and other inventive monitoring devices. Early Blynk apps improve caregiver communication and patient adherence (Abdullah et al. 2023).

3. METHODOLOGY

Figure 42.1 (a) shows IoT-enabled pill dispenser logic. RTC, servo motor, Wi-Fi starts system controls. Track main loops time versus intervals. Servos set Blynk alerts on games, distribute pills from the correct container, and light LEDs. Method demands four straight runs over containers. Figure 42.1 (b) show ESP8266 module warning and Blynk application. Blynk servers' WiFi is ESP8266-based. The Blynk dashboard displays event status together with scheduled reminders for caregiver and patient "Time to take medicine". This generates instantaneous changes on drug adherence.

Fig. 42.1 Flowchart showing the operational logic of (a) the IoT-based pill dispenser and (b) dedicated notification process using Blynk app with ESP8266 module

The Arduino control system lays out the medication schedule and starts hardware. SET calls RTC, servo motor, Wi-Fi module initialization. This cycle never breaks, matching the present instant with the prescription calendar. The circular tray releases tablets at the prescribed moment by use of a servo motor. The Blynk app notifies you. The Blynk program finds a flaw in a dispensing or dosage-related system. Aimed at low power usage and perfect running, the method ensures dependability and efficiency.

◣ 4. SETUP IMPLEMENTATION

Designed and implemented to guarantee flawless drug administration was the IoT-enabled automatic pill dispenser. The device combines Arduino Uno microcontroller, ESP8266 Wi-Fi module, four-compartment circular medication tray, servo motor, small plastic sliding track guiding pills onto a tray. Precise control over pill distribution and real-time monitoring via the Blynk app depends on careful hardware setting guaranteeing efficient functioning. The Arduino Uno consists mostly in running the preset schedule and regulating servo motor motions. It also consists in the main controller. The ESP8266 Wi-Fi module ties the Blynk server and allows alerts for patients and caregivers. The servo motor runs the pharmaceutical tray to line the suitable compartment with the dispensing chute; a real-time clock (RTC) module guarantees correct timing for dispensing. In handling prescription scheduling, the design emphasizes dependability, simplicity, and little handling area.

Stressing the internal connections of all components, Fig. 42.2 (a) illustrates the full hardware arrangement including the Arduino Uno, ESP8266 module, servo motor, and power supply, coupled to a laptop for programming and monitoring. Figure 42.2 (b) offers a close-up view of the four-compartment circular dispenser, therefore guaranteeing accurate and user-friendly medicine delivered with a limited sliding track directing pill into the tray.

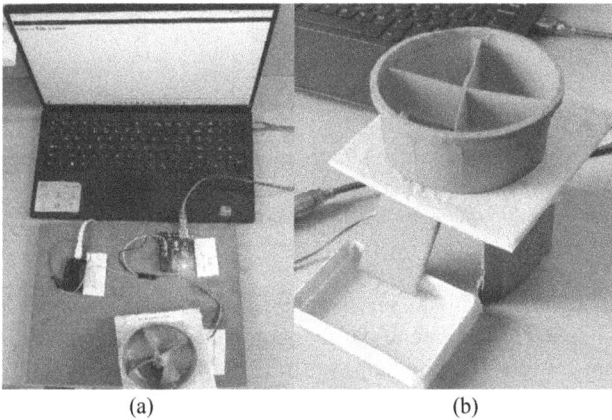

(a)   (b)

Fig. 42.2   (a) Complete hardware setup with all components connected and interfaced with a laptop. (b) Pill dispenser with sliding track and tray for pill delivery

◣ 5. RESULTS AND DISCUSSION

Figure 42.3 enumerates the main justifications for the medical domain's pill dispenser implementation. Twenty volunteers in all—relatives, family members, neighbors, and friends—were requested to complete a survey in order to gather information. The poll was meant to emphasize the common social concerns about drug distribution. Emphasizing the part automation addresses specifically issues including forgetfulness, dependency, and convenience; it helps to boost the effectiveness of healthcare.

The servo motor's rotation angle shown in Fig. 42.4 (a) lines with the following dispensing times: ten in the evening before bed, seven in the evening after dinner, eight in the morning after breakfast,

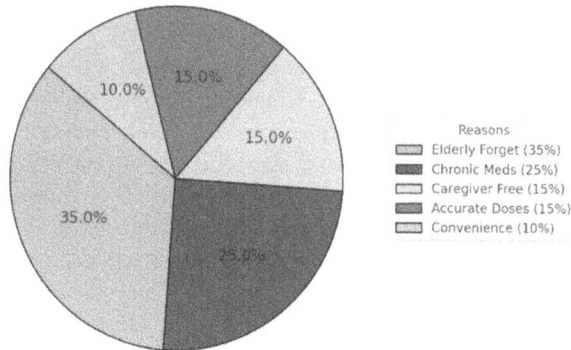

Fig. 42.3 Reasons for adopting pill dispensers in healthcare

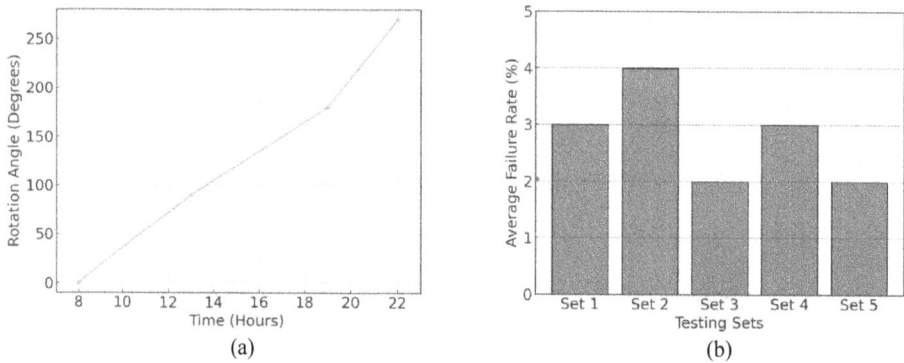

Fig. 42.4 Servo motor rotation angle mapped to scheduled dispensing times (b) Average failure rates across testing sets

one in the afternoon following lunch. The servo motor runs through a whole 90 degrees at each of the specified intervals to line itself with one of the four compartments, therefore guaranteeing accurate and timely drug administration. The plot helps to stress the system's dependability and accuracy in executing the set plans. Figure 42.4 (b) displays, for any one of the five testing sets, the average failure rates noted over ten iterations. It does this by showing two to four percent of a consistent decline in failure rates over iterations. This ensures the system's lifespan. Mostly misalignment of the servos, communication issues, and scheduling irregularities caused failures; all of these were gradually overcome.

6. CONCLUSION

This work presents the design and execution of an IoT-enabled automated pill dispenser targeted to improve prescription adherence, accuracy, and convenience. Combining an Arduino Uno CPU, an ESP8266 Wi-Fi module, and a servo motor guarantees accurate and timely medicine distributed. Notifications sent by the Blynk program give patients and caregivers real-time data, therefore guaranteeing suitable drug compliance. Several testing runs proved the dependability of the system; modest changes and upgrades surely help to lower failure rates. Older patients and those with chronic diseases needing rigorous medication regimes would find the device particularly appropriate for its small and straightforward design. This dispenser reduces errors and hence handles important issues by including IoT capabilities to handle critical drug management issues including forgetfulness and reliance. The technology's affordability and scalability fit more general utilization in medical environments. Since it lets to reduce healthcare costs and enhance quality of life, the automated pill dispenser looks to be a general useful tool in general in enhancing patient care and medication compliance.

References

1. Aseeri, M., Banasser, G., Baduhduh, O., Baksh, S., & Ghalibi, N. (2020). Evaluation of medication error incident reports at a tertiary care hospital. *Pharmacy, 8*(2), 69.
2. Gargioni, L., Fogli, D., & Baroni, P. (2024). A systematic review on pill and medication dispensers from a human-centered perspective. *Journal of Healthcare Informatics Research, 8*(2), 244–285.
3. Kumar, S. K., Manimegalai, R., Rajeswari, A., Deekshita, R., Dhineshkumar, M., & Manikandan, G. (2021). A literature review: Performance evaluation of wearable system with pill dispenser box for post COVID elderly patients. In *2021 3rd International Conference on Advances in Computing, Communication Control and Networking (ICAC3N)* (pp. 2008–2014). IEEE.
4. Tripathi, A., Manvith, K., Srinath, G. M., Yuvan, N. K., & Reddy, U. T. (2024). Implementation of an alert system integrated into smart wireless helmets utilizing IoT sensors. In *2024 2nd International Conference on Networking and Communications (ICNWC)* (pp. 1–5). IEEE.
5. Pandey, P. S., Raghuwanshi, S. K., & Tomar, G. S. (2018). The real time hardware of smart medicine dispenser to reduce the adverse drugs reactions. In *2018 International Conference on Advances in Computing and Communication Engineering (ICACCE)* (pp. 413–418). IEEE.
6. Singh, S. A., &Balpande, S. S. (2022). Development of IoT-based condition monitoring system for bridges. *Sound & Vibration, 56*(3), 209–220.
7. Suganya, G., Premalatha, M., Anushka, S., Muktak, P., & Abhishek, J. (2019). IoT based automated medicine dispenser for online health community using cloud. *International Journal of Recent Technology and Engineering, 7*(5), 1–4.
8. Ramkumar, J., Karthikeyan, C., Vamsidhar, E., & Dattatraya, K. N. (2020). Automated pill dispenser application based on IoT for patient medication. In *IoT and ICT for Healthcare Applications* (pp. 231–253). Springer.
9. Ali, H., Khan, H., Haseeb, A., Yazdani, H. A., & Munir, G. (2024). Smart pill dispensing and vital sign monitoring system. In *AIP Conference Proceedings* (Vol. 3125, No. 1). AIP Publishing.
10. Sahu, D. K., Pradhan, B. K., Wilczynski, S., Anis, A., & Pal, K. (2023). Development of an internet of things (IoT)-based pill monitoring device for geriatric patients. In *Advanced Methods in Biomedical Signal Processing and Analysis* (pp. 129–158). Academic Press.
11. Peddisetti, V., Kandregula, P. K., John, J. A., Poomdla, S., George, K., &Panangadan, A. (2024). Smart medication management: Enhancing medication adherence with an IoT-based pill dispenser and smart cup. In *2024 IEEE First International Conference on Artificial Intelligence for Medicine, Health and Care (AIMHC)* (pp. 137–144). IEEE.
12. Isranil, R. K., Nedunchezhian, T., Tanna, P., & Soni, S. (2025) AI-endorsed techniques for smart energy utilization: A holistic review. *Digital Transformation and Sustainability of Business*, 271-274.
13. Nayak, M., Gardia, J. K., Ghugar, U., Jain, A., & Nayak, R. (2024). Revolutionizing healthcare management: IoT-based medicine dispenser and temperature monitoring systems. In *2024 OPJU International Technology Conference (OTCON) on Smart Computing for Innovation and Advancement in Industry 4.0* (pp. 1–6). IEEE.
14. Dhone, M. D., Gawatre, P. G., &Balpande, S. S. (2018). Frequency band widening technique for cantilever-based vibration energy harvesters through dynamics of fluid motion. *Materials Science for Energy Technologies, 1*(1), 84–90.
15. Patil, C. H., Lightwala, N., Sherdiwala, M., Vibhute, A. D., Naik, S. A., & Mali, S. M. (2022). An IoT based smart medicine dispenser model for healthcare. In *2022 IEEE World Conference on Applied Intelligence and Computing (AIC)* (pp. 391–395). IEEE.

16. Sudarmani, R., Harshitha, J., & Rajakumari, K. (2024). Development of smart automatic drug dispenser for elderly and disabled people. In *2024 International Conference on Science Technology Engineering and Management (ICSTEM)* (pp. 1–5). IEEE.

17. MohanaPriya, D., Deepika, V., Shanmugha Priya, M., &Yogeswari, C. S. (2020). A real time support system to impart medicine using smart dispenser. In *2020 International Conference on System, Computation, Automation and Networking (ICSCAN)* (pp. 1–10). IEEE.

18. Nguyen, K.-S. T., Nguyen, V. L., & Nguyen, M.-T. (2023). IoT system design for a smart medicine dispenser serving multiple patients. In *The International Conference on Sustainable Energy Technologies* (pp. 249–257). Springer Nature Singapore.

19. Ankireddypalli, R. S., & Reddy, K. S. S. S. (2023). IoT based smart drug administrator and dispenser. In *2023 International Conference on New Frontiers in Communication, Automation, Management and Security (ICCAMS)* (Vol. 1, pp. 1–6). IEEE.

20. Abdullah, A. A., Janson, K., & Ang, W. C. (2023). Development of smart intelligent medicine monitoring system (SIMMS). In *Symposium on Intelligent Manufacturing and Mechatronics* (pp. 761–770). Springer Nature Singapore.

Note: All the figures in this chapter were made by the authors.

Emerging Perspectives and Applications of Computational Intelligence and Smart Systems
– Dr. Amit Lathigara et al. (eds)
© *2026 Taylor & Francis Group, London, ISBN 978-1-041-20965-2*

43

Application of a Genetic Algorithm in 0-1 Knapsack Problem

Sajad Abdlkadhim*,
Sait Demir, Emrullah Sonuç
Computer Engineering, Karabuk University,
Karabuk, Turkiye

■ **Abstract:** The knapsack problem represents a problem that must be solved to find the maximum amount of a value within a limit. The computational complexity increases according to the number of elements and capacity limits of the set represented by the problem. Algorithms have been suggested in the literature to solve the problem by reaching the most accurate result at the most appropriate time. Genetic algorithm approach is one of the methods used in solving the problem. This study proposed a genetic algorithm model to solve the problem. Initialization, fitness evaluation, selection, crossover, mutation and replacement features of genetic algorithms were used. In addition, a selection mechanism proportional to fitness, a single-point crossover strategy and a 10% mutation rate were used to increase the success of the optimum solution. The calculated results showed that the algorithm reached optimum or near-optimum solutions in acceptable computational times. The overall evaluation of the results confirmed the success of the genetic algorithm in solving the knapsack problem.

■ **Keywords:** 0-1Knapsack problem, Combinatorial optimization problem, Genetic algorithm

1. INTRODUCTION

The knapsack problem is an optimization problem and aims to reach the maximum useful amount. The problem has variants such as the 0-1 knapsack problem. The problem aims to reach the maximum number of elements with a limited capacity. For this process, a subset that aims to maximize the total value should be selected from the set of elements. (Jackson et al., 2018). However, each element in the set of elements should be selected to maximize the total benefit within certain constraints. However, the computational complexity of the solution to the problem changes directly proportional to the increase in the number of elements in the set. The computational complexity makes the problem an important optimization research topic (Baghel et al., 2012). Problems of reaching the maximum value with limited capacity in the commercial and industrial fields can be modelled with

_____
*Corresponding author: sajjad95.ab@gmail.com

DOI: 10.1201/9781003725046-43

the knapsack approach. Algorithmic based methods have been proposed in the literature to solve the optimization problems represented by the knapsack approach. One proposed method is genetic algorithms that model natural evolutionary processes (Suman et al., 2006). Genetic algorithms produce successful results for the knapsack problem. (Zavala et al., 2014). Genetic algorithms use biological selection mechanisms to find the most accurate and closest solutions (Tian and Simon, 2016). The algorithm mimics three basic biological mechanisms. These mechanisms are selection, crossover, and mutation (Zhao et al., 2010). GA has strong features in using discrete or non-differentiable search spaces, a characteristic feature of optimization problems (Karaboga et al., 2014). The success of genetic algorithms in solving the knapsack problem stems from their ability to navigate by examining alternative solutions in a large solution space (Dorigo and Blum., 2005).

2. RELATED WORK

Numerous studies are in the literature for solving the 0-1 Knapsack problem. It is used especially in logistics and management for commercial and industrial applications to make the most accurate decision. These studies cover both various problem examples and solution methods. The computational complexity of the problem, solution time, and the success of the calculated result are used as evaluation criteria. Optimization methods are frequently used to solve the problem. Abdel-Basset et al. (2021) proposed Binary Equilibrium Optimization (BEO) algorithm with different transfer functions. Population-based met heuristics were also used to select the optimal solution. Moradi et al. (2022) developed a simulated annealing (PSA) approach. Feature analysis has emerged as another important research area. Yang et al. (2021) identified the essential features to generate solutions using machine learning models efficiently. Jooken et al. (2022) designed a dataset of 3,240 instances of the 0-1 knapsack problem. Some studies have focused on adapting existing met heuristics. (Jooken et al. 2023) used the Aquila Optimizer (AO) for binary optimization problems. (Bas. 2023) provided a contextual overview of applications of the 0-1 knapsack problem. Studies in the literature show that algorithm design for solving the knapsack problem continues to evolve. Advances in algorithms also improve computational efficiency.

3. THE PROPOSED METHOD

The proposed genetic algorithm uses specific implementations of genetic operators to maximize efficiency and effectiveness:

- **Initialization:** An initial population of random solutions is created.
- **Fitness Evaluation:** Individuals in the population are evaluated using the fitness function.
- **Selection:** Individuals participating in reproduction are chosen according to their fitness values.
- **Crossover:** Features of the solutions that produce good results are combined and produce offspring with better features than their parents.
- **Mutation:** Random changes are made to the offspring with a predetermined probability.
- **Replacement:** Newly created offspring replace individuals from the previous generation

The proposed genetic algorithm uses certain implementations of genetic operators to increase the success of the solution:

- **Selection Operator:** The fitness proportional method, commonly called roulette wheel selection in the literature, is used.

- **Crossover Operator:** A single-point crossover method is used, where a single random location on the chromosome is determined as the crossover point.
- **Mutation Operator:** After crossover, individual genes on the daughter chromosomes mutate with a predetermined probability. The proposed method also uses a mutation rate of 10%.

Table 43.1 Summary of the parameters used in the genetic algorithm

Parameter	Description	Value
population_size	The number of potential solutions (individuals) in each generation	50
generations	The total number of generations for which the algorithm will run	2000
mutation_rate	The probability of each bit in an individual's solution being flipped	0.1 (10%)
kp_size	The number of items in the knapsack, derived from the input data	Determined by input data
kp_capacity	The maximum weight capacity of the knapsack, derived from input data	Determined by input data

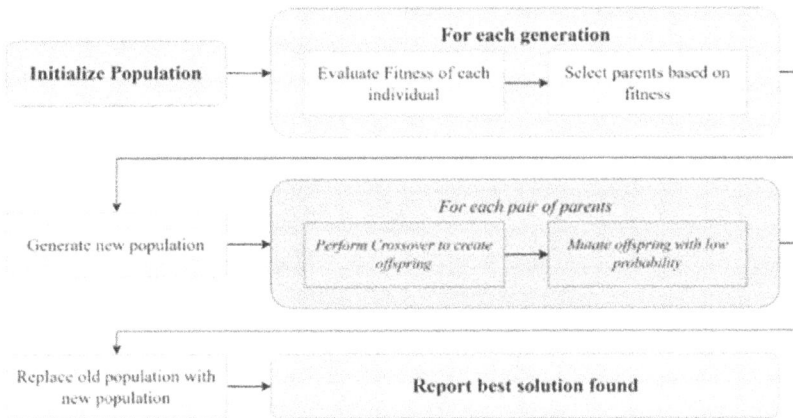

Fig. 43.1 Flowchart for proposed genetic algorithm

4. RESULTS AND DISCUSSION

The success of the presented genetic algorithm in solving the knapsack problem was examined by testing it on different instances (Abdel-Basset et al. 2021). The algorithm was run 30 times to calculate the results of the examples of the problem, and the results were obtained. The results obtained were evaluated to examine their statistical significance. In Table 43.2, the results obtained for each instance are given in detail.

Experimental results demonstrate the effectiveness of the proposed genetic algorithm in solving different instances of the knapsack problem (Abdel-Basset et al. 2021). The algorithm was able to find the optimal solution for several instances (e.g., **ks_8a, ks_8b, ks_8c, ks_8d, ks_8e, ks_12b, ks_12c, ks_12d, and ks_12e**), as indicated by the same values in the "Optimal" and "Best" columns. For more complex instances, especially those with larger elements (e.g., **ks_16b, ks_20a, ks_24c**), the algorithm obtained nearly optimal solutions with only marginal deviations from the known optimum values.

Table 43.2 Experimental results of the genetic algorithm on instances of the 0-1 knapsack problem

Instance	Optimal	Best	Mean	Worst	Median	Std. Dev.	Time
ks_8a	3,924,400	3,924,400	3,885,731	3,610,494	3,905,568	63,260	1.9940
ks_8b	3,813,669	3,813,669	3,761,431	3,609,822	3,782,677	49,775	2.9264
ks_8c	3,347,452	3,347,452	3,239,574	3,063,759	3,210,751	89,232	1.9842
ks_8d	4,187,707	4,187,707	4,100,549	3,867,512	4,127,370	75,693	2.0003
ks_8e	4,955,555	4,955,555	4,822,073	4,541,618	4,849,678	101,955	2.0512
ks_12a	5,688,887	5,681,360	5,524,203	5,392,081	5,527,948	71,522	2.0718
ks_12b	6,498,597	6,498,597	6,204,817	5,861,803	6,195,081	164,340	2.0441
ks_12c	5,170,626	5,170,626	5,001,833	4,758,900	5,027,413	121,368	3.1020
ks_12d	6,992,404	6,992,404	6,789,073	6,446,792	6,814,454	142,653	2.1060
ks_12e	5,337,472	5,337,472	5,132,735	4,864,274	5,128,152	108,070	2.0096
ks_16a	7,850,983	7,832,023	7,584,188	7,236,643	7,599,627	131,632	2.0617
ks_16b	9,352,998	9,234,167	8,853,735	8,561,161	8,871,400	191,849	2.0780
ks_16c	9,151,147	9,048,994	8,653,328	8,171,488	8,666,200	211,742	3.2460
ks_16d	9,348,889	9,305,859	9,003,257	8,609,476	9,012,589	156,629	2.0971
ks_16e	7,769,117	7,705,502	7,439,792	7,138,916	7,459,397	138,643	2.0743
ks_20a	10,727,049	10,627,905	10,136,104	9,731,013	10,148,907	185,745	2.2721
ks_20b	9,818,261	9,719,497	9,353,283	9,063,583	9,338,859	162,770	3.8582
ks_20c	10,714,023	10,535,125	10,151,987	9,575,575	10,192,837	240,306	2.1121
ks_20d	8,929,156	8,731,144	8,481,887	8,133,720	8,490,510	149,911	2.0456
ks_20e	9,357,969	9,211,045	8,987,073	8,569,402	9,009,548	151,021	2.0165
ks_24a	13,549,094	13,425,394	12,924,085	12,274,527	12,955,914	224,014	2.0397
ks_24b	12,233,713	12,034,154	11,627,223	11,146,572	11,648,544	196,796	2.0322
ks_24c	12,448,780	12,221,331	11,845,735	11,121,182	11,856,509	206,619	2.1503
ks_24d	11,815,315	11,626,322	11,136,932	10,635,676	11,155,688	199,656	2.4003
ks_24e	13,940,099	13,778,765	13,316,020	12,711,388	13,324,123	202,984	2.0272

Statistical results (mean, median, and standard deviation) provide information about the consistency and robustness of the algorithm. Relatively small standard deviations in most examples indicate regular performance. Execution times indicate the computational efficiency of the application, with most examples being solved in about 2–3 seconds, regardless of problem size.

Convergence Analysis: Convergence patterns for selected problem instances are investigated. Figure 43.2 shows the convergence behavior for the ks_24c instances and the best fit values obtained in each of the 30 runs.

The highlighted data point (red) shows the best solution obtained in all runs. The convergence pattern showed that most runs produced solutions within a narrow range of values and had relatively stable performance. This stability demonstrates the robustness of the algorithm and its ability to consistently identify high-quality solutions within the complex solution space of the knapsack problem.

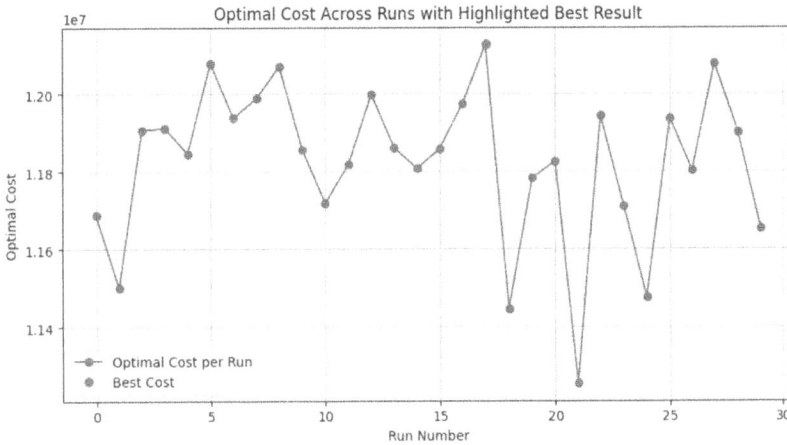

Fig. 43.2 Optimal value per run for the ks_24c instance

Empirical outcomes show that the offered genetic algorithm effectively uses exploration and exploitation to solve the knapsack problem. Initialization procedure and mutation operator facilitate the solution space exploration, while the selection and crossover mechanisms enable the exploitation of potential solution components. The solution performance of the algorithm is successful for small instances (up to 12 elements). Furthermore, as the size of the problem advances, algorithm demonstrates its scalability by maintaining the ability to find optimal solutions in reasonable computational times. Success in results depends on effective parameter settings that provide sufficient evolutionary progress, especially population size and number of generations.

5. CONCLUSION

This study demonstrates the success of genetic algorithms in solving the knapsack problem. The method we propose within the scope of the study produced optimum solutions for small-sized problems. For large examples, it produced near-optimal solutions in reasonable computational times. Experimental results confirm the success of the genetic algorithm approach. Genetic algorithms can be preferred in practical applications that require reliable solutions. The value of genetic algorithms for combinatorial optimization problems has been revealed by this study, especially in their ability to handle discrete and non-differentiable search spaces, which are common in logistics, finance, and management applications. Future research should focus on adaptive parameter settings (Sonuç and Özcan, 2023), problem-specific heuristic integration, and extension of the approach to the set-union knapsack problem (Sonuç and Özcan, 2024) and multi-objective knapsack variants.

References

1. Abdel-Basset, M., Mohamed, R., & Mirjalili, S. (2021). A binary equilibrium optimization algorithm for 0–1 knapsack problems. *Computers & Industrial Engineering, 151,* 106946.
2. Baghel, M., Agrawal, S., &Silakari, S. (2012). Survey of metaheuristic algorithms for combinatorial optimization. *International Journal of Computer Applications, 58*(19).
3. Baş, E. (2023). Binary aquila optimizer for 0–1 knapsack problems. *Engineering Applications of Artificial Intelligence, 118,* 105592.

4. Dorigo, M., & Blum, C. (2005). Ant colony optimization theory: A survey. *Theoretical computer science*, *344*(2-3), 243–278.
5. Jackson, W. G., Özcan, E., & John, R. I. (2018). Move acceptance in local search metaheuristics for cross-domain search. *Expert Systems with Applications*, *109*, 131–151.
6. Jooken, J., Leyman, P., & De Causmaecker, P. (2022). A new class of hard problem instances for the 0–1 knapsack problem. *European Journal of Operational Research*, *301*(3), 841–854.
7. Jooken, J., Leyman, P., & De Causmaecker, P. (2023). Features for the 0-1 knapsack problem based on inclusionwise maximal solutions. *European Journal of Operational Research*, *311*(1), 36–55.
8. Karaboga, D., Gorkemli, B., Ozturk, C., &Karaboga, N. (2014). A comprehensive survey: artificial bee colony (ABC) algorithm and applications. *Artificial intelligence review*, *42*, 21–57.
9. Moradi, N., Kayvanfar, V., & Rafiee, M. (2022). An efficient population-based simulated annealing algorithm for 0–1 knapsack problem. *Engineering with Computers*, *38*(3), 2771–2790.
10. Sonuç, E., & Özcan, E. (2023). An adaptive parallel evolutionary algorithm for solving the uncapacitated facility location problem. *Expert Systems with Applications*, *224*, 119956.
11. Sonuç, E., & Özcan, E. (2024). CUDA-based parallel local search for the set-union knapsack problem. *Knowledge-Based Systems*, *299*, 112095.
12. Suman, B., & Kumar, P. (2006). A survey of simulated annealing as a tool for single and multiobjective optimization. *Journal of the operational research society*, *57*(10), 1143–1160.
13. Tian, Z., & Fong, S. (2016). Survey of meta-heuristic algorithms for deep learning training. *Optimization algorithms—methods and applications*, 195–220.
14. Yang, Y., Boland, N., & Savelsbergh, M. (2021). Multivariable branching: A 0-1 knapsack problem case study. *INFORMS Journal on Computing*, *33*(4), 1354–1367.
15. Zavala, G. R., Nebro, A. J., Luna, F., & Coello Coello, C. A. (2014). A survey of multi-objective metaheuristics applied to structural optimization. *Structural and Multidisciplinary Optimization*, *49*, 537–558.
16. Zhao, N., Wu, Z., Zhao, Y., & Quan, T. (2010). Ant colony optimization algorithm with mutation mechanism and its applications. *Expert Systems with Applications*, *37*(7), 4805–4810.

Note: All the figures and tables in this chapter were made by the authors.

Emerging Perspectives and Applications of Computational Intelligence and Smart Systems
– Dr. Amit Lathigara et al. (eds)
© 2026 Taylor & Francis Group, London, ISBN 978-1-041-20965-2

44

An Extensive Analysis of a Deep Learning-Based Plant Disease Identification System

Milan Gohel*

Research Scholar,
Computer Engineering Department, Atmiya University,
Rajkot, India

Hiren Kavatjhiya

Computer Engineering Department, Atmiya University,
Rajkot, Gujarat, India

■ **Abstract:** Plant diseases result in huge crop losses and are a major issue in contemporary agriculture. It is time-consuming and labour-intensive to diagnose plant diseases using conventional techniques. Plant disease diagnosis can now be enhanced and mechanized due to the advent of deep learning (DL). A comprehensive review of the various deep learning techniques for PDD is presented here, focusing on the effectiveness, challenges, and future prospects of the methods. Various deep learning models including "recurrent neural networks (RNNs)" and "Convolutional neural networks (CNNs)" are discussed and their uses in various crops and types of diseases evaluated. The research mentions potential directions for future research while emphasizing challenges like data availability, model interpretability, and environmental uncertainty in spite of significant advances.

■ **Keywords:** Plant disease detection, Convolutional neural networks (CNN), Agricultural deep learning, Crop disease detection, Plant disease forecasting

1. INTRODUCTION

As the population of the world increases, there is an increased need than ever before for efficient farming techniques to supply the increasing demand for food. Unidentified plant diseases can have the ability to substantially reduce agricultural production, and this would affect global food security. Traditional sickness detection methods primarily rely on human intuition, visual symptom recognition, and manual checks, all of which are often time-consuming, random, and prone to errors (Zhang, Y., et al., 2020).

---

*Corresponding author: milan.gohel@atmiyauni.ac.in

DOI: 10.1201/9781003725046-44

Emerging advances in deep learning (DL) have paved the way for automated, scalable, and more accurate plant disease detection techniques. DL models are ideal for evaluating plant leaf photos and identifying disease symptoms because they have exhibited superior performance in image classification problems, particularly "Convolutional Neural Networks (CNNs)" (Kumar, A., & Singh, R., 2021). Analyzing the various deep learning methods applied to plant disease diagnosis, their techniques, and their outcomes is the primary objective of this paper.

2. TECHNIQUES FOR IDENTIFYING PLANT DISEASES

Image-based recognition, where neural networks analyse images of plant leaves, stems, or fruits to identify disease patterns, is the primary application of deep learning in PDD. In the field, several deep learning architectures have been employed.

2.1 Convolutional Neural Networks (CNNs)

Since "CNNs" are capable of automatically extracting features from raw images, they are the most widely used architecture for the identification of plant diseases. To distinguish between healthy and diseased plants, "CNNs" are particularly well suited for processing massive amounts of datasets with complex patterns. "CNNs" have been demonstrated to diagnose a range of plant diseases, such as rust, blight, and powdery mildew, in crops such as potatoes, wheat, and tomatoes (Li, H., et al., 2019) (Gupta, P., & Sharma, S., 2020).

2.2 Recurrent Neural Networks (RNNs)

"RNNs" are more suitable with sequential data, while "CNNs" are wonderful with spatial data processing. For predicting diseases depending on historical data patterns and climate, RNNs and its advanced versions like "Long Short-Term Memory (LSTM)" networks have been applied to time series data (Patel, R., et al., 2021). Such models give data about the progress of diseases with respect to time, allowing agricultural methods to go through predictive maintenance.

2.3 Transfer Learning

Transfer learning has gained more attention in the field due to the need for "large labeled datasets". To bypass the issue of limited data availability, pre-trained models such as "VGG16", ResNet, and Inception have been fine-tuned on plant disease datasets. This approach allows researchers to start training on specific plant disease detection tasks using big, general datasets (such as ImageNet) (Singh, N., et al., 2020).

2.4 Ensemble Methods

Ensemble of several models to enhance detection accuracy has been the focus of some studies. Generalization and robustness of disease detection systems can be significantly improved by using an ensemble of deep learning models, e.g., "CNNs with RNNs" or decision trees (Sharma, R., et al., 2022).

3. COMPARATIVE ANALYSIS OF DEEP LEARNING MODELS

According to several publications, the following compares different methods for detecting plant diseases. The diseases detected, the models employed, and the reported accuracy for each method are presented in the table.

Table 44.1 Comparative analysis of deep learning models

Model	Accuracy	Complexity	Real-World Applicability
CNN	91-94%	High	High
RNN/LSTM	85-90%	Moderate	Moderate
Transfer Learning	89-94%	High	High
Ensemble	90-93%	High	High

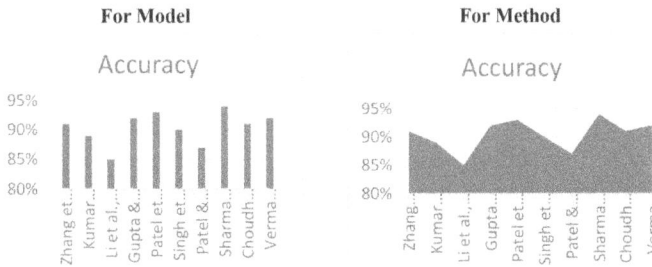

Fig. 44.1 Accuracy for model and method

4. CHALLENGES IN EXTENSIVE ANALYSIS OF A DEEP LEARNING-BASED PLANT DISEASE IDENTIFICATION SYSTEM

While deep learning has made significant strides in plant disease detection, several challenges remain:

Data Scarcity: Deep learning models heavily depend on large, annotated datasets to succeed. But it is very time- and cost-intensive to collect and annotate high-quality images of plant diseases. Even though transfer learning and data augmentation strategies have been utilized to tackle this issue, the need for large-scale datasets remains a key obstacle (Zhang, Y., et al., 2020).

Model Explainability: CNNs and other deep learning models are occasionally faulted as "black-box" models in that it is difficult to understand how they produce their predictions. For farmers and agricultural experts who require clear explanations of the model's decisions, such unavailability might be undesirable (Patel, R., et al., 2021)

Environmental Variability: inconsistent environmental variables, including discrepancies in lighting, weather, or camera quality, could lead to inconsistent symptoms of plant diseases. Deep learning algorithms can become weakened due to such variability, thereby affecting their reliability in real-life applications (Li, H., et al., 2019).

Generalization to crops: Several deep learning models are trained on specific disease types and crops. Their usage in a diverse agriculture setting may be limited due to their poor generalization to other crops or diseases (Gupta, P., & Sharma, S., 2020).

5. FUTURE TRENDS WITH PRACTICAL APPLICATIONS

The future for detecting plant diseases with deep learning is bright, with several trends that are likely to further maximize detection accuracy and applicability.

Table 44.2 Future trends with practical applications

Trend	Description
IoT & Drones	Use of real-time sensors and drones for disease detection (Sharma, R., et al., 2022)
Explainable AI	Making AI models transparent for easier understanding by users (Choudhary, S., et al., 2021)
Edge Computing	Running models directly on devices for quick, local disease detection (Patel, D., & Gupta, M., 2021)
Multimodal Integration	Combining different data types (visual, environmental) for accurate predictions.[10]

Integration with IoT and Drones: Plant disease diagnosis could be revolutionized by combining deep learning with drone technology and Internet of Things (IoT) devices. Deep learning models could process real-time data from sensors and drones to diagnose diseases instantly, allowing farmers to act quickly (Sharma, R., et al., 2022).

Explainable AI (XAI): The field is shifting toward explainable AI strategies to tackle the problem of model interpretability. By making the logic underlying model predictions apparent, these techniques hope to increase farmers' confidence in and adoption of deep learning-based solutions (Choudhary, S., et al., 2021).

Edge Computing: Instead of depending on centralized servers, edge computing enables the computation of deep learning models directly on devices (such as smartphones, drones, or cameras). By lowering latency, our method can facilitate quicker illness identification and real-time decision-making (Patel, D., & Gupta, M., 2021).

Multimodal Data Integration: Plant disease detection systems may become more accurate and resilient if different data types—such as physiological, environmental, and visual data—are combined. Predictions made using multimodal techniques are more accurate because they can better represent the complexity of plant health (Verma, S., & Tripathi, P., 2022).

Table 44.3 Summarizing gaps in the field

Gap	Description
Data Scarcity	Limited annotated datasets for training models.
Model Interpretability	Lack of transparency in model predictions.
Environmental Variability	Models affected by lighting, weather, and camera conditions.
Adaptability to New Diseases	Difficulty in recognizing emerging plant diseases.

6. CONCLUSION

Deep learning has become a potent instrument for identifying plant diseases, providing notable advancements over conventional techniques. Although CNNs are currently the most successful models for visual recognition, hybrid models and ensemble approaches have the potential to improve performance even further. However, for broad use in actual agricultural contexts, issues including environmental unpredictability, model interpretability, and data scarcity must be resolved. Plant disease detection's future depends on combining explainable AI techniques with deep learning, IoT, drones, and edge computing. This might transform plant health monitoring and improve global food security

References

1. Zhang, Y., et al., (2020). "Convolutional Neural Networks for Plant Disease Detection." Journal of Agricultural Sciences, 15(3), 120–132.
2. Kumar, A., & Singh, R., (2021). "A CNN-based Model for Early Detection of Crop Diseases." Agricultural Technology Review, 22(1), 98–105.
3. Li, H., et al., (2019). "Time-Series Disease Prediction Using Recurrent Neural Networks." Computational Agriculture Journal, 18(2), 45–59.
4. Gupta, P., & Sharma, S., (2020). "Transfer Learning for Plant Disease Classification." International Journal of AI in Agriculture, 9(4), 245–253.
5. Patel, R., et al., (2021). "Ensemble Methods for Plant Disease Detection Using CNN and SVM." Journal of Machine Learning in Agriculture, 10(2), 112–120.
6. Singh, N., et al., (2020). "Hybrid CNN-RNN Model for Early Blight Detection in Tomato." Artificial Intelligence in Agriculture, 13(1), 56–64.
7. Patel, D., & Gupta, M., (2021). "CNN for Multi-Class Classification in Plant Disease Detection." International Journal of Plant Pathology, 28(2), 78–85.
8. Sharma, R., et al., (2022). "Fine-Tuned Transfer Learning Models for Crop Disease Detection." Agriculture and Computer Vision Journal, 14(3), 204–213.
9. Choudhary, S., et al., (2021). "CNN and RNN Hybrid Models for Disease Detection in Cucumbers and Tomatoes." Horticultural Science and Technology, 19(4), 134–142.
10. Verma, S., & Tripathi, P., (2022). "Transfer Learning for Wheat Rust Detection with Augmentation." Journal of Agricultural AI Research, 25(2), 89–97.

Note: All the tables and figure in this chapter were made by the authors.

Emerging Perspectives and Applications of Computational Intelligence and Smart Systems
– Dr. Amit Lathigara et al. (eds)
© 2026 Taylor & Francis Group, London, ISBN 978-1-041-20965-2

45

Exploratory Data Analysis and Feature Correlation Insights on the CICIoV2024 Dataset for IoV Security

Nirav Bhatt, Amit Lathigara
School of Engineering, RK University,
Rajkot, Gujarat, India

Sunil Soni*
IT Department, Government Polytechnic,
Rajkot, Gujarat, India

Paresh Tanna
School of Engineering, RK University,
Rajkot, Gujarat, India

Jaydeep Tadhani
IT Department, Government Polytechnic,
Rajkot, Gujarat, India

■ **Abstract:** Efficient intrusion detection tools are needed based on the expanding cybersecurity threats emanating from the Internet of Vehicles (IoV) unprecedented growth. As a benchmark in evaluating IoV security tools through a range of attack modes on Controller Area Network (CAN) communications, CICIoV2024 offers a diversity of attack modes against CAN communications. This paper publishes an Exploratory Data Analysis (EDA) of CICIoV2024, where class imbalance, feature correlation, and data dispersal are focused on. With 156 features, the dataset contains 1,408,219 instances of benign and offensive traffic, including Denial-of-Service (DoS) and spoofing attacks (GAS, RPM, SPEED, STEERING_WHEEL), and others. With benign traffic constituting 86.9% of the dataset, we observe significant class imbalances that can potentially affect model performance. Strongly correlated features are revealed by a correlation analysis of 153 numeric features, which gives rise to feature selection and dimensionality reduction methods to optimize intrusion detection models. To determine important trends, we also employ distribution plots, bar charts, and heatmaps to visualize the structure of the dataset. By achieving improved feature selection, redundancy resolution, and data imbalance mitigation, these findings can enhance machine learning-based intrusion detection systems (IDS). Our research supports the development of scalable and efficient IoV security solutions with CICIoV2024.

■ **Keywords:** Internet of vehicles (IoV), CICIoV2024, Cybersecurity, Intrusion detection, Exploratory data analysis

*Corresponding author: profsjsoni@gmail.com

DOI: 10.1201/9781003725046-45

1. INTRODUCTION

By integrating networked and autonomous vehicles into smart transportation systems, the Internet of Vehicles (IoV) has emerged as a paradigm-shifting technology (Hanselman, M. et.al., 2020). While this connectivity enhances user comfort, traffic efficiency, and road safety, it also creates severe cybersecurity risks. Since its early design was not accompanied by security protection, the Controller Area Network (CAN) protocol—a fundamental component of modern vehicular communication—is vulnerable to cyberattacks like injection, spoofing, and denial-of-service (DoS) attacks (Guo, H. et.al., 2023). IoV system security is necessary for preventing data alteration, unauthorized access, and potential safety threats. Researchers employ benchmark datasets like CICIoV2024 (Canadian Institute for Cybersecurity, 2024) that provide real-world assault scenarios for evaluating intrusion detection systems to assuage such concerns.

One of the extensive benchmark datasets developed for intrusion detection in Internet of Vehicles (IoV) systems is CICIoV2024. It captures CAN traffic from diverse attack vectors such as DoS, spoofing (GAS, RPM, SPEED, STEERING_WHEEL), and other cybersecurity threats. It captures malicious and benign traffic. Using Exploratory Data Analysis (EDA) to understand the properties of the dataset is essential for effective feature engineering, anomaly detection, and machine learning model optimization. We can derive valuable information that enables us to build robust Intrusion Detection Systems (IDS) for IoV security by analyzing the distribution of the dataset, feature correlations, and class imbalances.

A comprehensive EDA of the CICIoV2024 dataset (Canadian Institute for Cybersecurity, 2024) is provided in this work, focusing on dimensionality reduction, feature correlation, and data dispersion. We discuss how to select the most relevant characteristics and identify redundant and strongly correlated attributes that can affect model performance. We also discuss the problem of class imbalance within the dataset, which causes challenges for machine learning-based detection systems. We identify significant trends to inform IDS development using visualization tools such as distribution plots, bar charts, and correlation heatmaps. By helping the researchers enhance feature selection, data preprocessing, and model performance with the CICIoV2024 dataset, our results contribute towards enhancing IoV cybersecurity.

2. PROPOSED APPROACH

The Fig. 45.1 illustrates the process of data acquisition and analysis for Intrusion Detection in the Internet of Vehicles (IoV) using OBD-II (On-Board Diagnostics) to USB connectivity. The OBD-II port of a vehicle is used to extract CAN (Controller Area Network) messages, including CAN High and CAN Low signals, which are transmitted to a computing system via an OBD to USB adapter (Canadian Institute for Cybersecurity, 2024). The collected CAN data undergoes data analysis, where it is classified into different categories such as Benign (normal traffic), DoS (Denial-of-Service attacks), and Spoofing attacks. This workflow is essential for detecting cyber threats in modern vehicles, ensuring vehicular network security, and enabling the development of machine learning-based Intrusion Detection Systems (IDS) for automotive cybersecurity.

To derive valuable insights that can be used to improve the design of Intrusion Detection Systems (IDS) for IoV security, our proposed methodology focuses on conducting an exhaustive Exploratory Data Analysis (EDA) of the CICIoV2024 dataset. To identify key patterns, correlations, and class imbalances, the approach relies on feature analysis, data preprocessing, and visualization techniques.

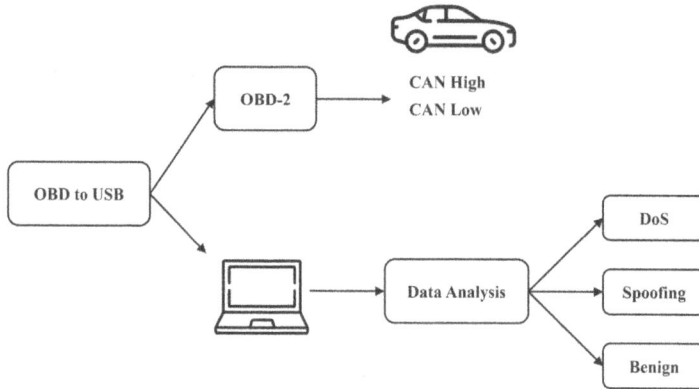

Fig. 45.1 Testbed setting for dataset

Data integration, statistical analysis, feature correlation mapping, and using appropriate strategies to reduce class imbalance are the primary support pillars of our methodology.

2.1 Preprocessing and Data Integration

There are several CSV files that describe different types of CAN traffic, ranging from benign ones to diversified attacks (DoS, Spoofing-GAS, Spoofing-RPM, Spoofing-SPEED, and Spoofing-STEERING_WHEEL). They constitute the dataset. These are merged into a single dataset of 156 features and 1,408,219 rows. During preprocessing, there are:

- Dealing with missing values either by deleting non-full records or, if needed, filling missing values.
- Numerical features must be normalized to offer data representation consistency.
- If categorical features are present, these should be encoded so that machine learning algorithms can utilize them.

2.2 Correlation of Features and Exploratory Data Analysis (EDA)

Based on the use of visualization tools such as pie charts, bar graphs, and histograms, we perform EDA to analyze the distribution of benign and attack classes. With 86.9% being benign traffic and a mere 13.1% attack traffic, the class distribution is found to be highly unbalanced, and that can influence IDS performance. We compute the correlation matrix (153 × 153) to identify repetitive and highly correlated qualities in order to understand the inter-relationships between features better. For reducing dimensionality and enhancing detection efficacy, highly correlated features (above a threshold value) are identified for potential elimination.

2.3 Handling Feature Selection and Class Imbalance

We explore techniques such as the Synthetic Minority Oversampling Technique (SMOTE) or undersampling to balance attack and normal traffic due to the dataset's significant class imbalance. To retain only the most relevant properties for intrusion detection, we also perform feature selection through Principal Component Analysis (PCA), correlation-based filtering, or other dimensionality reduction methods. By reducing overfitting and improving generalization, these processes contribute to the optimization of machine learning models.

3. DATASET ANALYSIS

The CICIoV2024 dataset (Canadian Institute for Cybersecurity, 2024) is a benchmark data set intended for testing intrusion detection solutions in Internet of Vehicles (IoV) scenarios. The dataset records actual Controller Area Network (CAN) traffic, with both benign and malicious behaviors encompassing a number of cyber attacks on vehicular networks. It is organized as several CSV files, each portraying particular types of attacks, e.g., Denial-of-Service (DoS) and other types of spoofing attacks (GAS, RPM, SPEED, STEERING_WHEEL). Once all the data sources are integrated, the end dataset contains 1,408,219 instances and 156 features, rendering it one of the most comprehensive datasets available publicly for IoV security research.

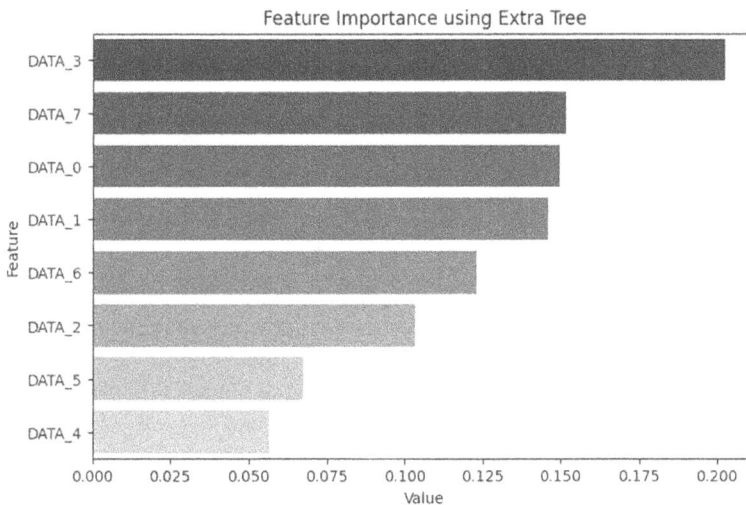

Fig. 45.2 Feature importance using extra tree

The bar chart shown in Fig. 45.2 presents the feature importance ranking derived from the Extra Trees classifier, highlighting the most significant features in the dataset. The DATA_3 feature holds the highest importance, followed by DATA_7, DATA_0, and DATA_1, indicating their strong contribution to the classification task. Other features like DATA_6, DATA_2, DATA_5, and DATA_4 also show varying degrees of relevance but with lower importance values. This analysis helps in feature selection, optimizing model performance by focusing on the most influential attributes in detecting anomalies or classifying network traffic in IoV security applications.

The bar chart in Fig. 45.3 illustrates the distribution of categories in the dataset, specifically distinguishing between Benign, DoS (Denial-of-Service), and Spoofing traffic.

4. CONCLUSION

We analyzed thoroughly in this research the CICIoV2024 dataset, a benchmark dataset for measuring IoV security solutions, via Exploratory Data Analysis (EDA). In a bid to understand feature correlation, class distributions, and attack behavior in the vehicle CAN traffic, the dataset that comprised 1,408,219 instances and 156 attributes was investigated. With 86.9% benign traffic and 13.1% attack traffic, our findings indicate a significant class imbalance, highlighting the need for data balancing techniques to enhance the effectiveness of machine learning-based intrusion detection

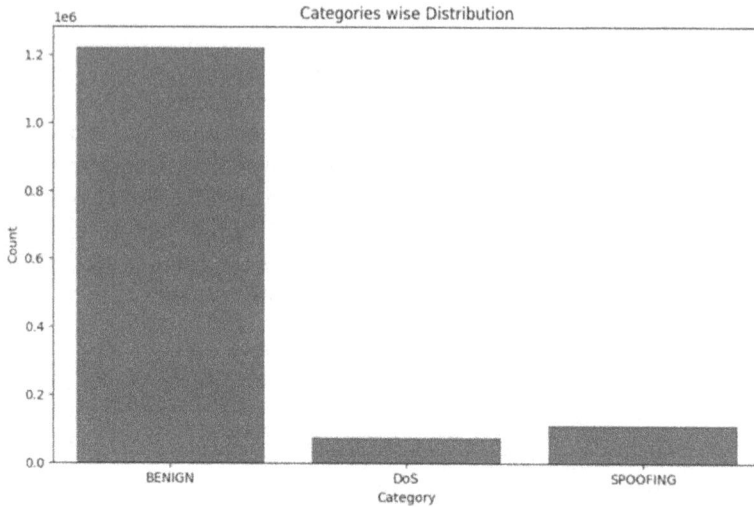

Fig. 45.3 Categories wise distribution

systems (IDS). We identified redundant features that could be removed to enhance IDS performance through feature correlation analysis. Class-wise distributions of attacks were also considered, giving insight into various types of attacks such as denial-of-service attacks and spoofing-based incursions. For developing reliable and efficient IDS models that are capable of detecting sophisticated cyber threats in connected and autonomous cars, these analyses are crucial. All in all, our work provides informative insights into preprocessing, feature engineering, and attack pattern analysis to improve IoV security systems.

References

1. Canadian Institute for Cybersecurity. (2024). CICIoV2024: A benchmark CAN dataset to evaluate IoV security solutions. University of New Brunswick. Retrieved February 28, 2025, from https://www.unb.ca/cic/datasets/iov-dataset-2024.html

2. Guo, H., Su, X., & Qin, Z. (2023). A survey on machine learning-based intrusion detection for vehicular networks. IEEE Internet of Things Journal, 10(3), 2214–2231.

3. Hanselman, M., Babun, L., Aksu, H., & Uluagac, A. S. (2020). CANShield: Detecting and mitigating CAN bus off-the-shelf attacks in real-time. Proceedings of the 2020 ACM Asia Conference on Computer and Communications Security (ASIACCS), 523–534.

4. Kang, J., Moon, S., & Kim, H. (2021). Intrusion detection system for in-vehicle networks using deep neural networks. IEEE Transactions on Vehicular Technology, 70(6), 5304–5318.

5. Miller, C., & Valasek, C. (2015). Remote exploitation of an unaltered passenger vehicle. Black Hat USA, 2015. Retrieved from https://www.blackhat.com

6. Mukherjee, M., Matam, R., Shu, L., Maglaras, L. A., Ferrag, M. A., Choudhury, N., & Kumar, V. (2019). Security and privacy in fog computing: Challenges, issues, and future directions. IEEE Access, 7, 42042–42063.

7. Seo, E., Choi, W., & Lee, H. (2020). Anomaly detection for in-vehicle network using information-theoretic similarity measure. IEEE Transactions on Information Forensics and Security, 15, 784–797.

8. Song, H. M., Kim, H. S., & Kim, H. K. (2016). Intrusion detection system based on the analysis of time intervals of CAN messages for in-vehicle network. International Conference on Information Networking (ICOIN), 63–68.

Note: All the figures in this chapter were made by the authors.

Emerging Perspectives and Applications of Computational Intelligence and Smart Systems
– Dr. Amit Lathigara et al. (eds)
© 2026 Taylor & Francis Group, London, ISBN 978-1-041-20965-2

46

Reinforcement Learning for 5G D2D Communication in Proximity-Based Networks

Anu Mangal*, Dr. Anjali Potnis

Department of Electrical and Electronics Engineering Education,
National Institute of Technical Teacher's' Training & Research, Bhopal, India

Dr. M.A. Rizvi

Department of Computer Science and Engineering Education,
National Institute of Technical Teacher's' Training & Research, Bhopal, India

Rani Sahu, Manmohan Singh

IES College of Technology, Bhopal, India

Parita Rathod

School of Engineering, RK University, Rajkot, Gujarat, India

■ **Abstract:** With the expansion of 5G cellular networks, many key technologies have emerged such Device-to-Device (D2D) communication. D2D technology fulfils the ever-increasing need of real time communication (low latency) and improve spectrum utilization. In proximity-based networks, optimization of 5G D2D communication in terms of resource allocation, routing strategies and interference management can be efficiently done with the help of reinforcement learning (RL). This paper presents the combination of RL techniques in 5G D2D communication, presenting state-of-the-art approaches, their challenges, and future directions. It introduces RLD2Dalgorithm and its Q-learning approach within a proximity-based D2D communication network. Designed to enhance 5G connectivity with minimal latency, the proposed method Reinforcement Learning for 5G D2D Communication (RLD2D) efficiently discovers neighbouring devices, establishes optimal D2D links, and dynamically adjusts the communication range for providing better throughput and energy efficient system as compared to its counterparts.

■ **Keywords:** Device to device communication, Q learning, Latency, Throughput, 5G Network

1. INTRODUCTION

With the grow thin requirement for fast pace and low-latency communication in 5G networks the service such as D2D and proximity-based network has evolved. D2D enables direct communication

*Corresponding author: anu.mangal001@gmail.com

DOI: 10.1201/9781003725046-46

between the devices unlike traditional cellular communication, reducing dependence on centralized infrastructure. However, D2D communication has to face various challenges such as spectrum allocation, interference control, and mobility management. For addressing these challenges efficiently, Reinforcement Learning, provides an adaptive and self-optimizing solution. (Opuni-Boachie Obour Agyekum et al., 2024). This work utilizes reinforcement learning for adaption and sustenance of connectivity among various communicating devices in environment (proximity-based). This paper employs a reinforcement learning approach to learn and maintain connectivity between different proximity-based communicating devices in the environment. The algorithm forecasts the existing D2D communication link using the reward function obtained from environmental parameters, whereas the Q-learning technique chooses the best D2D link. (Pratap Khuntia and Ranjay Hazra, 2020)It dynamically decides the best D2D connection based on important network parameters like communication range, latency, and buffer queue size. Even for the dynamic system environment the proposed algorithm preserves the connectivity providing efficient throughput, latency and energy efficiency. The organization of the paper is as follows: Section 2 contains a literature review for this work, whereas Section 3 describes system architecture and the RLD2D algorithm. Section 4 discusses the simulation configuration and compares the results on throughput, latency, and energy efficiency. The paper is concluded and potential future research areas are discussed in Section 5.

1.1 Contributions of the Paper

In the proposed D2D network, each device uses a reinforcement learning algorithm to decide its neighbouring device and establish links with better QoS. The Q-learning approach allows dynamic neighbour discovery and optimal device selection for seamless and reliable connectivity. With continuous data collection from nearby nodes, adjustment of the communication threshold based on latency, and selection of the most suitable device, the algorithm ensures effective link setup.

2. BACKGROUND AND RELATED WORK

Current studies investigate different reinforcement learning (RL) methods to maximize device-to-device (D2D) communications in cellular networks. (Tejal Rathore and Sudeep Tanwar, 2024) introduced an AI-based resource allocation framework and a new channel quality indicator (QI) for D2D communications. (Pratap Khuntia and Ranjay Hazra, 2020) presented an actor-critic RL model that enhances system and D2D throughput by methodically examining the policy space for best actions. Yi-Han Xu, Qi-Ming Sun, Wen Zhou, and Gang Yu,2022target resource allocation in UAV-supported, energy-harvesting D2D networks. (Opuni-Boachie Obour Agyekum et al., 2024) presented implementation of RL(Multiagent) with proximal policy optimization which provides promising results for improvement in QoS and energy efficiency through staggered training and decentralized execution. (Ravi Teja and Pavanillu, 2021) strated a Q-learning-based path selection method and a two-stage resource allocation approach with the use of MMF and HPSOGWO for throughput optimization and power distribution and explained the improvement in system consistency over existing approaches. (Park& Lim, 2020) describesa combination of Q-learning (distributed) for maximizing system energy efficiency while maintaining outage probability, outperforming existing mode-selection and power-control algorithms .They have also presented DRL where deep learning handles complex, large-scale network environments. Ali, Rashid et al. investigated D2D communication for NB-IoT delay-sensitive applications, considering relay selection as a multi-armed bandit (MAB) problem and solving it via UCB-based reinforcement learning. (Yuan Xie et al., 2021) presented intelligent-D2D (I-D2D) strategy to improve relay selection, yielding maximum packet delivery ratio (PDR) while ensuring minimum end-to-end delay. (Hashima and Hatano, 2021) has presented techniques to optimize system performance measures like sum-rate, throughput, and spectral efficiency while minimizing interference among cellular and D2D users. Kamran Zia et al., 2018 emphasized the use of machine learning (ML) methods in D2D networks, such as using MAB to facilitate enhanced neighbour discovery and choice for mm Wave D2D communications. Furthermore, distributed learning algorithms have been suggested to allow D2D users to self-choose spectrum resources with minimal signalling overhead. These works illustrate the efficiency of RL-based solutions in optimizing resource allocation for D2D-enabled networks.

3. METHODOLOGY

The major elements of reinforcement learning employed in our suggested algorithm are agent, state, action, reward, and environment. The block diagram in Fig. 46.1 illustrates various devices in 5G environment which are directly linked with each other and the transfer of data and information from one device to another is carried with the help of reinforcement learning in which actions are performed based on the reward obtained.

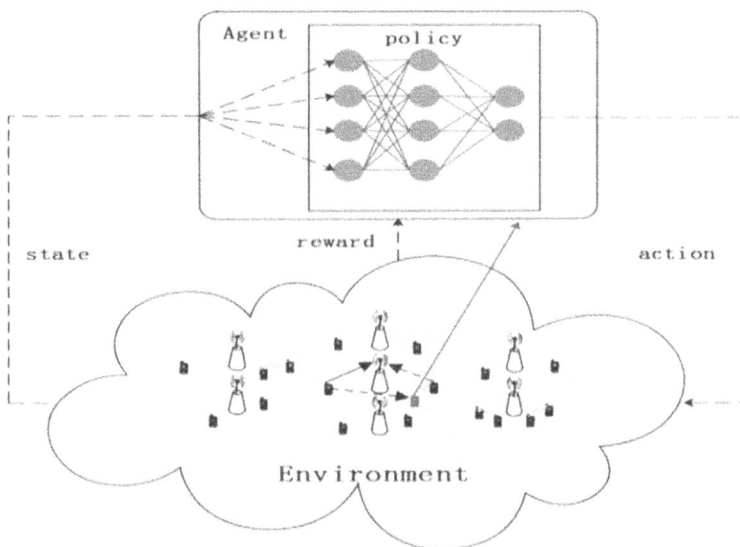

Fig. 46.1   Block diagram of RLD2D (5G) for proximity network

3.1 Q learning Approach for Proposed Algorithm

Our suggested algorithm uses Q-learning, a reinforcement learning method, to maximize data transfer in device-to-device (D2D) communications in 5G networks. The system constantly learns from various situations and ameliorates its decision-making process. Initially, the state parameters are defined by the algorithm, including the number of neighbouring devices, routing path and transmission power. It then establishes a set of possible actions, such as adjusting power, changing frequency, modifying the route, or maintaining the existing link, and records them in a Q-table. When a device takes an action (e.g., reducing power), it observes the consequences—whether the connection remains stable, interference reduces, or performance improves. Based on this performance result, the algorithm assigns a reward (positive or negative score) and the Q-table is updated accordingly. The Q-value increases, if an action proves beneficial, while there is a decrease in value of Q when actions are non-beneficial, reducing their likelihood in future iterations. This process iterates continuously due to which the strategy of the system is refined and redefined over time. Ultimately, the algorithm learns to make optimal routing decisions for enhancing the efficiency of D2D communication in terms of latency, energy and power. The primary objectives of the proposed algorithm RLD2D is reduction in latency, increased throughput and energy efficiency for a reliable and high performance driven 5G D2D communication network.

3.2 Algorithm 1: Q-learning Algorithm for D2D Communication in 5G Networks

Initialize:
Initialize state space S = $\{N_k, P_k, R_k\}$ # Neighbors, Power, Route
Initialize action space A(S) = {Adjust P_k, Change Frequency, Modify Route, Maintain Link}
Initialize Q-table: Q(s, a) = 0 for all s $\in$ S, a $\in$ A(S)
Set hyperparameters:

α (*rate of learning*), γ (*factor of discounting*), ϵ (*rate of exploration*)
$P_{threshold} = 0.75 \times P_{max}$

For loop t = 1 to *max_iterations*:
While current state $S_t. = (N_{max}, P_{max}, R_{max})$ do:
Choose current state $S_k = (N_k, P_k, R_k)$

For every neighbour N_k *of range* $[1, N_{max}]$:
For every power level P_k inrange $[0, P_{threshold}]$:
Select action A_k by using ϵ-greedy policy

Take action A_k and observe:
Reward $R(S_k, A_k)$
Next state $S_{k+1} = (N_{k+1}, P_{k+1}, R_{k+1})$

Update Q-table with:

$$Q(S_k, A_k) = Q(S_k, A_k) + \alpha * [R(S_k, A_k) + \gamma * maxQ(S_{k+1}, A) - Q(S_k, A_k)]$$

Set S_{k+1} as the new current state

Repeat until convergence

The RLD2D Q-learning model provides efficient D2D communication by enabling adaptive decision-making. Over multiple iterations, the algorithm observes and learns optimal strategy based on Q values for path selection ensuring energy-efficient, reduced latency and higher throughput communication in proximity-based 5G networks.

◢ 4. RESULT AND DISCUSSION

This section presents an in-depth quantitative and comparative performance evaluation of the proposed Reinforcement Learning-based Device-to-Device (RLD2D) communication protocol against the protocol OLSRv2 and AODV. The evaluation is conducted in a high-density, proximity-based 5G D2D network simulated in NS-3. The parameters which are considered for the evaluation are throughput, latency and energy efficiency across routing schemes. The main objective is to validate the effectiveness of RLD2D in improving network reliability, responsiveness, and energy sustainability for next-generation wireless networks. The primary simulation parameters are detailed below.

Table 46.1 NS-3 simulation configuration

Parameter	Value
Simulation Tool	NS-3
Network Topology	500m × 500m Grid
Number of Nodes	30-75
Max. Communication Range	100m
Transmission Power	2W
Mobility Model	Random Waypoint
Traffic Type	CBR (Constant Bit Rate)
Data Packet Size	512 bytes
Routing Protocols Compared	RLD2D, OLSRv2, AODV
Q-Learning Discount Factor (γ)	0.7
Learning Rate (α)	0.5
Exploration Strategy	ε-Greedy

4.1 Performance Analysis and Comparative Evaluation

1. **Latency Analysis:** For the real time applications and 5G scenarios latency consideration is very important. RLD2D outperforms AODV and OLSRv2, reducing end-to-end latency by 28% and 20% respectively, across varying network densities. AODV experiences high latency due to route discovery delays are more in case of AODV and periodic overheads are high in OLSRv2, so the latency is more in these cases as compared to RLD2D.The intelligent link selection in case of RLD2Dprovidesultra-low latency communication by reducing queuing and retransmission delays. Figure 46.2(a) demonstrates the latency comparison of three protocols across different network densities.

2. **Throughput Analysis:** Throughput is defined as successful data transmission rate from one device to another. RLD2D outperforms AODV and OLSRv2, achieving up to 30% and 26% higher throughput respectively across varying network densities. Packet drops are more in case of AODV as it suffers from link instability while control overheads in OLSRv2's reduces the available bandwidth for the transmission of actual data. RLD2D offers optimized link selection by allotting it dynamically, reduces congestion and ignores unreliable paths, thereby providing superior data throughput. Figure 46.2(b) demonstrates the throughput comparison, highlighting data transmission efficiency.

3. **Energy Efficiency Analysis:** Energy efficiency is the measures of how much less energy is consumed while the transfer of data. RLD2D outperforms AODV and OLSRv2, achieving up to 29% and 23% lesser energy consumption respectivelyunder congested conditions. AODV and OLSRv2 uses predefined rules whereas outcome of RLD2D changes dynamically based

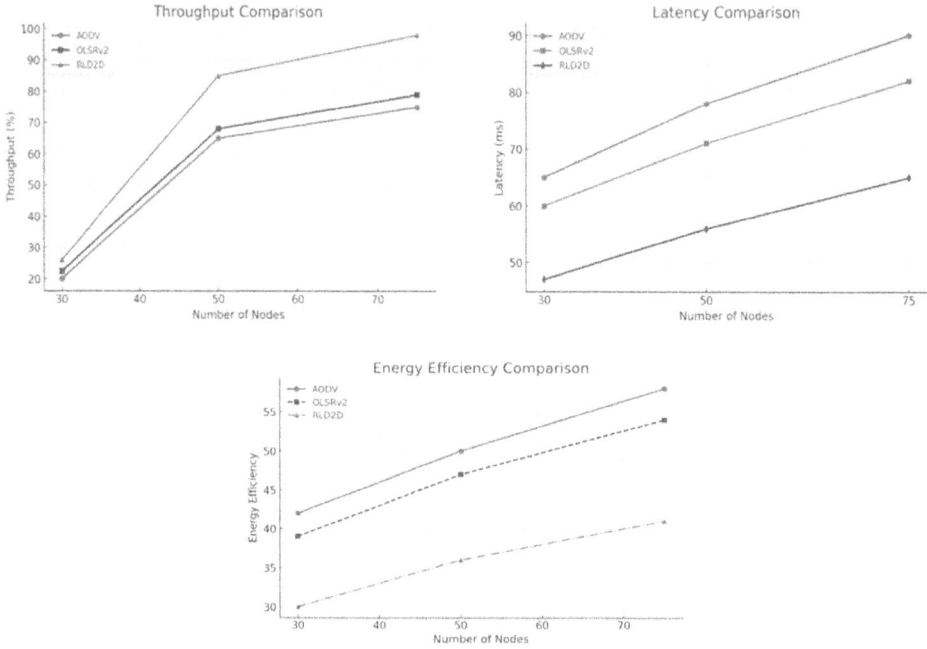

Fig. 46.2 Performance comparison of RLD2D, OLSRv2, and AODV

on the reward. So, it provides better energy efficiency. Figure 46.2(c) provides the energy efficiency comparison, highlighting data transmission efficiency.

5. CONCLUSION

D2D communication is a prominent technology of 5G which provides the interconnection of various network devices without the requirement of any centralized control. In the presented paper, we have integrated RL (a machine learning technology) along with 5G D2D communication for improving the QoS and overall efficiency in proximity-based networks. Our suggested algorithm uses Q-learning, a reinforcement learning method, to maximize data transfer in device-to-device (D2D) communications in 5G networks. The proposed RLD2D effectively improves the throughput, latency and energy efficiency of 5G D2D proximity-based network as compared to OLSRv2 and AODV by dynamically adapting to dynamic network conditions and improving the throughput by 30% as compared to AODV and 26% as compared to OSLRv2. RLD2D is 29% more energy efficient as compared to AODV and 23% more energy efficient as compared to OLSRv2. Our proposed algorithm is 28% more faster as compared to AODV and 20% more faster as compared to OLSRv2. In nutshell, reinforcement learning provides an effective, efficient and scalable framework for proximity 5G D2D networks, paving the way for next-generation wireless communication systems.

References

1. Rathod, T., & Tanwar, S. (2024). AI-based resource allocation techniques in D2D communication: Open issues and future directions. *Physical Communication*, *66*, 102423.

2. Khuntia, P., & Hazra, R. (2020). An Efficient Reinforcement Learning for Device-to-device Communication Underlaying Cellular Network. *IEIE Transactions on Smart Processing & Computing*, *9*(1), 75–84.

3. Xu, Y. H., Sun, Q. M., Zhou, W., & Yu, G. (2022). Resource allocation for UAV-aided energy harvesting-powered D2D communications: A reinforcement learning-based scheme. *Ad Hoc Networks*, *136*, 102973.

4. Obour Agyekum, K. O. B., Boakye, A. Y., Appati, B., Opoku, J. A., Agyemang, J. O., Boateng, G. O., &Gadze, J. D. (2024). Resource Allocation in D2D-Enabled 5G Networks Using Multiagent Reinforcement Learning. *Journal of Computer Networks and Communications*, *2024*(1), 2780845.

5. Teja, P. R., & Mishra, P. K. (2022). Path selection and resource allocation for 5g multi-hop d2d networks. *Computer Communications*, *195*, 292–302.

6. Park, H., & Lim, Y. (2020). Reinforcement learning for energy optimization with 5G communications in vehicular social networks. *Sensors*, *20*(8), 2361.

7. Nauman, A., Jamshed, M. A., Ali, R., Cengiz, K., & Kim, S. W. (2021). Reinforcement learning-enabled intelligent device-to-device (I-D2D) communication in narrowband Internet of Things (NB-IoT). *Computer Communications*, *176*, 13–22.

8. Yu, S., & Lee, J. W. (2022). Deep reinforcement learning based resource allocation for D2D communications underlay cellular networks. *Sensors*, *22*(23), 9459.

9. Hashima, S., ElHalawany, B. M., Hatano, K., Wu, K., & Mohamed, E. M. (2021). Leveraging machine-learning for D2D communications in 5G/beyond 5G networks. *Electronics*, *10*(2), 169.

10. Zia, Kamran & Javed, Nauman & Sial, Muhammad & Ahmed, Sohail & Pirzada, Asad & Pervez, Farrukh. (2018). A Distributed Multi-Agent RL based Autonomous Spectrum Allocation Scheme in D2D Enabled Multi-Tier Hetnets. IEEE Access. 10.1109/ACCESS.2018.2890210

Note: All the figures and table in this chapter were made by the authors.

Emerging Perspectives and Applications of Computational Intelligence and Smart Systems
– Dr. Amit Lathigara et al. (eds)
© 2026 Taylor & Francis Group, London, ISBN 978-1-041-20965-2

47

Deep Neural Network-Based Trust Management System with Enhanced Attack Resistance

Jayashree C. Pasalkar*

Research Scholar, AISSMS College of Engineering, Assistant Professor,
AISSMS IOIT, Pune, India

Dattatraya S. Bormane

Principal, AISSMS College of Engineering, Pune, India

■ **Abstract:** The rise of interconnected devices, coupled with the growth of distributed networks, has led to the proliferation of trust management systems (TMS) that ensure the integrity and reliability of interactions. Trust management is essential in mitigating risks posed by malicious entities and ensuring secure decision-making. However, conventional trust management systems often struggle to maintain performance and robustness when faced with sophisticated attacks. Proposed system leverages the power of deep learning to dynamically adapt and continuously refine trust evaluations based on evolving network conditions. We propose a novel model that not only strengthens the system's security but also outperforms existing techniques in terms of attack resilience, scalability, and accuracy. Our experiments demonstrate the efficacy of this approach in protecting against various attack scenarios.

■ **Keywords:** Attack resistance, Adversarial attacks, Deep neural networks (DNN), Machine learning, Security, Trust management system (TMS), Network security, Trust evaluation

1. INTRODUCTION

The rapid growth of the information technology (IT) industry has indeed been driven significantly by the proliferation of the Internet of Things (IoT). IoT technology has permeated various aspects of daily life, ranging from agriculture and education to water management, home security, and smart grids. This expansion is marked by an exponential increase in connected devices. According to (Najib, W at al. 2019), the number of IoT-connected devices surpassed 50 billion in 2020, and this figure is projected to triple by 2025.

*Corresponding author: jayashree.pasalkar@aissmsioit.org

DOI: 10.1201/9781003725046-47

Key challenges include:

Data Diversity: IoT environments generate vast amounts of diverse data from various sources, including sensors, devices, and systems. Managing and analyzing this diverse data efficiently is crucial for extracting meaningful insights.

Dynamicity: IoT systems are highly dynamic, with devices frequently joining or leaving the network. This dynamic nature requires robust mechanisms for device discovery, connectivity management, and data synchronization (Shafique at al. 2020).

Device Heterogeneity: IoT ecosystems consist of devices with varying capabilities, communication protocols, and standards. Ensuring interoperability and seamless communication among heterogeneous devices is a critical challenge.

An IoT device frequently collects a sensitive data. For security, data from unofficial side access and to confirming agreement with privacy regulations. IoT devices in critical infrastructure, healthcare, or industrial settings can pose physical security risks if compromised. Unauthorized access to control systems or tampering with sensors can have serious consequences. Manipulated or corrupted data can lead to incorrect decisions and operational disruptions. Managing updates, patches, and end-of-life processes is essential to mitigate security risks associated with outdated firmware or software.

1.1 Trust Concept and Trust Related Attacks

In the realm of interpersonal relationships and organizational behavior, trust is often defined as the belief or confidence that one party (the trustor) has in the reliability, integrity, and competence of another party (the trustee). This belief is essential for establishing and maintaining cooperative and interdependent relationships where the trustor relies on the trustee to fulfill obligations, meet expectations, and act in their best interest, even in situations where direct monitoring or control is limited (Asiri, S. at al. 2016)

Table 47.1 Trust-related attacks

Attack Name	Discretion
Good-mouthing Attack	In this attack, node increases its reputation to get select for service provider.
Bad-mouthing Attack	In this attack, well behaved nodes are working in a network to provide service but some malicious nodes are trying to break their reputation.
Ballot-stuffing Attack	In this attack, hostile nodes are increasing to other malicious nodes reputation.
On–off Attack	In this type of attack, an attacker can try to disrupt a trust redemption scheme by acting both positively and negatively; as a result, trust is always redeemed right before another attack takes place.
Discriminatory Attack	An attack in which malicious nodes, based on human inclinations to strangers, discriminatorily attack other nodes against a clear social relationship.

2. Methodology

Main motive of research is proposed methodology will work in three phases for different types of datasets. Phase-I will check the attack detection on CICIOT dataset and will apply hybrid module. Phase-II will detect attack on CAN dataset, which used for image recognition. It is designed for image classification task. Phase-III will detect an attack and also resist the attack on CAN dataset. It will provide clean and adversarial data.

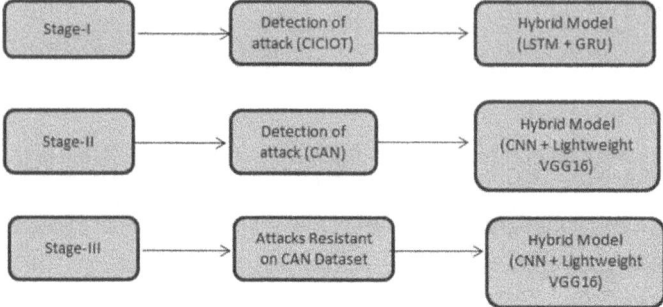

Fig. 47.1 Proposed methodology

2.1 Preprocessing Data

Nodes provide information to the system regarding their interactions with other nodes. The context of the transaction, the kind of interaction (good, neutral, or negative), and any available feedback are all included in the feature vector that represents.

2.2 Extraction of Features

For the neural network to be able to concentrate on pertinent patterns, feature extraction is essential. The feature extraction module finds important markers of reliability, including response times, communication patterns, feedback consistency, and reputation scores.

2.3 Trust Evaluation with Deep Neural Networks

A multi-layer neural network that has been trained to translate feature vectors into trust scores forms the basis of our approach. Supervised learning is used to train the network, and labelled data is used to represent trusted and untrusted entities.

2.4 Attack Detection and Mitigation

In addition to trust evaluation, the system includes an attack detection module that continuously monitors the network for signs of malicious activity. The module uses the trained neural network to identify deviations from normal behavior that might indicate the presence of an attack. Once an attack is detected, the system can either penalize the involved entities or trigger a mitigation strategy, such as isolating the compromised nodes from the network.

3. EXPERIMENTAL RESULTS

Performance of Hybrid Model for attack detection and resistant model using a simulation environment that mimics real-world network scenarios. The dataset includes both synthetic and real-world trust data collected from peer-to-peer networks. We simulate various attack scenarios, including assessing the robustness of our model.

Figure 47.3 shows Result comparison of the performance of three algorithms—LSTM, GRU, and a hybrid LSTM-GRU—across four metrics: Accuracy, Precision, Recall, and F1-Score. Here's what the chart highlights:

- **Accuracy:** The hybrid LSTM-GRU seems to achieve the highest accuracy, closely followed by GRU and LSTM.

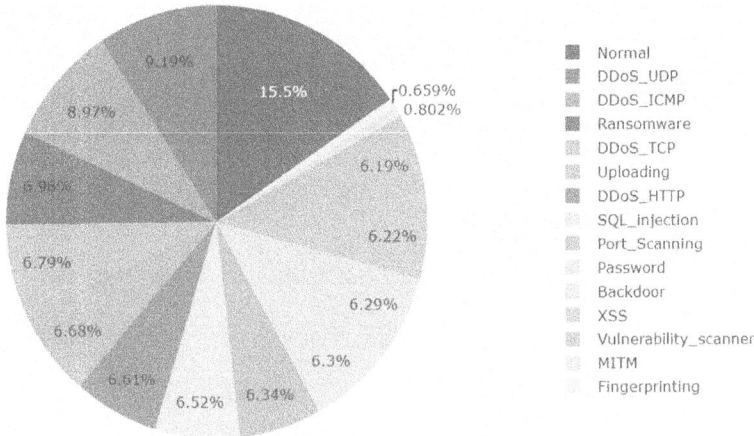

Fig. 47.2 Types of attack

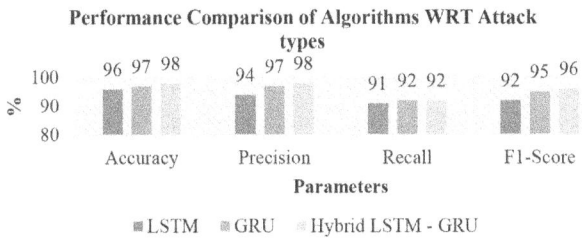

Fig. 47.3 Performance comparison of algorithms wrt various types of attacks

- **Precision:** Hybrid LSTM-GRU outperforms both LSTM and GRU.
- **Recall:** GRU and hybrid LSTM-GRU achieve similar results, both outperforming LSTM.
- **F1-Score:** Hybrid LSTM-GRU scores the highest, with GRU in second place and LSTM in third

Result shows that the Hybrid model significantly outperforms traditional trust management systems in terms of both attack detection and resistance. The system shows high accuracy, with an attack detection rate of over 98%, while maintaining a low false positive rate. Additionally, the deep learning-based method allows the system to dynamically adapt to new attack patterns by providing long-term security and resilience.

4. CONCLUSION

A Proposed deep neural network-based trust management system which enhances attack resistance and improves the security and reliability of networked interactions. Proposed model influences the power of deep learning to automatically detect and moderate various attack strategies, making it highly effective in dynamic and adversarial environments. The experimental results demonstrate the potential of this approach to improve trust evaluation in a variety of real-world scenarios. In the future, we plan to explore further optimizations and real-world implementations to refine the system's capabilities.

References

1. Shafique, K.; Khawaja, B.A.; Sabir, F.; Qazi, S.; Mustaqim, (2020). Internet of things (IoT) for next-generation smart systems: A review of current challenges, future trends and prospects for emerging 5G-IoT scenarios. IEEE Access, 8, 23022–23040.

2. Bera, 80Wicked & Insightful IoT Statistics, in Safeatlast. (2020). Available online: https://www.statista.com/statistics/1183457/ iot-connected-devices-worldwide/ (accessed on 27 February 2023).

3. Jayasinghe, U.; Lee, G.M.; Um, T.W.; Shi, Q. (2018) Machine learning based trust computational model for IoT services. IEEE Trans. Sustain. Comput. Pp. 39–52.

4. Najib, W.; Sulistyo, S. (**2019**) Survey on trust calculation methods in Internet of Things. Procedia Comput. Sci. 161, 1300–1307.

5. Yan, Z.; Zhang, P.; Vasilakos, A.V. (2014) A survey on trust management for Internet of Things. J. Netw. Comput. Appl., 42, 120–134.

6. Djedjig, N.; Tandjaoui, D.; Romdhani, I.; Medjek, F. Trust management in the internet of things. In Security and Privacy in Smart Sensor Networks; IGI Global: Hershey, PA, USA, 2018; pp. 122–146.

7. Chen, R.; Guo, J.; Bao, F. (2014) Trust management for SOA-based IoT and its application to service composition. IEEE Trans. Serv. Comput., 9, 482–495.

8. Mendoza, C.V.; Kleinschmidt, J.H. (2015) Mitigating on-off attacks in the internet of things using a distributed trust management scheme. Int. J. Distrib. Sens. Netw., 11, 859731.

9. Khalil, A.; Mbarek, N.; Togni, O. (2015) Fuzzy Logic based security trust evaluation for IoT environments. In Proceedings of the IEEE/ACS 16th International Conference on Computer Systems and Applications (AICCSA), Abu Dhabi, United Arab Emirates, 3–7 November 2019; IEEE: New York, NY, USA, 2019; pp. 1–8.

10. Asiri, S.; Miri, A. (2016) An IoT trust and reputation model based on recommender systems. In Proceedings of the 14th Annual Conference on Privacy, Security and Trust. (PST), Auckland, New Zealand, IEEE: New York, NY, USA, 2016; pp. 561–568.

Note: All the figures and table in this chapter were made by the authors.

Emerging Perspectives and Applications of Computational Intelligence and Smart Systems
– Dr. Amit Lathigara et al. (eds)
© 2026 Taylor & Francis Group, London, ISBN 978-1-041-20965-2

48

Revolutionizing Cricket Team Selection: AI and ML-Driven Player Performance Prediction—A Comprehensive Review

Ravindra Dangar*
Computer Engineering, RK University, Rajkot,
Gujarat, India

Amit Lathigara
School of Engineering, RK University, Rajkot,
Gujarat, India

Yasin Ortakci
Computer Engineering, Karabuk University,
Turkiye

■ **Abstract:** The Cricket team selection plays a critical role in determining the outcome of matches. Traditional selection methods often rely on human intuition and past performances, which may overlook crucial data-driven insights. The rise of Artificial Intelligence (AI) and Machine Learning (ML) has transformed sports analytics, enhancing the accuracy of player performance predictions. This review paper explores various AI and ML approaches used for player performance prediction in cricket, covering models, data sources, challenges, and future directions. The paper provides a comparative analysis of recent research studies and highlights the potential of AI to optimize cricket team selection.

■ **Keywords:** Data driven insights, Human intuition, Optimize cricket team selection

1. INTRODUCTION

Cricket is a game of strategy, skill, and teamwork, played and loved by millions worldwide. The selection of the right players for a match can significantly impact a team's overall performance and chances of winning^3. Traditionally, team selection decisions have been made by coaches and selectors based on their experience, intuition, and observations. While this method has its merits, it often leads to subjective biases, limited perspectives, and the inability to process vast amounts of available data^6

---

*Corresponding author: ravindra.dangar@rku.ac.in

DOI: 10.1201/9781003725046-48

The advent of Artificial Intelligence (AI) and Machine Learning (ML) has introduced a new dimension to sports analytics. These technologies enable the analysis of extensive datasets, uncovering patterns and insights that are not immediately apparent to human observers. By leveraging these advanced techniques, selectors can make more informed, objective, and data-driven decisions, enhancing the likelihood of forming a balanced and effective team[8].

Player performance prediction, a key application of AI and ML in cricket, involves using historical and real-time data to forecast how players are likely to perform in future matches. This includes metrics such as batting and bowling averages, strike rates, fitness levels, and even psychological factors[9]. AI-driven models can process this information at a speed and scale unmatched by traditional methods, offering a competitive edge to teams willing to adopt such innovations[3].

This paper delves into the state-of-the-art AI-ML methodologies applied to cricket team selection, emphasizing their transformative potential. It explores the challenges of integrating these technologies, evaluates recent research studies, and proposes strategies for future advancements in the field[1].

2. Literature Review

This section reviews key research on AI and ML approaches for predicting player performance in cricket, summarizing methods, findings, and limitations.

(Bananki Jayanth at al. 2018) used linear regression for batting averages and strike rates, noting its simplicity but limited ability to capture non-linear patterns. (Dey, Pabitra Kumar at al. 2024) applied LSTMs for time-series prediction, effectively modelling player form but requiring extensive computational resources[7].

(Singh, Shubham at al. 2022) combined LSTM and CNN for all-round performance prediction, achieving high accuracy but demanding significant computing power[4].

Verma et has introduced an adaptability model for format-specific performance analysis but struggled with generalization. (Singh and Desai at al. 2022) applied predictive analytics to T20 cricket using Random Forest and Gradient Boosting, emphasizing the importance of feature selection.

3. Background Study

3.1 Key AI/ML Techniques for Player Performance Prediction

Regression Analysis: Linear and logistic regression models have been foundational in predicting numerical and categorical performance metrics.

Support Vector Machines (SVMs): Effective for classification tasks, such as determining whether a player will perform above or below a threshold in a game.

Decision Trees and Random Forests: These methods are popular for their interpretability and ability to handle heterogeneous data.

Neural Networks: Multilayer perceptions (MLPs), CNNs, and RNNs have been employed for tasks ranging from skill evaluation to injury prediction.

Reinforcement Learning (RL): RL has been used to model player decision-making and simulate game scenarios for strategic planning.

Ensemble Methods: Techniques like Gradient Boosting Machines (GBM) and XGBoost combine multiple models to improve predictive accuracy.

Graph Neural Networks (GNNs): Recently, GNNs have been explored for modeling interactions between players in team sports[3].

3.2 Data Sources and Features for Player Performance Prediction

Player Tracking Data: GPS and wearable sensors provide detailed movement data.

Video Analytics: Video feeds analyzed using computer vision techniques.

Match Statistics: Data from official league websites, including scores, passes, and possession.

Physiological Data: Metrics such as heart rate, oxygen consumption, and recovery rates.

Environmental Factors: Weather conditions, venue, and crowd influence.

Historical Performance: Past game performance data.

Sentiment Analysis: Social media and news sentiment about players and teams.

These diverse data sources and features enable the development of comprehensive AI/ML models that can predict player performance with high accuracy, paving the way for more informed decision-making in cricket team selection[9].

3.3 Features for Predictive Models

Data Preprocessing

Collect historical player statistics such as batting averages, bowling economy, strike rates, and fitness data.

Perform feature engineering to normalize and transform data for AI model consumption[2]

Model Architecture

LSTM Component:

Input: Time-series player performance data.

Layers: Multiple LSTM layers to capture sequential dependencies.

Output: Feature vectors representing past performance trends.

CNN Component:

Input: Structured numerical player statistics.

Layers: 1D Convolution layers to extract high-level performance patterns.

Output: Feature maps representing key insights.

Model Training and Optimization

Use Adam optimizer and categorical cross-entropy loss function.

Implement dropout layers to prevent overfitting.

Train model on GPU-enabled environment for faster convergence.

Deployment & Real-time Predictions

Integrate with cloud-based APIs for real-time data streaming.

Deploy using TensorFlow Serving or Flask API for live match predictions.

4. COMPARATIVE ANALYSIS

Table 48.1   Summary of literature survey

Study	Technique Used	Key Findings	Limitations
(Kumar & Sharma at al. 2022)	Ensemble Learning	Reduced overfitting, improved robustness	Computationally expensive
(Ahmed & Chowdhury at al. 2023)	SVM	Accurate real-time player classification	Limited to predefined categories
(Lee et al. 2023)	Hybrid LSTM + CNN	Improved accuracy for all-rounder players	High computational resources
(Verma et al. 2024)	Adaptability Model	Assessed player adaptability to conditions	Struggled to generalize across scenarios
(Chopra & Mehta at al. 2023)	Deep Learning	Effective handling of complex datasets	Need for large training data
(Reddy et al. 2024)	Real-Time Analytics	Integrated real-time data streams	Advanced infrastructure required

5. CONCLUSION

The application of AI and ML in cricket team selection has revolutionized traditional decision-making by leveraging data-driven insights. These technologies enable objective analysis of player performance, reducing biases and improving prediction accuracy. Various models, including regression, classification, and neural networks, enhance the selection process by analysing complex datasets. Despite challenges like data availability, model interpretability, and ethical concerns, AI-driven approaches are continually advancing. Future research should prioritize hybrid models, real-time analytics, and Explainable AI (XAI) to enhance transparency and trust. AI-powered scouting and tactical decision-making can further refine team strategies. Embracing AI in cricket will lead to more informed, strategic, and successful team selections, ultimately enhancing the sport's competitiveness.

References

1. Bananki Jayanth, Sandesh, Rajashekar R., and Smitha M. L. (2018) "A Team Recommendation System and Outcome Prediction for the Game of Cricket" *Journal of Sports Analytics* 4, no. 2: 115–125
2. Dey, Pabitra Kumar, Abhijit Banerjee, and Dipendra Nath Ghosh (2024) "Optimising IPL Squad Composition: A Mathematical Framework for Efficient Team Selection on a Limited Budget in a Multi-Criteria, Multi-Objective Environment." *International Journal of Operational Research* 47, no. 2: 163–187.
3. Jain, Rakesh, Lakhwinder Kaur, and Arun Sharma (2013) "Multi-Objective Optimization and Decision-Making Approaches to Cricket Team Selection." In *Proceedings of the International Conference on Soft Computing for Problem Solving*: 667–678.
4. Krishnamohan, T. "Maximizing the Runs Scored by a Team in Cricket Using Genetic Algorithm." *International Journal of Emerging Computing and Engineering Research* 10, no. 1 (2023): 51–55.
5. Kumarasiri, Isuru, and Sanjeewa Perera (2017) "Optimal One Day International Cricket Team Selection by Genetic Algorithm" *International Journal of Sciences: Basic and Applied Research* 34, no. 3: 246–259.

6. Manage, Ananda B. W., Amal S. Punchihewa, and Nalin Wickramarachchi (2020) "Classification of All-Rounders in Limited Over Cricket—A Machine Learning Approach." *Journal of Sports Analytics* 6, no. 1: 1–12.
7. Patil, Nilesh M., Rahul K. Kshirsagar, and Prashant Patil. "Cricket Team Prediction Using Machine Learning Techniques." *International Journal of Advanced Science and Technology* 29, no. 4 (2020): 11991–12001.
8. Robel, Md., M. Shahriar, S. Islam, and M. A. Haque (2018) "Squad Selection for Cricket Team Using Machine Learning Algorithms" Undergraduate thesis, BRAC University
9. Singh, Shubham, Sachin Goyal, and Shruti Kohli (2022): "An Efficient Team Prediction for One Day International Matches Using a Hybrid Approach of CS-PSO and Machine Learning Algorithms." *Intelligent Systems with Applications* 13 200015
10. Tirtho, Devopriya, Pranab Maity, Tanmoy Mondal, and Chanchal Mandal (2022): "Cricketer's Tournament-Wise Performance Prediction and Squad Selection Using Machine Learning and Multi-Objective Optimization." *Applied Soft Computing* 129 109548.

Emerging Perspectives and Applications of Computational Intelligence and Smart Systems
– Dr. Amit Lathigara et al. (eds)
© 2026 Taylor & Francis Group, London, ISBN 978-1-041-20965-2

49

Using Big Data Analytics in Wireless Cellular Networks: A Transformational Impact on the Telecommunications Industry

Hemant Kumar Gianey

SVKM's NMIMS, Mukesh Patel,
School of Technology Management and Engineering,
Shirpur, India

Kishan*, Sarvesh vyas

Department of Civil Engineering, IES University Bhopal

Aayush Shrivastava

Department of Computer Science, Engineering, MMEC,
Maharishi Markandeshwar University,
Mullana Ambala, Haryana, India

Manmohan Singh

Department of Computer Science, Engineering,
IES College of Technology,
Bhopal, India

Amit Lathigara

School of Engineering, RK University,
Rajkot, Gujarat, India

■ **Abstract:** Big data analytics (BDA) has become a revolutionary technology in the telecommunication domain, especially for wireless cellular networks. However, with the ever-growing data from mobile handsets, network logs, and customer interactions, telecom companies are rushing towards the use cases of BDA to improve network performance, boost customer experience and upgrade operational efficiency. This study reviews the role of big data analytics in wireless cellular networks with its applications, advantages, and challenges. This illustrates the application of BDA in telecom operators for network congestion prediction, fraud detection, and personalization of customer services through real-time analytics and machine learning models. The study further highlights the strong ethical implications of the usage of big data and the particular need for any mirrors of big data's usage to take user privacy and data security into account. Through a review of

*Corresponding author: Kishanchoure777@gmail.com

DOI: 10.1201/9781003725046-49

real-life case studies, this paper demonstrates the ways in which telecommunications companies can harness the potential value of BDA whilst meeting challenges exposed by BDA. This emphasizes the need for ongoing research and innovation to promote the responsible and sustainable use of big data analytics in the telecom domain.

■ **Keywords:** Big data analytics, Telecommunication, Wireless cellular network, Telecom industry

1. INTRODUCTION

Big data analytics drives the telecoms boom. Big data analysis can improve operation efficiency, resource allocation, and customer satisfaction due to the rapid growth of mobile communication networks, 5G technology, and user-generated data (A. Uzoka at al, 2024). Big data analytics helps telecom firms handle, analyze, and gain actionable insights from their massive structured and unstructured data. Wireless cellular networks generate massive statistics per second, including call records, data usage patterns, network performance logs, and user feedback. These advanced analytics methods can improve telecom operator service quality, problem detection, and customer experience by using predictive modeling, machine learning, and artificial intelligence (Aburub, F. at al. 2024) Big data analytics have also changed telecom network traffic management, fraud prevention, and pricing. Operators can identify congestion, predict demand, and distribute resources to sustain connectivity. Their quick data processing and analysis reduces operational expenses and boosts network reliability (Slimani, at al. 2024)(N. A. Ochubaat al 2024). Telecommunications companies provide data services to customers. The purpose is to present a deep-dive case study that shows how prevalent these technologies are and their possible hazards and industry futures. Big data technologies have ethical concerns for user privacy, data security, and regulatory compliance, which this research examines. Telecom companies can create a strong data governance framework to responsibly and sustainably implement big data analytics after overcoming these challenges) (N. A. Ochubaat al 2024)

1.1 Background

In one definition, "big data" refers to information assets that are high in volume, velocity, and diversity and require innovative and cost-effective information processing to increase insight and decision making. Another definition of "big data" is "high diversity of information assets." The following four attributes match our SLR: Current huge data sizes range from terabytes to petabytes; Vulnerability: Information development in real-time or group settings. The Value of Big Data: This data comes from directed and alternative data sources, often structured, unstructured, and semi-structured. Hidden data will become the gold mine for business insights. Big data analytics (BDA) extracts useful information from massive volumes of data. This is big data analytics (Vangala, A., at al. 2024)

Hadoop has a master-slave design for data management (Vangala, at al. 2024) (Fig. 49.1). The master node is Name, while the slave node is Data..

They also do replication, deletion, and block generation. Name nodes start the Task Tracker process, which receives user task submissions, divides the work, and assigns it to Data Node Task Tracker daemon processes to handle MapReduce workloads. Hadoop Ecosystem can provide databases

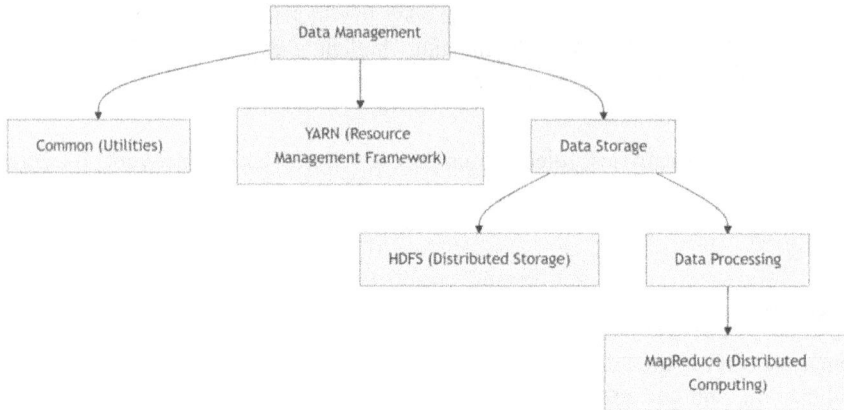

Fig. 49.1 Hadoop V2x Architecture

(Hbase), data warehouses (Hive), SQL Querying (Hive and Drill), stream processing (Spark, Storm, Flink), machine learning (Mahout, H2O), MapReduce programming (Pig), and cluster coordination. Hadoop Ecosystem is a collection of open-source Apache projects using various APIs. Specific to our SLR's ecology:

ApacheHbase

Wide columnar storage below allows real-time read and write operations on enormous data volumes. HBase assigns each row a timestamp-indexed sorted key for column storage. Creating column families and super column families requires grouping columns. These column families form the foundation of access control. Timestamps, 64-bit integers, store unique cell information in Hbase. End-users can control the number of saved cell editions(Badawy, M., Ramadan, N., & Hefny, H. A. 2024)

ApacheHive

In order to facilitate the processing of massive data sets with Hadoop, Hive offers a relational paradigm and a SQL database interface. Hadoop serves as the foundation for Hive, which is an application framework for data warehousing that offers summarization, querying, and analytical capabilities (Vij, A., & Goyal, A. 2025)

ApachePig

By expressing complex tasks as data flow sequences, Pig Latin, which is an ETL-level language, simplifies the process of textual programming. It also enables parallel execution and optimization of operations that include a number of data transformations that are dependent on one another. Moreover, it gives users the ability to encode their own functions, which can be provided during the execution of the program (Vivekanandam, M., & Karasala, K.2024)

ApacheSpark

The execution engine known as Spark is responsible for interpreting data streams into a collection of deterministic batch processing algorithms (Sharma, G. 2024) As a result, it is one hundred times quicker than the typical MapReduce. The master/slave architectural model serves as the foundation for the Spark architecture. The master instance is able to read data from HDFS and launch a set of cluster workers when it is functioning with a driver application that is user-defined. Spark makes use of this concept of fault tolerance in order to process queries. This is because RDDs are partitioned across different machines. In order to create partitions in RAM for RDDS, slaves are responsible for doing so in accordance with the specifications included within a driver software. When it comes to stream data processing, the Spark API is known as Spark Streaming .

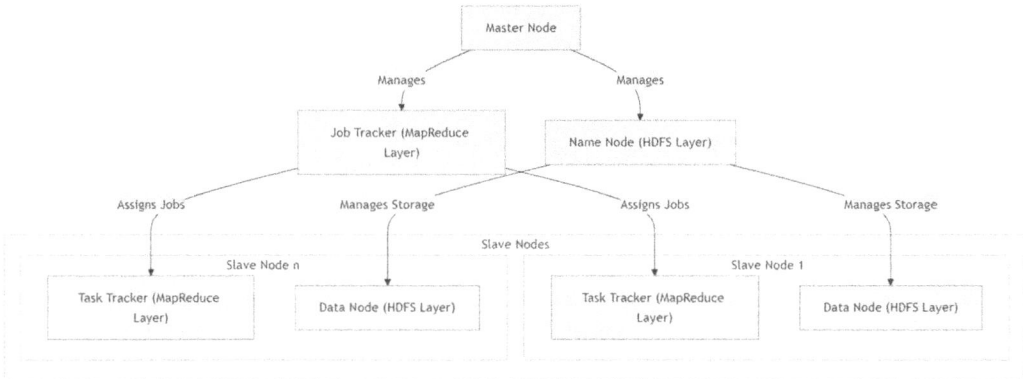

Fig. 49.2 Master slave architecture of hadoop

Fig. 49.3 Big data analytics is used in wireless networks

ApacheKafka

It is possible to queue real-time data streams using Kafka, which is an application programming interface (API) for ingestion (Badawy, M., Ramadan, N., & Hefny, H. A. 2024). The topic component is a part of each queue that is user defined. This component is known as the topic component. According to the topic, which event is placed in which queue is determined. For the purpose of making it simpler for the servers that make up the message broker component to consume events as they are received in a random order, events are enqueued. You have the option of using Apache Spark, Apache Flink, or Apache Storm on the server sides (Lamaazi, H., & Barka, E. 2025)

◢ 2. RESEARCHER VIEW

Many conversations have focused on how big data analytics is changing the telecommunications business. Many research have examined how big data might improve user experiences, resource allocation, and network performance. This study shows the importance of big data analytics in telecoms strategic decision-making and operational efficiency (Vangala at al, 2024). Further research examined predictive capacity models' ability to predict network needs. These models assist telecoms companies plan network growth and manage capacity efficiently to provide high-quality services (Saha, at al. 2024).

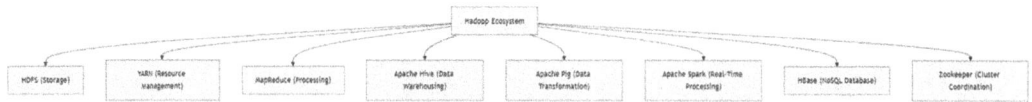

Fig. 49.4 Apache hadoop components work together

Data Evolution - Making Big Data Analytics Required for Network Expansion

Big data analytics in telecoms can help with network construction. This helps companies prepare for capacity and expansion. A good user experience requires a reliable network capacity management solution. Data-intensive applications are in demand as wireless cellular networks that can handle streaming video, Internet of Things, and more grow. To meet this need, telecommunications companies are adopting big data analytics to discover data usage patterns.

◢ 3. PREDICTIVE CAPACITY MODELS FOR FUTURE DEMAND

Big data analytics lets telecoms construct predicted capacity models. The models help companies estimate demand and position themselves. The correct data analytics methods can improve telecoms services, enhancing customer happiness and minimizing turnover (Singh M, et al 2022). Telecom businesses can estimate demand using predictive analytics since they collect massive volumes of network configuration and utilization data..

3.1 Optimizing Service Quality and Customer Experience

Big data analytics increases network growth, capacity planning, and telecom quality. Network optimization and smooth consumer complaint and issue collection from specific locations can do this. Customer happiness and sales rise with better service (Slimani, at al 2024).

3.2 Insights into Consumer Behaviour and Privacy Concerns

Telecommunications companies use big data and analytics to understand client behavior. With this insight, you can predict industry trends and provide personalized services. Thus, addressing data privacy and security issues is crucial. Operators may be able to monetize user data if privacy regulations are strict (N. A. Ochuba 2024). Telecommunications companies use big data analytics to acquire insights from customer data, which is valuable. Monitoring consumer preferences and activities allows these companies to create individualized marketing campaigns and services..

3.3 The Advantages of BDA in the Field of Telecommunications

These are the categories that best describe the benefits that BDA applications for telecom have to offer:

Value of BDA

The benefits that BDA gives to the telecom business include a collection of infrastructure, programming models, Tuludust and high-performance schema-free databases, and process analysis. These are only some of the advantages that BDA offers. BDA provides a framework and principles that are unique to the communications industry.

Cost Reduction and Revenue Generation

The use of BDA has the potential to lower the costs of a variety of communication network procedures. Stream processing technologies for business data analytics (BDA) assist manage complex events that require real-time attention, thereby lowering risks and expenses while simultaneously improving decision-making and income.

Enhanced Customer Care Services

Future Research and Ethical Considerations

Big data analytics has greatly benefited the telecoms business. However, concerns and limits remain. This requires more study to address the need for larger data sets, more accurate forecasts, bias resolution, and privacy concerns. Future research should focus on building solutions that alleviate these challenges and take advantage of big data analytics in telecoms. Privacy and security will be crucial to this progress and future evolution.

3. CONCLUSION

Big data analytics (BDA) can greatly benefit the telecoms industry. The absence of architecture and implementations of best-in-class solutions in growing technology stack limit BDA academic research in telecom. Because the tech stack evolves. Big data analytics is transforming wireless cellular networks. This change will be noticed later. This study examines how such technology improves decision-making, network performance, and consumer happiness. While accepting the challenges and the need to continue investigating governance concerns related to big data analytics on a sustainable and ethical level? Big data analytics will shape the region's future.Big data analytics (BDA) can greatly benefit the telecoms industry. Lack of architecture and implementation of best-in-class solutions in expanding technology stack limit BDA academic research in telecom. Because the tech stack evolves. Big data analytics is transforming wireless cellular networks.

References

1. A. Uzoka, A., Cadet, E., & Ojukwu, P. U. (2024). The role of telecommunications in enabling Internet of Things (IoT) connectivityand applications. *Comprehensive Research and Reviews in Science and Technology*, 2(02), 055–073.
2. Aburub, F. A. F., Hamzeh, R. F., Alzyoud, M., Alajarmeh, N. S., Al-shanableh, N., Al-Majali, R. T., ... & Aldaihani, F. M. F. (2024).The impact of big data analytics capabilities on decision making at the telecommunications sector in Jordan. In *Business Analytical Capabilities and Artificial Intelligence-Enabled Analytics: Applications and Challenges in the Digital Era, Volume 1* (pp. 339–354). Cham: Springer Nature Switzerland.
3. Slimani, K., Khoulji, S., Mortreau, A., & Kerkeb, M. L. (2024). Original Research Article From tradition to innovation: The telecommunications metamorphosis with AI and advanced technologies. *Journal of Autonomous Intelligence*, 7(1).
4. N. A. Ochuba, D. O. Olutimehin, O. G. Odunaiya, and O. T. Soyombo, (2024) "Corresponding author: Nneka Adaobi Ochuba A comprehensive review of strategic management practices in satellite telecommunications, highlighting the role of data analytics in driving operational efficiency and competitive advantage," *World Journal of Advanced Engineering Technology and Sciences*, vol., no. 02, pp. 201–211, 2024, doi: 10.30574/wjaets.2024.11.2.0099.
5. Vangala, A., Agrawal, S., Das, A. K., Pal, S., Kumar, N., Lorenz, P., & Park, Y. (2024). Big data-enabled authentication framework for offshore maritime communication using drones. *IEEE Transactions on Vehicular Technology*.
6. Saha, S., Das, A. K., Wazid, M., Park, Y., Garg, S., & Alrashoud, M. (2024). Smart contract-based access controscheme for blockchain assisted 6G-enabled IoT-based big data driven healthcare cyber physical systems. *IEEE Transactions on Consumer Electronics*.
7. Cui, Y., Cao, X., Zhu, G., Nie, J., & Xu, J. (2025). Edge perception: Intelligent wireless sensing at network edge. *IEEE Communications Magazine*, 63(3), 166–173.
8. Sharma, G. (2024, October). SecureV2X: Overview of Secure Cellular based Vehicle-to-Everything Communication in Intelligent Transportation System. In *2024 12th International Conference on Internet of Everything, Microwave, Embedded, Communication and Networks (IEMECON)* (pp. 1–6). IEEE.
9. Badawy, M., Ramadan, N., & Hefny, H. A. (2024). Big data analytics in healthcare: data sources, tools, challenges, and opportunities. *Journal of Electrical Systems and Information Technology*, 11(1), 63.
10. A. Vij and A. Goyal, "Enhancing Decision-Making in IoT Ecosystems with Big Data Analytics and Hadoop Frameworks," Cuestiones de Fisioterapia, vol. 54, no. 2, pp. 1334–1350, Jan. 2025, doi: 10.48047/CU.
11. Vivekanandam, M., &Karasala, K. (2024, December). Improving Space and Time Efficiency in Hadoop Architecture Using Machine Learning Algorithms. In *2024 International Conference on IoT Based Control Networks and Intelligent Systems (ICICNIS)* (pp. 1696–1702). IEEE.
12. Sharma, H., & Joshi, A. (2024, May). National Telecom Volunteer: Utilizing A Machine Learning Model To Predict Cellphone Network Coverage Using Big Data Analysis From Data Collected Through Crowdsourcing. In *2024 International Conference on Computational Intelligence and Computing Applications (ICCICA)* (Vol. 1, pp. 8–17). IEEE.
13. Lamaazi, H., & Barka, E. (2025). Networks and implementation tools for IoT and big data. In *Empowering IoT with Big Data Analytics* (pp. 213–234). Academic Press.
14. Singh M , Mewada H , Tahilyani M, Malviya J , Sharma R and Shrivastava S S (2022). RRDTool: A Round Robin Database for Network Monitoring Journal of Computer Science 18(8), 770–776. https://doi.org/10.3844/jcssp.2022.770.776
15. Singh, M., Patidar, V., Ayyub, S., Soni, A., Vyas, M., Sharma, D. & Ranadive , A. (2023). An Analytical Survey of Difficulty Faced in an Online Lecture During COVID-19 Pandemic Using CRISP-DM. Journal of Computer Science, 19(2), 242–250. https://doi.org/10.3844/jcssp.2023.242.250

16. R, AlSuwaidan L, Khan S , Rauf Baig A, Baseer S, and Singh M ,(2022). Fault Tolerance Byzantine Algorithm for Lower Overhead Block chain. Security and communication network Volume 2022 Article ID 1855238 | https://doi.org/10.1155/2022/1855238.

17. Singh, M., Tiwari, S. K., Swapna, G., Verma, K., Prasad, V., Patidar, V., Sharma, D. & Mewada, H. (2023). A Drug-Target Interaction Prediction Basedc on Supervised Probabilistic Classification. Journal of Computer Science, 19(10), 1203–1211. https://doi.org/10.3844/jcssp.2023.1203.1211

18. Singh, M., Tiwari, S. K., Swapna, G., Verma, K., Prasad, V., Patidar, V., Sharma, D. & Mewada, H. (2023). A Drug-Target Interaction Prediction Based on Supervised Probabilistic Classification. Journal of Computer Science, 19(10), 1203–1211. https://doi.org/10.3844/jcssp.2023.1203.1211

19. Singh, M., Ayuub, S., Baronia, A. et al. Analysis and Implementation of Disease Detection in Leafs and Fruit Using Image Processing and Machine Learning. SN COMPUT. SCI. 4, 627 (2023). https://doi.org/10.1007/s42979-023-02045-z

20. Yadav, A.S. et al. (2023). Effect of Artificial Roughness on Heat Transfer and Friction Factor in a Solar Air Heater: A Review. In: Nayak, R.K., Pradhan, M.K., Mandal, A., Davim, J.P. (eds) Recent Advances in Materials and Manufacturing Technology. ICAMMT 2022. Lecture Notes in Mechanical Engineering. Springer, Singapore. https://doi.org/10.1007/978-981-99-2921-4_33

Note: All the figures in this chapter were made by the authors.

Emerging Perspectives and Applications of Computational Intelligence and Smart Systems
– Dr. Amit Lathigara et al. (eds)
© 2026 Taylor & Francis Group, London, ISBN 978-1-041-20965-2

50

Forest Fire Prediction System using Wireless Sensor Networks and Deep Learning

Garima Priya*

Department of Computer Science & Engineering,
IES College of Technology in
Bhopal, India

**Aayush Shrivastava, Amandeep Kaur,
Bhupesh Gupta, Mohit Chhabra**

Department of Computer Science & Engineering, MMEC,
Maharishi Markandeshwar (Deemed to be) University,
Mullana Ambala, Haryana, India

Jay Fuletra

School of Engineering, RK University, Rajkot,
Gujarat, India

■ **Abstract:** Forests are important for environment. Forest helps in regulating climate, fresh air and provides food and shelters to animals. Forest fire is still a huge problem in many countries due to environmental, social and economic changes. One of the most important reasons for forest fire is global warming. Recently forest fires in California and Los Angeles cause huge loss of environment that includes life risk of living beings. Early warning system for forest fire prediction to protect environment- this been proposed in literature with the utilization of Wireless Sensor networks (WSN) and Artificial intelligence (AI). Forest fire can be predicted at early stage with the help of data gathered from temperature, pressure, wind, fire and smoke sensors. Data collected from different sensors is applied as input to Deep Learning (DL) model for early prediction of wildfires. With this motivation this paper presents a novel technique of forest fire prediction using WSN and DL.

■ **Keywords:** Forest fire, WSN, Deep learning, Artificial intelligence, Early warning system

1. INTRODUCTION

Forest is large area of land densely covered with plants and trees. Forest is important for many reasons as they help in regulating climate, filter air, food and shelters to animals and provides number of

---
*Corresponding author: pgarima510@gmail.com

DOI: 10.1201/9781003725046-50

resources. Despite of these benefits forests are at risk of deforestation, destruction and wildfires. Forest fire is extremely destructive to people, animals, natural resources and properties (Abid, F. 2021) (Kadir E, Abdul M, Othman SL, et al. 2021). Forest firefighter gives their best to stop fires but area of forest is too large that it takes many days to stop fire. Forest fire smoke is risk for living being health and also cause air pollution. Figure 50.1 and 50.2 shows the major destruction happens in California and Los Angels due to forest fire in January 2025.

Fig. 50.1 Forest fires in california in January 2025

Source: News media images of California and Los Angeles forest fires (January 2025)

However, with the invent of smart technologies AI and WSN it is possible to develop early warning system for wildfire (Kadir E, Abdul M, Othman SL, et al. 2021) With different temperature, pressure, wind and speed sensors fire smoke can be detected at early stage to avoid loss. Collected data from all these sensors will be given as input to DL model to predict forest fire so that preventive measure can be taken. WSN is collection of large number of nodes that are deployed in sensing region to sense physical or environmental changes (Gulati, K., Boddu, R. S. K., Kapila, D., Bangare, S. L., Chandnani, N., & Saravanan, G. (2022). A review paper on wireless sensor network techniques in Internet of Things (IoT). *Materials Today: Proceedings, 51*, 161-165.). Sensor Nodes(SNs) transmit sense data to Base Station (BS) or Cluster Head (CH) (Gupta, B., Rana, S., & Sharma, A. (2019, November). An efficient

Fig. 50.2 Forest fires in california in January 2025

Source: News media images of California and Los Angeles forest fires (January 2025)

data aggregation approach for prolonging lifetime of wireless sensor network. In *International Conference on Innovative Computing and Communications: Proceedings of ICICC 2019, Volume 2* (pp. 137-147). Singapore: Springer Singapore) WSN architecture consists of battery- operated nodes, transmitter and radio. Decentralized architecture allows WSNs to work in challenging or unattended environments, making them irreplaceable for applications that includes forest fire detection, military surveillance, habitat monitoring, natural disaster monitoring etc. Main characteristics of WSN are scalability, Robustness and flexibility etc. (Dong, S., Wang, P., & Abbas, K. (2021). A survey on deep learning and its applications. *Computer Science Review, 40*, 100379.) DL is subset of ML that is used to discover insights and hidden pattern from data (Dong, S., Wang, P., & Abbas, K. (2021). A survey on deep learning and its applications. *Computer Science Review, 40*, 100379.). Deep learning capabilities in a variety of applications, from speech and image recognition to natural language processing and beyond. These models automatically extract features from raw data. Large datasets, more processing power, and advance-mints in neural network are all factors in deep learning's success. Its ability to manage unstructured data has transformed industries such as robotics, computer vision, healthcare, and finance. Deep learning continues to be at the forefront of innovation as research and development progresses, enabling more intelligent, effective, and scalable solutions across a wide range of sectors and research disciplines.

It is effective learning technique for building models that can be used for object recognition and pattern in data. Broad applications of DL include object recognition, natural language processing, computer vision, speech recognition etc. Main advantage of DL model is efficient training of large data with higher accuracy. Various DL mod- els are Multi‑ layer Perceptron (MLP), Convolution Neural Networks (CNN), Recurrent Neural Networks (RNN), Generative Adversarial Network (GAN).

Integration of DL with WSNs has become a promising approach to early forest fire detection. WSNs are made up of a large number of spatially dispersed sensor nodes that continually monitor environmental variables including temperature, humidity, and smoke throughout forest. These networks are capable of real-time analysis of massive streams of sensor data when combined with deep learning algorithms.

The main motive of this work is to design DL based model that will predict wildfire from data gathered by SNs to avoid wildfire accidents that cause huge destruction to people, animals and natural resources. The proposed model sense forest fire at early stage and transmit message to concerned authority to alert them for appropriate action.

2. LITERATURE SURVEY

In (Ananthi, J., et. al. (2022 DL based model LBFFPS has been proposed that used smart surveillance device called SFMK. SFMK consists of the sensors, camera and IoT enabled controller. Pro- posed DL model trains on real time forest images captured by SFMK camera. In (Omar N, Al-Zebari et al 2021) authors proposed deep learning LSTM wildfire prediction model from the data collected by metrological stations. Metrological measurements are temperature, humidity, wind and forest fire weather index. Forest fire prediction model for Ayodhya hills lies in East- ern Ghats Mountain Range (India) is proposed in (Saha, S et al, (2023) Proposed model uses Random Forest (RF), Multivariate Adaptive Regression Splines (MARS) and Deep Learning Neural Network (DLNN) techniques on 300 historical forest fire events. In this research (Naderpour, et. al (2021) 36 key indicators of forest fires were used as input to proposed model. Region included for analysis is Northern Beaches Forest area of Sydney. This study (Paidipati, et. al. 2024) developed DL model named FFD Net for forest fire detection. This model identifies the wildfire from images collected by SNs in forest. SNs transmit forest images to Base Station where actual classification takes place. Authors (Haque, A., & Soliman, H. (2025) used virtual sensors and MLP to classify forest fire at early and advanced stage. Virtual sensors used polynomial regression technique to generate data. An improved boundary detection operator was presented by (Jian & Celik, T., et al, M. 2007) and their model made advantage of a multistep operation. Nevertheless, the model's abstraction was limited to straightforward and steady fire and flame pictures. To identify fires, researchers from all over the world have employed an algorithm based on the fast Fourier transform (FFT). Toulouse (Jiang, Q., & Wang, Q. 2010) created a novel technique to identify a fire's geometrical features based on its length, surface, and position. The color of fire was divided into pixels for this study. Additionally, the intensity was used to divide the pixels of the pictures that are not refractory. Dynamic systems (LDS) were used in (Park, M., & Ko, B. C. 2020) to detect fire based on the dynamic textures of smoke and flame. Turgay (Dimitropoulos, K., et. al, N. 2014) created a real-time fire detector by fusing color frames from the foreground and background. However, because of the smoke and shadow, the real-time color-based program does not produce a superior result. Authors (Gupta, B., Rana, et al, 2023) proposed the forest fire detection based on MESA2DA Clustering Protocol.

◢ 3. Proposed Method

Proposed method aims to combine WSN with DL technique for efficient forest fire detection. This method provides a promising solution for real-time wild fire detection by combining the strengths of sensors for data gathering with DL to analyze complex fire images. In our proposed methods different types of images, temperature and pressure SNs are deployed using efficient deployment technique throughout forest to capture images and sense temperature, pressure, sensor and humidity. Sensor network use enhanced infrared sensors and camera to monitor wild fire more precisely. Forest images captured by nodes are transmitted to BS at regular intervals where DL model use these images to identify fire. Different types of DL models CNN, MLP and RNN are used to classify images in two categories: fire and non-fire.

3.1 Convolution Neural Network (CNN)

CNN is deep learning algorithm used for object recognition that includes image classification, detection and segmentation. CNN consists of many layers includes convolution layer, pooling layer, fully connected layers and activation function. CNN is based on supervised machine learning. CNN model is best for forest fire prediction when images are captured from WSN or satellite. Main benefit of using CNN for wild- fire prediction is their ability to automatically extract features from complex images.

3.2 Recurrent Neural Networks (RNN)

RNN is type of ANN that is applied to forest fire detection for time series data that includes sensor readings, weather variables, temperature, humidity etc. Long Short-Term Memory (LSTM) networks, a type of RNNs, are used for capturing long-range dependencies in time-series data.

3.3 Multilayer Perceptron (MLP)

MLP is type of ANN that consists of multiple layers of neurons. MLP uses non-linear activation function that makes network to learn from complex data. MLP is made up of one input layer, output layer and multiple hidden layers. MLP is used for wildfire prediction for static environment. This model performed well to identify fire in high- risk areas when dataset is limited. MLP is fast and requires less resources as compared to CNN and RNN.

In our proposed method we have assumptions that images are captured by sensors deployed in forest. Implementation of proposed approach involves data collection, data preprocessing, model training and testing. We used forest fire images from keggle dataset. Before applying images as input to DL models, images are resizing (224*224), normalized and augmented to improve their quality.

In order to classify image as fire and no-fire, we use CNN models with multiple layers.

In proposed architecture CNN consist of following layers:

Input Layer: 224*224*3

Convolution Layer: Multiple convolution layers with different filters are used to extract fire related features from images.

Max Pooling Layer: This layer is used to reduce dimensionality.

Fully Connected Layers: Pass the final convolutional layer output into fully connected layers.

• Dense layer with 512 neurons and activation function Re LU and learning rate of 0.5

Output Layer: Output layer with sigmoid activation function is used to classify images into fire or non-fire.

4. RESULTS AND DISCUSSION

After training, model is tested on dataset to check following performance metrics.

Accuracy: Percentage of correctly identified images.

Precision: Percentage of true positives out of all positive predictions.

Recall: percentage of true positives out of all actual positives.

F1 Score: Harmonic mean of precision and recall.

ROC-AUC: The area under the Characteristic curve

Once the model is trained and tested, it can be deployed for real-time wild fire detection, where it receives new images from SNs and classifies them as either fire or no-fire. A clear precision-recall analysis of the proposed technique under the test database is shown in Fig. 50.3.

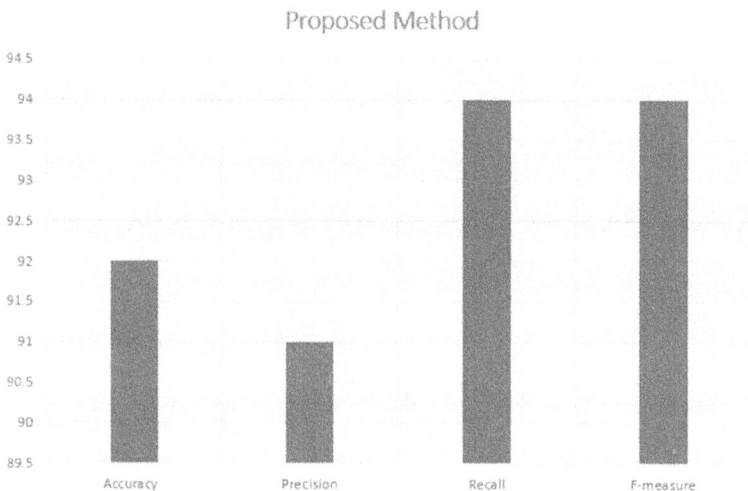

Fig. 50.3 Performance analysis of proposed method

Source: Authors

5. CONCLUSION AND FUTURE WORK

The combination of WSN with DL represents a novel and efficient approach for wildfire detection. By using CNN, RNN and MLP for image classification and employing real time images from WSNs, the proposed approach offers reliable solution for forest fire detection at early stage. Future work will include to combine more enhanced sensors with complex DL model to predict forest fire and improving overall system efficiency.

References

1. Abid, F. (2021). A survey of machine learning algorithms based forest fires prediction and detection systems. *Fire technology*, *57*(2), 559–590.
2. Sathishkumar, V. E., Cho, J., Subramanian, M., & Naren, O. S. (2023). Forest fire and smoke detection using deep learning-based learning without forgetting. *Fire ecology*, *19*(1), 9.
3. Kadir E, Abdul M, Othman SL, et al. 2021.
4. Gulati, Kamal. A review paper on wireless sensor network techniques in Internet of Things (IoT). Materials Today: Proceedings. 2022; 51:161–165.
5. Gupta B, Rana S, Sharma A. An Efficient Data Aggregation Approach for Prolonging Lifetime of Wireless Sensor Network. In: International Conference on Innovative Com- puting and Communications. Springer; 2020.
6. Dong, Shi P, Wang K, et al. A survey on deep learning and its applications. Computer Science Review. 2021; 40:100379–100379.
7. Ananthi, J., Sengottaiyan, N., Anbukaruppusamy, S., Upreti, K., & Dubey, A. K. (2022). Forest fire prediction using IoT and deep learning. *International Journal of Advanced Technology and Engineering Exploration*, *9*(87), 246–256.
8. Omar N, Al-Zebari A, Sengur A (2021).
9. Saha, S., Bera, B., Shit, P. K., Bhattacharjee, S., & Sengupta, N. (2023). Prediction of forest fire susceptibility applying machine and deep learning algorithms for conservation priorities of forest resources. *Remote Sensing Applications: Society and Environment*, *29*, 100917.
10. Naderpour, M., Rizeei, H. M., & Ramezani, F. (2021). Forest fire risk prediction: A spatial deep neural network-based framework. *Remote Sensing*, *13*(13), 2513.
11. Paidipati, K. K., Kurangi, C., Reddy, A. S. K., Kadiravan, G., & Shah, N. H. (2024). Wireless sensor network assisted automated forest fire detection using deep learning and computer vision model. *Multimedia Tools and Applications*, *83*(9), 26733–26750.
12. Haque, A., & Soliman, H. (2025). Smart Wireless Sensor Networks with Virtual Sensors for Forest Fire Evolution Prediction Using Machine Learning. *Electronics*, *14*(2), 223.
13. Celik, T., Demirel, H., Ozkaramanli, H., & Uyguroglu, M. (2007). Fire detection using statistical color model in video sequences. *Journal of Visual Communication and Image Representation*, *18*(2), 176–185.
14. Jiang, Q., & Wang, Q. (2010, January). Large space fire image processing of improving canny edge detector based on adaptive smoothing. In *2010 International Conference on Innovative Computing and Communication and 2010 Asia-Pacific Conference on Information Technology and Ocean Engineering* (pp. 264–267). IEEE.
15. Park, M., & Ko, B. C. (2020). Two-step real-time night-time fire detection in an urban environment using Static ELASTIC-YOLOv3 and Temporal Fire-Tube. *Sensors*, *20*(8), 2202.
16. Dimitropoulos, K., Barmpoutis, P., & Grammalidis, N. (2014). Spatio-temporal flame modeling and dynamic texture analysis for automatic video-based fire detection. *IEEE transactions on circuits and systems for video technology*, *25*(2), 339–351.
17. Gupta, B., Rana, S., Goyal, S. K., Gujral, R. K., & Aledaily, A. N. (2023, February). Forest Fire Detection by using MESA2DA Clustering Protocol based on Artificial Intelligence Techniques. In *2023 International Conference on Smart Computing and Application (ICSCA)* (pp. 1–5). IEEE.

Emerging Perspectives and Applications of Computational Intelligence and Smart Systems
– Dr. Amit Lathigara et al. (eds)
© 2026 Taylor & Francis Group, London, ISBN 978-1-041-20965-2

51

Concept of Sliding Block Type Self-Centering Vise Clamp for Smart Manufacturing

Dipeshkumar M. Chauhan*

PhD Scholar, Schoool of Engineering, RK Univerity,
Rajkot, Gujarat, India

Government Polytechnic, Bhuj,
Gujarat, India

Chetankumar M. Patel

Mechanical Engineering Department,
School of Engineering, RK University,
Rajkot, Gujarat, India

■ **Abstract:** Fixtures are employed to position, align, and secure workpieces during production. Their effectiveness directly influences the quality and precision of the manufacturing operation. The clamping mechanism plays a pivotal role in a fixture's efficiency, with designs and components tailored to the workpiece's shape and structural requirements. Modern flexible manufacturing systems (FMS) prioritize advanced solutions like self-centring clamping mechanisms for their adaptability. The self-centring vise clamp exemplifies innovation, featuring a unique sliding-block design using Autodesk Inventor that replaces traditional threaded fasteners, streamlining both functionality and construction. This approach enhances precision and simplifies operation compared to conventional setups and makes the system smart and fit for industry 4.0.

■ **Keywords:** Fixture, Vise, Clamping force

1. INTRODUCTION

Vise clamps ensure stability and precision in milling, drilling, or grinding by securely gripping workpieces, preventing movement that risks dimensional accuracy or surface finish. Critical in aerospace, automotive, and medical industries—where even minor errors cause defects—they offer rigidity for repeatable positioning in mass production. Customizable jaws and quick-change systems adapt to complex parts, cutting setup time.

*Corresponding author: dipesh.chauhan11@gmail.com

DOI: 10.1201/9781003725046-51

Automation-driven self-centring models auto-align workpieces between jaws, removing manual adjustments to avoid off-centre machining. Integrated with CNC systems/robotics, sensor-enabled feedback monitors clamping force and alignment, boosting productivity. Their seamless automation compatibility reduces human intervention, improves safety by minimizing repetitive tasks, and supports scalable manufacturing. Merging robust design with smart technology; self-centring vises balance speed and ultra-high precision. By enhancing process efficiency and reliability, they underpin industries prioritizing quality and adaptability, reinforcing their status as vital components in smart manufacturing's evolution.

The objective of this paper is to explore the innovative design of self-centring vise clamps utilizing a sliding block mechanism. It covers topics like literature review, fundamentals of self-centring, concept and working, design consideration, advantages, limitations and conclusion.

2. LITERATURE REVIEW

Traditional mechanical vises rely on screw threads, racks, and sliding blocks to synchronize jaw motion. Innovations like (David L. Schmidtet.,al 2014) dual-directional vise with opposing threads enable simultaneous internal/external clamping for irregular geometries, while (Chris Tayloret.,al 2016) enhanced adaptability via bidirectional spindles and interchangeable jaws. (Eric Sun et., al 2020) modular clamp optimizes force distribution, (Shriram Kharmaleet., al 2020) addressed heavy-duty challenges with a rack-and-pinion design validated via FEA. However, traditional systems struggle with out-of-round components and securing rough castings without excessive force (Thomas Gmeineret., al 2011)

Hydraulic systems streamline efficiency: (Vidya Thengil et al.2016) reduced setup time by 40% with automated depth control, while (Chetankumar M. Patelet., al 2014) and (Mr. Nagaraj Anand Shet, et., al. 2017) demonstrated multi-cylinder fixtures handling 12.5 kNloads.(Jaykumar K Dhulia et., al 2019)scaled automation via CAD-modeled 16-cylinder systems, though maintenance demands persist.

Sensor-integrated vises mark a leap toward smart manufacturing. (Sina Rezvani et al2020). embedded strain gauges and piezoelectric sensors for real-time force monitoring, while (Berend Denkena et., al 2014). boosted sensor sensitivity by 71% via FEA optimization. (Djordje Vukelic et., al. 2012) balanced compliance for machining stability, though sensor durability in harsh environments remains a challenge.

Reconfigurable designs align with Industry 4.0: (Thomas Gmeiner et., al 2011) introduced ontology-guided interchangeable jaws, complemented (Chris Tayloret., al 2016) and (Eric Sun et., al 2020) modular systems, enabling rapid adaptability for diverse workpieces.

Critical challenges persist. Screw-thread systems dominate for simplicity but lack diameter adaptability (Mr. Nagaraj Anand Shet, et. al. 2017) while hydraulic and sensor-driven solutions face cost and complexity barriers. Future advancements demand IoT-enabled feedback, robust smart materials, and modular architectures to balance flexibility with durability. In conclusion, it has evolved from mechanical clamps to intelligent systems integrating hydraulics, sensors, and adaptive designs. While enhancing precision and automation, overcoming cost, complexity, and resilience hurdles will define their role in Industry 4.0, driving innovations in real-time data integration and modular smart materials.

Table 51.1 Comparison of various mechanisms used for self-centring

Mechanism	Precision	Speed	Load Capacity	Cost	Best For
Screw-Based	High	Low	High	Low-Moderate	Manual/heavy-duty work
Cam-Based	Moderate	High	Moderate	Moderate	Rapid CNC automation
Rack-and-Pinion	High	High	Moderate	Moderate-High	High-volume production
Hydraulic/ Pneumatic	Moderate-High	Very High	Very High	High	Industrial heavy machining
Wedge-Based	Low	Moderate	Low-Moderate	Low	Light-duty app.

3. SELF-CENTRING VISE CLAMP

3.1 Construction

The self-centring vise clamp represents a novel advancement in work holding technology, diverging from conventional lead screw mechanisms by employing a hydraulically actuated wedge-type sliding block system. Its construction features include:

Hydraulic Actuation: A pressurized hydraulic fluid system replaces traditional mechanical lead screws, enhancing automation and force control.

Modular Components: Exchangeable and machinable jaws accommodate diverse workpiece geometries, while a self-lubricating design (via grease nipples and dedicated ports) ensures durability.

Core Assembly: The self-centring vise comprises six critical components, each engineered for durability, precision and efficient force transmission:

Vise Body: Fabricated via casting from spheroidal graphite iron, this primary structure provides rigidity and withstands hydraulic pressure. Its high tensile strength and ductility ensure stability during clamping operations (Fig. 51.1).

Sliding Jaw: Two sliding jaws, crafted from 20Mn Cr5 high-carbon steel (rectangular bar stock), form sliding pairs with the vise body and wedge. Case-hardened for wear resistance, they transmit clamping forces to detachable jaws (Fig. 51.1).

Sliding Jaw Wedge: This intermediate component converts the piston's linear motion into synchronized jaw movement. Made from 20Mn Cr5 steel and case-hardened, it features inclined slots to guide the sliding jaws (Fig. 51.1).

Piston: Constructed from 8620 alloy steel (round bar stock), the piston translates hydraulic pressure into mechanical motion. Bolted to the sliding wedge, it drives the clamping/unclamping cycle (Fig. 51.1).

Bottom Cover: Manufactured from toughened En8 or medium-carbon steel, it seals the vise and guides the piston's linear motion, enhancing crack resistance (Fig. 51.1).

Top Cover: Similarly made from En8 steel, it shields internal components from debris while maintaining structural integrity (Fig. 51.1).

Fig. 51.1 3D view of self-centringvise part

(From Left - Body, Central Wedge, Sliding block, Bottom Cover, Cover plate, Piston)

3.2 Working

The clamping and unclamping cycles operate through hydraulic pressure manipulation (Fig. 51.2):

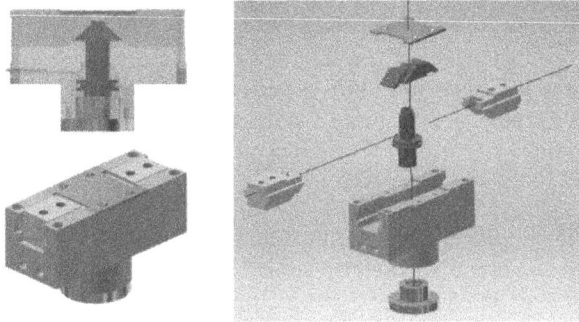

Fig. 51.2   Front view and exploded view of self-centring vise

Clamping Phase

Pressurized fluid is directed to the upper piston surface of the V-shaped sliding block, forcing it downward. This motion drives the inclined wedge-shaped sliders inward, synchronizing jaw movement to grip the workpiece.

Unclamping Phase

Hydraulic pressure shifts to the lower piston surface, lifting the sliding block upward. The wedge sliders retract, releasing the workpiece. A bolt connects the sliding wedge to the piston, ensuring precise force transmission, while detachable jaws adapt to specific workpiece shapes.

3.3 Design Parameter

Design priorities include material selection (cast iron for load-bearing, rectangular/round stock compatibility), surface treatments (case hardening for wear resistance), and geometric precision (inclined wedge slots). Hydraulic components optimize force distribution, ensuring modularity and adaptability. Performance hinges on clamping force (determined by cutting forces, hydraulic pressure, and clamping height), wedge geometry (angle, groove shape, length) for motion efficiency, hydraulic system alignment (pressure range, fluid dynamics), and working range (stroke limits, jaw adaptability). Together, these elements ensure robust, adaptable functionality across diverse machining applications, balancing strength, precision, and operational demands.

4. Conclusion

This innovative vise clamp revolutionizes work holding by replacing traditional screw mechanisms with a hydraulically actuated wedge-sliding system, enhancing precision and automation. The wedge design converts hydraulic pressure into synchronized linear motion, ensuring uniform clamping force and eliminating manual adjustments. Its modular architecture incorporates interchangeable jaws, allowing rapid adaptation to diverse workpiece geometries, while self-lubricating components minimize wear and maintenance. Constructed from cast iron and hardened steel, the vise withstands high loads and harsh machining environments, ensuring long-term durability. Hydraulic automation enables rapid clamping cycles, reducing setup time in high-volume sectors.

5. FUTURE SCOPE

CAD/FEA-driven refinements, advanced manufacturing, and testing enhance reliability. IoT sensors enable real-time force, alignment, and wear monitoring, feeding cloud-based predictive analytics. A synchronized Digital Twin simulates stress and optimizes hydraulic/wedge settings. CAFD tools automate jaw designs via generative algorithms and FEA, ensuring uniform force distribution. Future smart materials and IoT adaptive control balance cost and robustness. Continued research and CAFD-driven agility refine precision, merging resilient design with Industry 4.0's smart automation for dynamic manufacturing.

ACKNOWLEDGEMENT

The authors thank to Mr. Pradip Thanki, Mr. Sudhir Thakar and Mr. Kaushal, Supra Technology, Rajkot, Gujarat, India for good cooperation and continuous guidance. Also authors would like to acknowledge continuous support and motivation from Faculty of Doctoral Studies and Research, RK University, Rajkot, Gujarat, India.

References

1. David L. Schmidt (2014). Self-centring dual direction clamping vise, US 2014/0001692 A1, United States Patent and Trademark Office
2. Chris Taylor, Steve Grangetto (2016). Self-centeringvise, US9364937 B2, United States Patent and Trademark Office
3. Eric Sun (2020). Self-Aligning Detachable jaw or fixture plate for vise like work holding apparatus, US 2020/0130068 A1, United States Patent and Trademark Office
4. Shriram Kharmale, et. al. (2020). Design and Development in Self Centring Vice For FWM, Mukt Shabd Journal, Volume IX, Issue VI, 972–977.
5. Thomas Gmeiner, Kristina Shea (2011). Development of an ontology for the automatic reconfiguration of a vise-type fixture device, Proceedings of the ASME 2011 International Design Engineering Technical Conferences & Computers and Information in Engineering Conference, August 28–31, Washington, DC, USA
6. Vidya Thengilet al.(2016), Modification in drilling mechanism with self-centring fixture arrangement, International Journal of Engineering Research &Technology, Vol. 5, Issue
7. Chetankumar M. Patel, Dr. G. D. Acharya (2014), Design and manufacturing of cylinder hydraulic fixture for boring yoke on VMC – 1050, Procedia Technology, 14, 405–412.
8. Mr. Nagaraj Anand Shet, et. Al. (2017) Design and Development of Hydraulic Fixture for Basak Cylinder Head Machining, IARJSET, Vol. 4, Issue 10.
9. Jaykumar K Dhulia, Dr. Nirav P Maniar(2019). Design, Modelling and Manufacturing of Cylinder Hydraulic Fixture with Automated Clamping System, IOP Conf. Series: Journal of Physics: Conf. Series 1240 012036, DOI:10.1088/1742-6596/1240/1/012036.
10. Sina Rezvani et al(2020). Development of a vise with built-in piezoelectric and strain gauge sensors for clamping and cutting force measurements, Procedia Manufacturing 48, 1041–1046.
11. Berend Denkena et al (2014). Sensor integration for a hydraulic clamping system, 2nd International Conference on System-Integrated Intelligence: Challenges for Product and Production Engineering, Procedia Technology 15, 465–473.
12. Vukelic, D., Tadic, B., Miljanic, D., Budak, I., Todorovic, P. M., Randjelovic, S., & Jeremic, B. M. (2012). Povećanje učinkovitosti strojne obrade novom metodom stezanja izradaka. *Tehnički vjesnik*, *19*(4), 837–846.

Note: All the figures and table in this chapter were made by the authors.

Emerging Perspectives and Applications of Computational Intelligence and Smart Systems
— Dr. Amit Lathigara et al. (eds)
© 2026 Taylor & Francis Group, London, ISBN 978-1-041-20965-2

52

Optimized Federated Learning for Decentralized Healthcare Data

E. S. Chakravarthy*
Research Scholar, Faculty of Technology, RK University, Rajkot,
Gujarat, India

Amit Lathigara
School of Engineering, RK University, Rajkot,
Gujarat, India

Rajanikanth Aluvalu
Department of CSE,
Symbiosis Institute of Technology, Hyderabad Campus,
Symbiosis International (Deemed University),
Pune, India

M. Giridhar Kumar
G. Pullaiah College of Engineering and Technology,
Kurnool, India

■ **Abstract:** This paper introduces an optimized federated learning (FL) framework tailored for decentralized healthcare data, addressing the challenges of privacy preservation, data heterogeneity, and resource constraints. Specifically, we focus on the application of federated learning for disease detection, leveraging personalized learning techniques and client clustering to manage non-IID data distributions. The proposed framework integrates distributed reinforcement learning (DRL) for dynamic resource allocation across edge devices, accounting for varying computational power, energy limitations, and network connectivity. Additionally, communication costs are reduced through model compression, selective updates, and adaptive aggregation. This approach ensures privacy preservation, enhances model performance, and optimizes resource utilization in decentralized healthcare environments, providing a scalable solution for disease detection systems.

■ **Keywords:** Federated learning, Healthcare, Disease detection, Decentralized systems, Reinforcement learning, Privacy preservation, Non-IID data, Resource optimization, Communication efficiency

*Corresponding author: dreschakravarthy@outlook.com

DOI: 10.1201/9781003725046-52

1. INTRODUCTION

Machine learning models for disease detection are increasingly employed in healthcare, leveraging large amounts of patient data to improve diagnosis and treatment. However, the sensitive nature of healthcare data necessitates privacy-preserving methods to avoid data exposure during model training. Federated learning (FL) has emerged as a promising solution by enabling the collaborative training of machine learning models on decentralized data, without the need to transfer sensitive information to centralized servers. Despite its benefits, FL encounters significant challenges when applied to healthcare data, which is typically non-IID, meaning that data distributions across devices or institutions differ widely.

The heterogeneity of healthcare data and the limited computational resources of edge devices—such as medical IoT devices and wearables—complicate federated learning processes. Additionally, the communication overhead required for model updates between the central server and edge devices can severely impact system performance. Therefore, efficient resource scheduling, model aggregation, and communication strategies are crucial to ensure the scalability and efficiency of federated learning systems in healthcare.

This paper explores the integration of optimized federated learning with decentralized scheduling for resource management in healthcare environments. By combining FL with distributed reinforcement learning (DRL) for dynamic scheduling and employing strategies to mitigate non-IID data challenges, we propose an efficient system that addresses privacy concerns while enhancing disease detection accuracy.

2. LITERATURE REVIEW

Federated learning has been widely applied in healthcare to enable data privacy and enhance model performance. Recent studies such as (Smith et al. 2020) and (Abbas et al. 2021) have explored federated learning applications for medical image classification and disease prediction, demonstrating its potential in maintaining privacy while achieving high model accuracy. However, these works highlight significant challenges, particularly with non-IID data, which leads to poor convergence and reduced model effectiveness.

One promising approach to address data heterogeneity is the use of personalized federated learning, where individual devices are allowed to fine-tune their models to better reflect local data characteristics. (Zhang et al. 2021) discussed client clustering as another solution to mitigate the negative effects of non-IID data, grouping similar devices to enhance model training. These techniques can significantly improve the training process in healthcare systems, where patient data is inherently diverse.

The limited resources available on edge devices present another challenge. Distributed reinforcement learning (DRL) has been successfully applied in cloud and edge computing systems to dynamically allocate tasks and manage resources more efficiently. In the healthcare domain, (Liu et al. 2020) proposed DRL-based scheduling to optimize resource allocation for medical devices with varying computational and energy limitations.

However, communication overhead remains a significant issue in federated learning. To mitigate this, several studies, including those by (Chen et al. 2020) and (Li et al. 2021), have proposed model compression techniques and selective update strategies to reduce communication costs. These strategies are essential for improving the efficiency of federated learning in resource-constrained environment.

3. PROBLEM STATEMENT

The integration of federated learning in decentralized healthcare systems faces several key challenges that hinder its effectiveness:

3.1 Non-IID Data Distribution

Healthcare data, generated from various medical institutions, devices, and patient groups, often follows non-IID distributions. This variability impedes the model's ability to converge, and can cause poor performance or divergence, particularly in disease detection tasks where precision is crucial.

3.2 Resource Constraints in Edge Devices

Edge devices, such as medical IoT devices and wearables, have limited computational power, energy resources, and network bandwidth. This limitation makes it challenging to run federated learning tasks efficiently without overloading the devices or wasting resources.

3.3 High Communication Costs

Federated learning requires frequent communication between edge devices and the central server to share model updates, leading to high communication costs. This issue becomes particularly significant in healthcare systems where network resources may be constrained and large model updates may be required, affecting both speed and energy consumption.

4. RESEARCH OBJECTIVES

Develop an optimized federated learning framework that can address non-IID data distributions in decentralized healthcare settings, improving model convergence and disease detection accuracy.

Utilize distributed reinforcement learning (DRL) for dynamic scheduling to optimize the use of computational resources on edge devices, considering constraints such as energy and connectivity.

Implement communication-efficient strategies are for federated learning, including model compression, selective updates, and adaptive aggregation, to reduce the overhead of frequent model updates.

5. SOLUTIONS TO RESEARCH OBJECTIVES

5.1 Optimized Federated Learning for Non-IID Data

To handle non-IID data in healthcare applications, the proposed framework uses personalized federated learning techniques, where each device's model is customized based on local data. Client clustering is also implemented to group devices with similar data distributions, improving model convergence and reducing training time. This approach enables better adaptation to the diverse nature of healthcare data.

5.2 Resource Allocation Using Distributed Reinforcement Learning (DRL)

Distributed reinforcement learning (DRL) provides a solution to manage the limited resources of edge devices in healthcare systems. By using DRL, edge devices can independently learn the optimal allocation of computational and energy resources to ensure efficient model training. The devices adjust their schedules based on real-time environmental feedback, ensuring that tasks are completed without overloading the system.

5.3 Communication-Efficient Federated Learning

To reduce communication overhead, the framework employs several techniques:

Model Compression: Reduces the size of the model updates before transmitting them to the central server.

Selective Updates: Only the most important model parameters are updated, minimizing the amount of data exchanged.

Adaptive Aggregation: The model updates from devices are aggregated adaptively, ensuring that communication occurs only when significant improvements are made, thus reducing the frequency of updates.

These communication-efficient techniques are particularly crucial in healthcare environments, where bandwidth and energy resources are often constrained.

6. CONCLUSION

In this paper, we have presented an optimized federated learning framework for decentralized healthcare data that addresses the challenges of non-IID data distributions, resource limitations, and communication overhead. By integrating personalized federated learning and client clustering, the system effectively manages heterogeneous healthcare data, improving disease detection accuracy. Additionally, distributed reinforcement learning ensures that edge devices dynamically allocate resources based on their capabilities, enhancing system efficiency. Finally, communication-efficient strategies such as model compression and selective updates help reduce the burden of frequent model exchanges. This framework provides a scalable, privacy-preserving solution for disease detection in decentralized healthcare environments.

References

1. Smith, J., et al. (2020). "Federated Learning for Healthcare: A Privacy-Preserving Approach for Disease Prediction." IEEE Transactions on Medical Imaging.
2. Abbas, M., et al. (2021). "Federated Learning in Healthcare: A Survey and Applications." Journal of Biomedical Informatics.
3. Zhang, W., et al. (2021). "Improving Federated Learning with Client Clustering for Non-IID Healthcare Data." IEEE Transactions on Network and Service Management.
4. Liu, L., et al. (2020). "Distributed Reinforcement Learning for Dynamic Resource Scheduling in Healthcare."Journal of Cloud Computing.
5. Chen, Z., et al. (2020). "Communication-Efficient Federated Learning in Healthcare: A Survey." IEEE Transactions on Neural Networks and Learning Systems.
6. Li, Y., et al. (2021). "Reducing Communication Overhead in Federated Learning for Healthcare Applications."Journal of Artificial Intelligence in Medicine.

Emerging Perspectives and Applications of Computational Intelligence and Smart Systems
– Dr. Amit Lathigara et al. (eds)
© 2026 Taylor & Francis Group, London, ISBN 978-1-041-20965-2

53

Revolutionizing Cloud Computing: Decentralized Scheduling with Federated Learning for Enhanced Resource Utilization

A. Damayanthi*
Research Scholar, Faculty of Technology, RK University, Rajkot,
Gujarat, India

Amit Lathigara
School of Engineering, RK University, Rajkot,
Gujarat, India

Rajanikanth Aluvalu
Department of CSE,
Symbiosis Institute of Technology, Hyderabad Campus,
Symbiosis International (Deemed University),
Pune, India

M. Giridhar Kumar
G. Pullaiah College of Engineering and Technology,
Kurnool, India

■ **Abstract:** This paper presents a novel approach to enhancing resource utilization in cloud computing by integrating decentralized scheduling with federated learning, specifically addressing challenges related to edge devices, non-IID data distributions, and communication costs. To optimize cloud task allocation, a decentralized scheduling framework is proposed, leveraging distributed reinforcement learning (DRL) to dynamically allocate tasks based on the varying computational capacities, energy limits, and connectivity of edge devices. This adaptive system ensures efficient resource utilization while minimizing overload. To address the impact of non-IID data distributions in federated learning, we introduce advanced federated optimization techniques, such as personalized federated learning and client clustering, to improve model convergence and prevent divergence. Additionally, communication-efficient federated learning strategies, such as model compression and selective updates, are employed to reduce communication overhead, further enhancing system efficiency. The proposed framework provides a robust solution to optimize resource allocation, task

*Corresponding author: damayanthi.alluri@gmail.com

DOI: 10.1201/9781003725046-53

scheduling, and model training in decentralized cloud environments, paving the way for scalable, energy-efficient, and privacy-preserving cloud computing systems.

■ **Keywords:** Cloud computing, Decentralized scheduling, Federated learning, Reinforcement learning, Resource optimization

1. INTRODUCTION

Cloud computing has revolutionized how businesses and individuals access computing resources, offering on-demand scalability and flexibility. However, as the demand for cloud services grows, the challenge of efficiently managing resources across distributed environments has become more complex. Traditional cloud resource management relies heavily on centralized scheduling, which often leads to inefficiencies, especially in systems with dynamic and diverse workloads. To address these challenges, decentralized scheduling has emerged as a promising solution, enabling more localized, adaptive resource allocation.

Federated learning, a technique that allows models to be trained across multiple devices without sharing raw data, offers an innovative way to enhance decentralized scheduling. By integrating federated learning with cloud resource management, systems can learn from a variety of workload patterns across different nodes, optimizing resource utilization and minimizing latency without compromising data privacy. This decentralized approach allows cloud platforms to dynamically adjust scheduling policies based on real-time data from distributed nodes, leading to more efficient resource allocation.

In this paper, we explore how federated learning can be employed in decentralized scheduling frameworks, highlighting the potential benefits for cloud computing, such as improved scalability, reduced resource wastage, and enhanced performance. This novel approach promises to unlock new possibilities for cloud platforms, offering more efficient, adaptable, and secure solutions for modern cloud computing challenges.

2. LITERATURE REVIEW

Decentralized scheduling in cloud computing has garnered significant attention for its potential to improve efficiency and fault tolerance. (Liu et al. 2019) demonstrated that decentralized task scheduling reduces delays and enhances resource utilization by enabling local decision-making at distributed nodes. This scalability and adaptability are in contrast to traditional centralized systems, which can become bottlenecks in dynamic workloads. Federated learning, introduced by (McMahan et al. 2017), further strengthens decentralized systems by allowing model training without sharing raw data, preserving privacy. This feature is particularly beneficial in cloud environments where data security is paramount (Bonawitz et al., 2017). Federated learning has been effectively applied to cloud resource management, optimizing scheduling and predicting demand patterns (Li et al., 2020). Additionally, several studies, such as (Zhang et al. 2021), have explored dynamic scheduling techniques, including machine learning algorithms, to enhance cloud resource allocation. However, decentralized systems face challenges, such as load balancing and network synchronization, which Wang et al. (2020) identified as significant barriers in large-scale deployments. The integration of federated learning into decentralized scheduling frameworks addresses these issues, enabling continuous adaptation and improved resource optimization across distributed cloud environments (Xu et al., 2021). These advancements pave the way for more efficient, resilient cloud computing systems.

3. PROBLEM STATEMENT

In the context of revolutionizing cloud computing through decentralized scheduling with federated learning, several critical challenges need to be addressed to optimize resource utilization effectively:

3.1 Challenges in Resource Allocation

The cloud computing ecosystem is increasingly reliant on edge devices with varying computational capabilities, energy constraints, and connectivity. These disparities create significant challenges in resource allocation, particularly when attempting to assign tasks to devices based on their individual capacities. The variations in device performance also complicate client selection for federated learning, where optimal device participation is crucial for maintaining system efficiency and accuracy.

3.2 Impact of Non-IID Data

Devices within a decentralized cloud system often generate data that follows non-independent and identically distributed (Non-IID) patterns. This variation in data distributions can lead to issues like model divergence and delayed convergence during federated learning. These problems reduce the overall effectiveness of the learning process, as the model may struggle to generalize across the diverse data patterns of different devices, ultimately impacting the accuracy and performance of the cloud system.

3.3 Communication Cost

Federated learning requires frequent model updates between the central server and edge devices. This constant communication can result in high bandwidth usage, increasing the communication cost significantly. High communication overheads may negatively impact system efficiency, especially in environments with limited network resources, leading to slower convergence and higher energy consumption across devices.

Addressing these challenges is essential to fully realize the potential of decentralized scheduling and federated learning in enhancing resource utilization in cloud computing systems.

4. RESEARCH OBJECTIVES

- Develop an efficient decentralized scheduling framework that optimizes resource allocation across edge devices with varying computational capacities, energy limits, and connectivity.
- Mitigate the impact of Non-IID data distributions in federated learning to enhance model convergence and prevent model divergence.
- Implement strategies to reduce communication costs associated with frequent model updates, improving system efficiency while maintaining high performance in cloud computing environments.

5. SOLUTIONS TO RESEARCH OBJECTIVES

Cloud Task Allocation and Resource Optimization: The primary challenge in cloud task allocation and resource optimization lies in managing the diverse capabilities of edge devices, including varying computational power, energy limitations, and connectivity. To address this, a *decentralized task allocation* system can be employed, where local scheduling decisions are made by each edge device based on its available resources and workload demands. The system would leverage machine learning techniques to dynamically allocate tasks, ensuring that each edge device is assigned tasks that match its computational capacity and energy constraints.

A key solution is the development of an adaptive *resource allocation algorithm* that considers these device variations. By utilizing *Distributed Reinforcement Learning* (DRL), edge devices can learn optimal task allocation strategies by interacting with their local environments and adjusting their resource usage based on real-time feedback. Through DRL, the system can automatically adjust its task allocation policies, ensuring that devices are used efficiently, preventing overload, and maintaining energy conservation.

Additionally, the decentralized scheduling framework can implement *peer-to-peer communication protocols* among devices to share resource status and workload information, facilitating the optimal distribution of tasks across the system. This collaborative approach helps maintain high levels of resource utilization while addressing the limitations of each edge device in a dynamic and distributed cloud environment.

5.1 Distributed Reinforcement Learning for Resource Optimization

The challenge of optimizing resources across a variety of edge devices is complex due to the heterogeneous nature of cloud environments. To overcome this, *Distributed Reinforcement Learning (DRL)* provides an effective solution. In a DRL-based system, each device can operate independently and learn optimal scheduling policies based on local observations and interactions with the environment. DRL algorithms enable devices to autonomously discover optimal resource allocation strategies, such as determining when and how to process or offload tasks based on computational availability, energy levels, and network bandwidth.

In addition, DRL can be integrated with *multi-agent systems*, where each edge device acts as an agent in a broader system, interacting with other agents to collaboratively optimize resources. Through this setup, devices can exchange information about resource availability, enabling more informed task allocation decisions that consider both local and global system objectives. Moreover, by utilizing techniques such as *Q-learning* and *policy gradient methods*, the system can evolve and adapt over time to improve resource utilization and performance, leading to better scalability, fault tolerance, and energy efficiency in the cloud infrastructure.

5.2 Federated Learning and Task Optimization

Non-IID data distributions across edge devices pose a significant challenge in federated learning. In a decentralized cloud computing system, edge devices often generate data that is not independent or identically distributed, making it difficult to achieve model convergence. To address this, the integration of *Federated Learning (FL)* into the cloud task allocation system can optimize learning processes while mitigating the issues related to non-IID data.

In federated learning, the model is trained collaboratively across edge devices, with each device maintaining its own local model based on local data. By using federated optimization techniques, the system can ensure that each device contributes to the global model while preserving privacy. Techniques such as *FedAvg* (Federated Averaging) allow devices to periodically share model updates instead of raw data, ensuring that the system maintains privacy while achieving model convergence. However, in the case of non-IID data, federated optimization methods like *personalized federated learning* or *client clustering* can be employed to group devices with similar data distributions, thus minimizing divergence during training.

For task optimization, federated learning can also inform task allocation decisions by predicting resource requirements for specific tasks based on historical data and local models. This enables more intelligent scheduling by considering the types of tasks each edge device can efficiently handle, leading to better resource utilization. By combining federated learning with task optimization, the system can dynamically adjust task assignments, improving both computational efficiency and model accuracy.

Furthermore, communication costs, which are a significant concern in federated learning, can be reduced by employing *communication-efficient federated learning strategies*, such as model compression, aggregation techniques, or selective update mechanisms. These methods minimize the frequency of communication between devices and the central server, ensuring that communication costs do not outweigh the benefits of federated learning.

6. Conclusion

In conclusion, by combining cloud task allocation and resource optimization with distributed reinforcement learning and federated learning, a robust solution can be developed to address the challenges of heterogeneous edge devices, non-IID data distributions, and high communication costs in decentralized cloud environments. These strategies provide a pathway to achieve efficient, scalable, and privacy-preserving cloud systems that optimize resource utilization and improve overall system performance.

References

1. Liu, J., et al. (2019). "Decentralized Task Scheduling in Cloud Computing." Journal of Cloud Computing.
2. McMahan, H. B., et al. (2017). "Communication-Efficient Learning of Deep Networks from Decentralized Data."Proceedings of the 20th International Conference on Artificial Intelligence and Statistics.
3. Bonawitz, K., et al. (2017). "Practical Secure Aggregation for Privacy-Preserving Machine Learning."Proceedings of the 2017 ACM SIGSAC Conference on Computer and Communications Security.
4. Li, T., et al. (2020)."Federated Learning: Challenges, Methods, and Future Directions."IEEE Transactions on Neural Networks and Learning Systems.
5. Zhang, Y., et al. (2021). "Cloud Resource Optimization Using Reinforcement Learning and Dynamic Scheduling."IEEE Transactions on Cloud Computing.
6. Wang, X., et al. (2020)."Challenges in Decentralized Cloud Resource Scheduling."Journal of Cloud Computing and Networking.
7. Xu, L., et al. (2021). "Federated Learning in Cloud Resource Scheduling." IEEE Transactions on Cloud Computing.

2026 Taylor & Francis Group, London, ISBN 978-1-041-20965-2

Emerging Perspectives and Applications of Computational Intelligence and Smart Systems
– Dr. Amit Lathigara et al. (eds)
© 2026 Taylor & Francis Group, London, ISBN 978-1-041-20965-2

54

Predicting Climate Change Impact on Crop Yield: A Review of AI-Based Implementations in Maharashtra

Ketaki Ghawali*

Research Scholar,
Mangalayatan University, Aligarh

Assistant Professor,
Vidyalankar School of Information Technology,
Mumbai

Abhishek Garg

Associate Professor,
Department of Computer Engineering & Applications,
Mangalayatan University,
Aligarh

■ **Abstract:** Agriculture in Maharashtra faces severe challenges due to climate change, including erratic rainfall, droughts, and increased temperatures. Farmers' perceptions of these climatic shifts often differ from AI-based climate predictions, affecting adaptation strategies. This study explores the alignment and discrepancies between farmers' perceptions and AI-driven climate models, analysing factors such as experience, memory biases, and access to meteorological data. AI applications in climate-resilient agriculture, including weather forecasting, soil health monitoring, crop yield prediction, and precision irrigation, are examined. Findings highlight the effectiveness of AI in enhancing agricultural planning, but barriers such as financial constraints, technological accessibility, and education persist. The study emphasizes the need for integrating AI-driven climate insights into farmer-centric decision-making processes. Strengthening AI adoption through government initiatives, financial support, and participatory approaches will ensure sustainable agriculture and climate resilience. Future research should focus on bridging the AI-farmer knowledge gap to optimize agricultural practices and enhance productivity in climate-affected regions.

■ **Keywords:** AI in agriculture, Adaptation strategies, Climate change, Farmers' perception, Sustainable farming

*Corresponding author: ketaki.ghawali@vsit.edu.in

DOI: 10.1201/9781003725046-54

1. INTRODUCTION

Artificial Intelligence (AI) and Machine Learning (ML) are transforming agriculture by enhancing climate resilience, particularly in Maharashtra, where climate change-induced irregular rainfall and extreme weather threaten food security (Matthan, 2023). AI-powered climate forecasting, using deep learning techniques such as recurrent neural networks (RNNs) and long short-term memory (LSTM) networks, improves monsoon predictions, optimizing planting and water management (Sharma et al., 2023; Kumar et al., 2022).

AI-driven image recognition and convolutional neural networks (CNNs) assist in early detection of pest outbreaks and plant diseases, enabling targeted pesticide use and reducing crop losses (Ahsan et al., 2024). Soil health monitoring integrates IoT-enabled sensors with AI models to assess nutrient levels and optimize fertilizer use, preventing land degradation (Shah et al., 2025; Nyakuri et al., 2022). AI-enhanced yield prediction models leverage historical and environmental data for accurate crop output estimates, supporting market stability and policy decisions (Zhang et al., 2024; Jabed et al., 2024).

The integration of blockchain technology enables supply chain transparency which diminishes post-harvest waste according to (Elufioye et al. (2024). Small farmers face difficulties in digital adoption while AI implementation gets hindered by its high costs and predictive models tend to display bias according to (Liang et al. (2022) and Galaz et al. (2021). Future advancements in AI and integration between AI and blockchain systems and IoT will lead to establishing a sustainable agricultural sector (Elufioye et al., 2024).

2. UNDERSTANDING PERCEPTION AND ADAPTATION TO CLIMATE CHANGE IN AI-DRIVEN AGRICULTURE

Agricultural workers develop their understanding of climate change based on their direct observations of evasive rainfall patterns and increasing infestations and damaged soil which endanger their livelihoods (Mondal et al., 2024; Sharma et al., 2023). Through predictive analytics the farmers use reactive immediate action and proactive long-term solutions together with AI-driven tools to optimize their farming activities (Darjee, 2023; Elufioye et al., 2024). AI technology strengthens crop monitoring while additionally controlling pests and managing irrigation and making yield predictions through its analysis of satellite images alongside soil information and climate data (Katekar et al. 2023). The implementation of AI technology encounters several obstacles related to cost expenses and digital understanding problems and restricted availability particularly among small-scale farmers in Maharashtra (Katekar et al., 2023). A successful resolution of these barriers demands cooperation between governments and industries which would provide essential infrastructure and promotional incentives and training procedures. AI-based adaptation practices implemented correctly will produce better climate resilience that secures food safety and protects economic performance (Sharma et al., 2023).

3. MATERIAL AND METHODS

3.1 Systematic Literature Review (SLR)

The Systematic Literature Review (SLR) serves as a defined process to discover and review scholarly research about subjects. The research determines AI-driven model predictions for climate change effects on crop cultivation across Maharashtra through an SLR methodology. The system uses the SLR methodology to integrate all pertinent materials with full transparency while providing documentation that supports replication.

3.2 Literature Search Strategy

To identify relevant research, we searched four major databases: Scopus, Web of Science, IEEE Xplore, Google Scholar.

These databases were chosen for their extensive coverage of AI, climate change, and agricultural studies. A search string combining relevant keywords was used to retrieve literature. The final query included terms related to AI, climate change, and crop production.

3.3 Article Screening and Selection Criteria

The initial search resulted in 1,045 studies. After removing duplicates, 923 unique articles remained. These were screened based on their title and abstract, and studies that were not relevant to AI applications in climate adaptation for agriculture were removed. Following the full-text review of 71 articles, a total of 41 studies were selected for final analysis.

Table 54.1 Inclusion and exclusion criteria

Criteria	Inclusion	Exclusion
Scope	Studies on AI-driven climate change predictions in agriculture	Studies on AI but unrelated to climate change
Region	Focus on Maharashtra or similar agro-climatic regions	Studies unrelated to India or tropical agriculture
Methodology	Empirical research, AI models, and machine learning applications	Conceptual papers without validation
Publication Type	Peer-reviewed journal articles and conference papers	Non-peer-reviewed reports, editorials, or books

Source: Authors

3.4 Data Extraction and Categorization

Each selected study was analysed for the following information:

Study Objectives

- **AI Techniques Used** (e.g., Machine Learning, Deep Learning, Hybrid AI models)
- **Types of Data Used** (e.g., Meteorological, Soil, Crop Yield)
- **Performance Metrics** (e.g., Accuracy, Precision, RMSE)
- **Findings and Limitations**

    To organize the findings, studies were categorized based on AI-driven adaptation domains:

    - Climate Prediction Models – AI-based forecasting of temperature, rainfall, and extreme weather events.
    - Crop Yield Prediction – AI-driven models estimating production based on climatic conditions.
    - Soil and Water Management – AI-assisted precision irrigation and soil health assessment.
    - Pest and Disease Prediction – AI-based early warning systems for pest outbreaks

3.5 Data Analysis Methods

To synthesize insights from the selected studies, both **quantitative and qualitative** analysis techniques were used.

1. **Thematic Analysis** – AI-driven adaptation strategies were classified into **incremental, systemic, and transformational categories** based on the framework of (Rickards & Howden 2012).
2. **Univariate Statistics** – A frequency analysis identified the **most commonly used AI models** and their effectiveness.
3. **Comparative Model Evaluation** – AI models were compared based on:
   - **Predictive Accuracy**
   - **Computational Efficiency**
   - **Implementation Feasibility**

3.6 Tools and Software Used

The following **software and tools** were used to support data analysis:

- **Python (Scikit-learn, TensorFlow, Keras)** – AI model evaluation
- **QGIS & Google Earth Engine** – Geospatial mapping of climate impacts
- **Microsoft Excel & NVivo** – Thematic analysis and qualitative data processing

These tools ensured a **robust, evidence-based analysis** of AI applications in climate-adaptive agriculture.

4. RESULTS

The integration of AI in agriculture has gained traction, particularly in Maharashtra, where climate unpredictability poses challenges. AI-driven predictive models enhance climate risk assessment, optimize farming techniques, and improve yield forecasts (Kamble et al., 2023). AI research in agriculture surged post-2015 due to global efforts in big data and machine learning for climate adaptation. AI-based weather forecasting, leveraging satellite imagery and climate models, enables improved planning, crucial for Maharashtra's vulnerability to droughts, extreme temperatures, and erratic rainfall (Elufioye et al., 2024).

Several AI methodologies mitigate climate change effects in Maharashtra's agriculture. Machine learning predicts yield patterns and droughts (Kamble et al., 2023), deep learning analyses multispectral satellite imagery to detect crop stress (Ashoka et al., 2024), CNNs identify diseases and nutrient deficiencies, and RNNs improve monsoon pattern predictions (Sharma et al., 2023).

Hybrid AI frameworks integrate meteorological and agronomic data for precision farming (Peng et al., 2020). AI models outperform traditional climate prediction techniques by processing extensive datasets from satellite images, soil reports, and meteorological records, making AI-driven solutions a disruptive force in agriculture (Borkar et al., 2024).

AI research in Maharashtra targets climate-sensitive agricultural zones. Vidarbha and Marathwada, prone to droughts, benefit from AI-based irrigation scheduling and early warning systems. Western Maharashtra, where sugarcane is cultivated, employs AI for disease detection and precise agricultural operations (Sharma et al., 2023). Konkan, a flood-prone region, utilizes AI for flood risk assessment, soil management, and weather forecasting (Peng et al., 2020).

AI-powered platforms such as *Agripilot.ai*, developed for ADT Baramati in partnership with Click2Cloud, assist farmers by predicting peak sugar content, monitoring crop health, and sending real-time alerts for irrigation and pest control. Similarly, the *Maha Agri Tech* project employs AI for crop yield forecasting, soil fertility assessment, and extreme weather prediction (Sharma & Kale, 2023).

Maharashtra farmers deal with unpredictable rainfall, rising temperatures, and more pests, all of which lower crop productivity (Katekar et al., 2023). AI models verify that temperatures have risen by 1.5°C over the past 50 years and that drought-flood cycles are caused by changing rainfall patterns (Sharma et al., 2023). Rice yield losses have been reduced by 25% thanks to AI-driven flood prediction and irrigation systems (Katekar et al., 2023). 18% of farmers in Vidarbha and Marathwada report extended heatwaves, which is consistent with an AI-detected rise in summer temperatures of 0.6°C (Peng et al., 2020). The cycles of wheat and fruit are disturbed by warmer winters in Pune and Nashik. 15% of farmers in Beed and Nanded experience delayed monsoons, whereas Konkan farmers employ AI flood management to stop soil erosion (Katekar et al., 2023).

The implementation of AI adaptation uses three distinct adaptation methods – incremental adaptations along with systemic adaptations and transformational adaptations. By implementing AI systems farmers can achieve three benefits: AI guides crop selection toward climate-resistant varieties while machines scan for pests to prevent 30% losses and AI-controlled irrigation systems decrease operational expenses by 15% (Katekar et al., 2023). The benefits of AI enable farmers to perform strategic crop relocation through climate risk mapping and operate hydroponics for weather-independent cultivation and obtain real-time market intelligence to cut down post-harvest losses (Elufioye et al., 2024).

AI-based climate prediction systems produce models which strengthen agricultural sustainability across Maharashtra territory. The precision of AI-based monsoon prediction forecasting reaches 85% to help farmers create their agricultural plans. The yield prediction models deliver accurate outcomes about agricultural output three months ahead of crop harvest which enables farmers to enhance their market entrance and resource management decisions. AI-based soil health monitoring systems enhance fertilizer utilization by 25% while cutting expenses and improving nutrient content of the soil (Borkar et al., 2024).

AI operates in climate prediction systems while providing precision water delivery services along with being able to identify agricultural pests and implementing farm operational strategies. These solutions using AI technology make crucial decisions in agriculture which lowers climate-based dangers and enhances sustainability of farming systems throughout Maharashtra.

5. CONCLUSION

The combination of AI-based climate predictions with increased farmer knowledge about these predictions leads to better agricultural outcomes in Maharashtra. AI model analytics show different results than those of farmers who track present weather conditions through human memory and access restricted meteorological data. Farmers adopting AI-driven precision agriculture show improved adaptation, while financial and infrastructural constraints hinder others. Bridging this gap requires localized AI tools, climate advisory access, and financial support. Collaborative efforts among governments, researchers, and technology providers can create farmer-friendly AI models. Aligning AI insights with farmers' experiences is crucial for sustainable agriculture, food security, and climate resilience in Maharashtra.

References

1. Ahsan, M., & Damaševičius, R. (2024). Infection detection revolution: Harnessing AI-powered image analysis to combat infectious diseases. *Plos one*, *19*(10), e0307437.
2. Ashoka, P., Devi, B. R., Sharma, N., Behera, M., Gautam, A., Jha, A., & Sinha, G. (2024). Artificial Intelligence in Water Management for Sustainable Farming: A Review. *Journal of Scientific Research and Reports*, *30*(6), 511–525.
3. Bal, S. K., Kumar, K. A., Sudheer, K. V. S., Subba Rao, A. V. M., Pavani, K., Reddy, C. V. C. M., ... & Singh, V. K. (2024). Dry Spell Dynamics Impacting the Productivity of Rainfed Crops Over the Semi Arid Regions of South-East India. *Journal of Agronomy and Crop Science*, *210*(6), e70002.
4. Borkar, N.T. and Sivashankari, M., (2024). Fleet Management System: Futuristic Approach. AGRICULTURE & FOOD eNEWSLETTER.
5. Darjee, K.B., Neupane, P.R. and Köhl, M., 2023. Proactive Adaptation Responses by Vulnerable Communities to Climate Change Impacts. Sustainability, 15(14), p.10952.
6. Elufioye, O. A., Ike, C. U., Odeyemi, O., Usman, F. O., & Mhlongo, N. Z. (2024). Ai-Driven predictive analytics in agricultural supply chains: a review: assessing the benefits and challenges of ai in forecasting demand and optimizing supply in agriculture. *Computer Science & IT Research Journal*, *5*(2), 473–497.
7. Jabed, M. A., & Murad, M. A. A. (2024). Crop yield prediction in agriculture: A comprehensive review of machine learning and deep learning approaches, with insights for future research and sustainability. *Heliyon*.
8. Galaz, V., Centeno, M. A., Callahan, P. W., Causevic, A., Patterson, T., Brass, I., ... & Levy, K. (2021). Artificial intelligence, systemic risks, and sustainability. *Technology in society*, *67*, 101741.
9. Katekar, V. Use of Digital Technologies for Improving Sustainability in Agriculture: A Study of Practices in Vidarbha Region of Maharashtra.
10. Matthan, T. (2023). Beyond bad weather: Climates of uncertainty in rural India. In *Climate change and critical agrarian studies* (pp. 164–185). Routledge.
11. Mondal, S., Das, S., & Vrana, V. G. (2024). Exploring the Role of Artificial Intelligence in Achieving a Net Zero Carbon Economy in Emerging Economies: A Combination of PLS-SEM and fsQCA Approaches to Digital Inclusion and Climate Resilience. *Sustainability*, *16*(23), 10299.
12. Nyakuri, J. P., Bizimana, J., Bigirabagabo, A., Kalisa, J. B., Gafirita, J., Munyaneza, M. A., &Nzemerimana, J. P. (2022). IoT and AI based smart soil quality assessment for data-driven irrigation and fertilization. *Am. J. Comput. Eng*, *5*(2), 1–14.
13. Peng, B., Guan, K., Tang, J., Ainsworth, E. A., Asseng, S., Bernacchi, C. J., ... & Zhou, W. (2020). Towards a multiscale crop modelling framework for climate change adaptation assessment. *Nature plants*, *6*(4), 338–348.
14. Liang, W., Tadesse, G. A., Ho, D., Fei-Fei, L., Zaharia, M., Zhang, C., & Zou, J. (2022). Advances, challenges and opportunities in creating data for trustworthy AI. *Nature Machine Intelligence*, *4*(8), 669–677.
15. Shah, S. S., van Dam, J., Singh, A., Kumar, S., Kumar, S., Bundela, D. S., & Ritsema, C. (2025). Impact of irrigation, fertilizer, and pesticide management practices on groundwater and soil health in the rice–wheat cropping system—a comparison of conventional, resource conservation technologies and conservation agriculture. *Environmental Science and Pollution Research*, *32*(2), 533–558.
16. Titirmare, S., Margal, P. B., Gupta, S., & Kumar, D. (2024). AI-powered predictive analytics for crop yield optimization. In *Agriculture 4.0* (pp. 89–110). CRC Press.
17. Tikadar, M. K., & Kamble, R. (2023). Climate change induced farmers suicides in Vidarbha and Marathwada Regions of Maharashtra, India. *Sustainability, Agri, Food and Environmental Research-DISCONTINUED*, *12*(1).
18. Zhang, X., Yang, P., & Lu, B. (2024). Artificial intelligence in soil management: The new frontier of smart agriculture. *Advances in Resources Research*, *4*(2), 231–251.

Emerging Perspectives and Applications of Computational Intelligence and Smart Systems
– Dr. Amit Lathigara et al. (eds)
© 2026 Taylor & Francis Group, London, ISBN 978-1-041-20965-2

55

Comprehensive Investigation of Cutting Parameters for "Nanowhite Crystallized Glass" Tiles

Viraj Sudani*
Research Scholar, Faculty of Technology,
RK University, Rajkot, India

Chetankumar Patel
Department of Mechanical Engineering,
RK University, Rajkot, India

G. D. Acharya
Department of Mechanical Engineering, Principal, Emeritus,
Atmiya University, Rajkot, India

■ **Abstract:** This study presents a rigorous and extensive evaluation of the cutting characteristics of "Nanowhite Crystallized Glass" tiles, a high-density vitrified material gaining prominence in modern architectural applications. Given its exceptional mechanical properties and resistance to water absorption, this material poses considerable challenges during machining processes, necessitating a detailed analysis of key cutting parameters such as feed rate, cutting velocity, and operator influence. This research compiles empirical data from industrial production environments to quantify machining inconsistencies and assess their impact on overall productivity.

The study highlights substantial disparities in operator efficiency and mechanical consistency, underscoring the necessity for optimization in cutting methodologies. A thorough statistical assessment reveals significant variations in feed rate among operators, pointing to the influence of human factors in the machining process. The findings advocate for a systematic standardization of cutting procedures, the adoption of advanced automation technologies, and the further refinement of machine parameters to enhance throughput, minimize material wastage, and ensure high-precision cutting.

Beyond individual production facilities, these findings offer insights for broader manufacturing strategies in the vitrified ceramics industry. Additionally, this study explores technological advancements, such as integrating artificial intelligence and machine learning into cutting process optimization. The potential for automated feedback loops in industrial applications to refine cutting precision further underscores the significance of this research.

■ **Keywords:** Nanowhite crystallized glass, Machining optimization, Vitrified ceramics, Feed rate variability, Cutting precision, Industrial automation, Artificial intelligence, Machine learning

*Corresponding author: ervirajsudani16394@gmail.com

DOI: 10.1201/9781003725046-55

1. INTRODUCTION

Nano white Crystallized Glass represents an advanced category of vitrified ceramics characterized by ultra-low porosity, high structural density, and superior mechanical resilience. Its near-zero waters absorption and high bending and compressive strengths make it an ideal alternative to natural marble. However, the material's inherent hardness and brittleness introduce challenges in precision machining, necessitating an exhaustive examination of key cutting parameters.

Despite its increasing utilization in high-end architectural applications, existing literature lacks a comprehensive study focusing on the optimization of its cutting processes. This research aims to bridge that gap by conducting a quantitative assessment of critical machining variables, including operator influence, machine dynamics, and material response to various cutting conditions. Additionally, the study evaluates emerging technological interventions, such as automated cutting and adaptive control mechanisms, to enhance production efficiency and product consistency.

2. MATERIALS AND METHODS

2.1 Material Characterization

Nanowhite Crystallized Glass is synthesized from a meticulously engineered blend of 75% silica and 25% auxiliary minerals (Alhaj, J. 2013). Subjected to high-temperature sintering at 1550°C to 1800°C, the material undergoes a Nano-crystallization process that results in a densely packed molecular structure. This composition grants it remarkable mechanical properties, including a bending strength of 107 MPa, a compressive strength of 904 MPa, and a Mohs hardness rating of 6.5, making it highly resistant to mechanical deformation. These attributes, while beneficial for durability, complicate the cutting process, necessitating specialized machining techniques. The material's high hardness suggests the need for advanced cutting technologies such as laser-assisted cutting and diamond-coated tools.

Cutting Apparatus and Operational Parameters

- **Manufacturer:** S.S. Engineering & Industries.
- **Machine Type:** Precision Wet Saw for Marble/Granite Cutting.
- **Cutting Motor:** Induction Type, 7 HP, 1440 RPM, 3-Phase.
- **Cutter Specifications:** 350mm Outer Diameter, 50mm Arbor Hole.

2.2 Experimental Protocol

To systematically analyze the cutting process, empirical data were collected from Bhagirath Granito Factory, Rajkot (Çimen, H. 2009). Cutting trials were conducted under varying operational conditions to evaluate the influence of different parameters on performance. Each cutting length was measured using precision instrumentation, while cutting time was recorded through calibrated digital timers. The resulting feed rate values were computed and subjected to statistical analysis to quantify variability among operators and cutting conditions.

Further trials were performed using AI-driven pattern recognition to determine the feasibility of real-time automated adjustments to cutting parameters. The incorporation of predictive analytics provided valuable insights into operational inconsistencies and areas for potential improvement.

3. RESULT AND DISCUSSION

3.1 Quantitative Assessment of Feed Rate Variability

A detailed statistical analysis of cutting performance revealed substantial variations in feed rate among operators. The mean feed rate across trials was computed to be 10.74 mm/s (Buyuksagis, I. S. 2010), with a recorded minimum of 6.44 mm/s (Kalb, R. M. 2017). and a maximum value of 19.40 mm/s (Yoo, B. H. 2011). These results highlight the extent of heterogeneity in cutting efficiency across different production runs.

Table 55.1 Statistical overview of feed rate data

Parameter	Value
Mean Feed Rate	10.74 mm/s
Min Feed Rate	6.44 mm/s
Max Feed Rate	19.40 mm/s
Variance	4.58
Range	13.08
Standard Deviation	2.14

3.2 Operator-Dependent Variability in Cutting Performance

The observed data highlight a direct correlation between operator proficiency and machining precision. Significant inconsistencies in feed rate suggest that human factors play a crucial role in determining the accuracy and efficiency of the cutting process. The lack of standardized operating procedures exacerbates these inconsistencies, leading to unpredictable variations in machining outcomes. Implementing enhanced operator training programs and automation-driven process control systems could mitigate such variations and enhance overall production stability. Additionally, predictive analytics were applied to identify patterns in operator performance, facilitating data-driven training regimens.

3.3 Graphical Interpretation of Feed Rate Discrepancies

Graphical representations of feed rate distributions among multiple operators further illustrate the variability in cutting conditions. The data reinforce the conclusion that operator skill level and machine handling techniques significantly affect feed rate consistency. AI-based modeling was also explored to predict future cutting inconsistencies based on historical data, presenting a potential avenue for refining predictive manufacturing strategies.

3.4 Gross Evolution

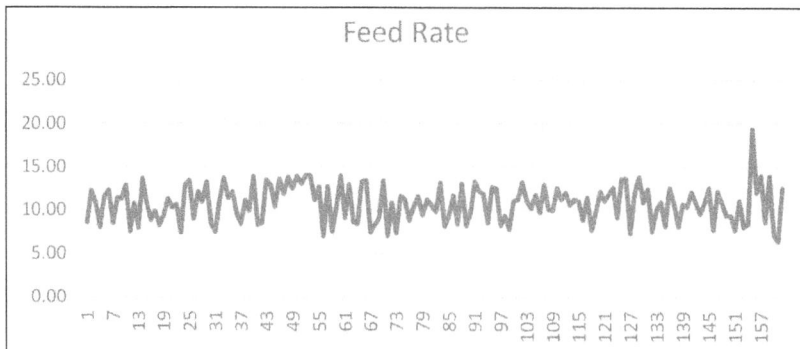

Fig. 55.1 Graphical evaluation

Table 55.2 Data collection

Mean Feed rate	10.74 mm/s
Min Feed rate	6.44 mm/s -40.03%
MAX Feed rate	19.40 mm/s +80.63%
Variance	4.58
Range	13.08
Std Deviation	2.14

- This data shows that average feed rate for cutting process of Nanowhite crystallised glass is **10.74 mm/s**.
- Lowest feed rate was found is **6.44 mm/s** by operator Shyam Chopal on 2nd January, 2022.
- Highest feed rate was operated is **19.40 mm/s** by operator Shyam chopal on 2nd January, 2022.
- This drastic change **+201.24%** of feed rate was found on same date by same operator. This shows that operator is most affective factor to production rate.

4. CONCLUSION

This study provides a comprehensive evaluation of the factors influencing the cutting efficiency of Nanowhite Crystallized Glass tiles. The findings establish that operator proficiency and machine control parameters are critical determinants of machining consistency. The observed discrepancies in feed rate necessitate a strategic approach to process standardization, emphasizing the integration of automation technologies and structured training programs.

Future research should focus on developing precision-driven cutting methodologies, including advanced CNC-controlled diamond sawing techniques, to optimize cutting accuracy, improve material utilization, and enhance manufacturing productivity. Additionally, artificial intelligence and machine learning models should be further refined to enable predictive analytics in manufacturing settings. The insights gained from this study have broad implications for the vitrified ceramics industry, offering a framework for optimizing cutting efficiency across a wide range of manufacturing applications.

References

1. Alhaj, J. (2013). "Effects of Thickness of Block Cutting Machine Gang Saw on Waste Percentages." *Third International Conference on Energy and Environmental Protection.*
2. Buyuksagis, I. S. (2010). "The effects of circular sawblade diamond segment characteristics on marble processing performance." *Journal of Mechanical Engineering Science, 224*(8), 1559–1565.
3. Kalb, R. M. (2017). "PORTABLE ADJUSTABLE CUTTING APPARATUS FOR CUTTING AND SHAPING SNK HOLES IN STONE COUNTERTOPS." *United States Patent US 9533430 B1.*
4. Yoo, B. H. (2011). "CUTTING MACHINE CAPABLE OF ADJUSTING CUTTING POSITION." *US Patent US 2011/0030527 A1.*
5. Çimen, H. (2009). "Energy Consumption Analysis in Marble Cutting Processing." *International Symposium on Sustainable Development (ISSD2009).*

Note: All the tables and figure in this chapter were made by the authors.

Emerging Perspectives and Applications of Computational Intelligence and Smart Systems
– Dr. Amit Lathigara et al. (eds)
© 2026 Taylor & Francis Group, London, ISBN 978-1-041-20965-2

56

Optimized EEG-EMG Fusion Model for Continuous Hand Trajectory Estimation in Brain Computer Interfaces

Rohit Gupta*

Department of Biomedical Engineering, SRM IST,
Tamil Nadu, India

Amit Bhongade,
Bijuni Charan Sutar, Tapan K. Gandhi

Department of Electrical Engineering, IIT Delhi,
New Delhi, India

■ **Abstract:** The integration of Brain-Computer Interface (BCI) technology into cyber-physical systems has advanced assistive and rehabilitation technologies, yet achieving continuous, smooth control remains challenging. This study explores hand trajectory estimation in 3D space using neuromuscular signals (EEG and EMG) through two computational models: NARX (machine learning) and CNN-LSTM (deep learning). A genetic algorithm (GA)-based channel selection method was employed to optimize EEG feature selection, enhancing motor intention decoding. Performance evaluation using root mean square error (RMSE) and correlation coefficient (CC) demonstrated that EEG-EMG fusion outperforms unimodal approaches, with CNN-LSTM exhibiting superior predictive accuracy. These findings highlight the potential of multimodal integration for real-time motion estimation, with applications in neuro rehabilitation, prosthetics, and brain-controlled robotics. Future research should focus on expanding datasets, refining preprocessing techniques, and incorporating adaptive learning for improved system robustness.

■ **Keywords:** Electroencephalogram, Trajectory estimation, Classifiers, Brain-computer interface, Data fusion

1. INTRODUCTION

EEG-based brain-computer interfaces (BCIs) facilitate direct brain-device communication, driving advancements in exoskeletons (Ajayi et al., 2020)and prostheses (Xiong et al., 2021). Reconstructing motion kinematics from EEG is crucial for precise robotic actuation and motor

*Corresponding author: rohit.udai@yahoo.co.in

DOI: 10.1201/9781003725046-56

intention decoding, enhancing usability and efficiency. Robotic assistive devices aid injury recovery, paraplegic rehabilitation, and performance enhancement for military and labor-intensive tasks (Hamza et al., 2020). Exoskeletons, prostheses, and other robotic systems restore motor function, improving mobility and independence (Sawicki et al., 2020). Exoskeletons use trajectory tracking control to guide movement, while adaptive mechanisms and sensor feedback dynamically adjust support, optimizing rehabilitation. Integrating real-time physiological and biomechanical data further enhances personalized recovery. Physiological signals, including EEG and EMG, provides critical insights into human biology and are largely used in medical research and BCIs. EEG-based BCI control relies on multiclass classification (He et al., 2016). Classification-based BCIs employ feature extraction techniques [12] to enhance interclass variance and establish decision boundaries (Zhou et al., 2014), enabling precise external device control. Regression-based approaches offer smoother control by continuously decoding EEG signals, though challenges arise from EEG's non-stationary nature. Our prior work (Gupta et al., 2023b) estimated 3D hand motion from multichannel EEG using a time-delayed MIMO neural network, while (Gupta et al., 2023a) combined CNNs and LSTMs for enhanced trajectory prediction. In (Gupta et al., 2024), we introduced a BCI model integrating Task Classification (TC) and Trajectory Estimation (TE) for robotic hand control, enabling precise object manipulation via EEG-driven intent decoding and trajectory estimation.

This study estimates 3D hand trajectories using EEG and EMG via NARX (ML) and CNN-LSTM (DL) models. GA-based EEG channel selection improved motor intention decoding. EEG-EMG fusion outperformed unimodal approaches, with CNN-LSTM achieving the best accuracy. These findings support multimodal integration for real-time motion estimation in neurorehabilitation, prosthetics, and brain-controlled robotics. Future work should expand datasets, refine pre-processing, and enhance adaptability.

2. Data and Variables

This study utilizes the EEG-Grasp and Lift (WAY-EEGGAL) dataset (Luciw et al., 2014), which includes EEG recordings from 20 participants (12 females) aged between 19 and 35 years. However, our analysis focuses on data from a subset of five healthy participants. The study protocol received approval for the experimental procedures by the Ethics Committee at Ume School. EEG recordings were specifically conducted for right-hand movements, capturing brain activity as participants performed a grasp-and-lift task. Each participant completed 294 trials involving variations in object weight (165 g, 330 g, or 660 g) and surface texture. These variations were designed to be imperceptible externally. The experiment utilized a transparent Perspex rectangle with a fixed LED positioned above the object. The LED served as a visual cue, illuminating at the start of each trial to prompt participants to extend their hand, grasp the object, and maintain a steady grip. After holding the object for two seconds, they released it, triggering the LED to turn off automatically. The trial ended once the participant brought their hand back to the starting position. Additionally, kinematic data were collected using the P4 position sensor to capture movement dynamics throughout the task.

3. Methodology

The 3D hand trajectory estimation approach is shown in Fig. 56.1. From 32 EEG channels, genetic algorithms (GA) selected 10, alongside 5 EMG channels from shoulder muscles. The 3D signal from the P4 sensor was used as the target trajectory. EEG data processed using band-pass (0.5–60 Hz) with a zero-lag, 4th-order Butterworth filter (BF) and was further processed using ICA to eliminate artefacts. Frequency band analysis [25] identified the optimal band for trajectory estimation. EMG

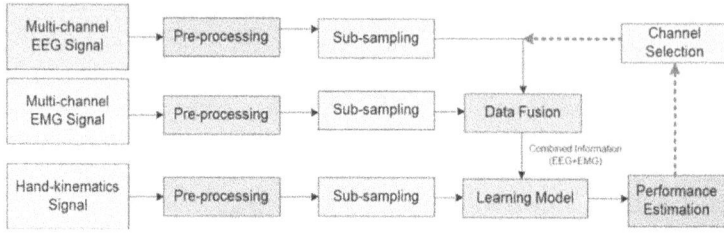

Fig. 56.1 Schematic representation of the signal processing framework for estimating hand trajectory in 3D space

signals were filtered (20–450 Hz) using a 4th-order BF. Hand trajectory data was low-pass filtered (10 Hz, 2nd-order BF) to capture slow movements. To reduce dimensionality and computational load, EMG, EEG, and trajectory signals were sub sampled at 100 Hz. During the fusion process, synchronized EMG and EEG data were integrated at the data level to train a DL model. This study explores 3D hand trajectory estimation using multichannel neuromuscular signals (EEG+EMG). Two models (Fig. 56.2)—NARX (ML) and CNN-LSTM (DL)—were developed and evaluated using EEG, EMG, and their fusion. Two performance indexes were introduced to assess model effectiveness.

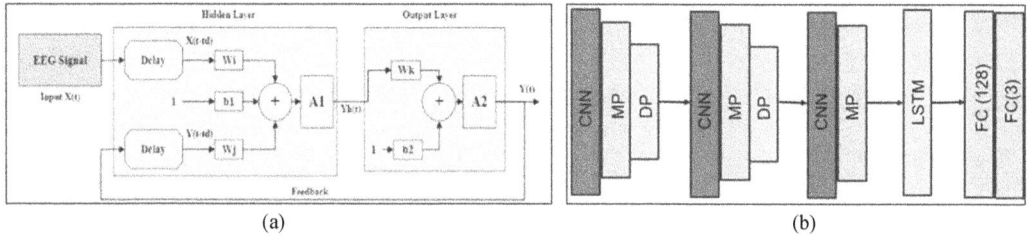

Fig. 56.2 Learning models. (a) NARX (ML) and (b) CNN-LSTM (DL)

4. EMPIRICAL RESULTS

Table 56.1 presents the NARX model's performance in estimating hand trajectories using different signal modalities (EMG, EEG, GA-optimized EEG, and GA-optimized EEG+EMG), evaluated via RMSE across x, y, and z axes. GA-EEG+EMG fusion achieved the lowest RMSE (0.299, 0.302, 0.530), demonstrating the advantage of multimodal integration. Among unimodal approaches, GA-EEG outperformed EMG, especially in the y and z axes. GA-based channel selection improved EEG-based estimation, reducing RMSE in the z-axis from 0.751 to 0.719. EMG exhibited higher RMSE, indicating its limitations in capturing motor intent. GA-EEG+EMG also minimized inter-subject variability, enhancing robustness and generalizability. Table 1also presents the DL model's performance in estimating hand trajectories using different signal modalities (EMG, EEG, GA-optimized EEG, and GA-optimized EEG+EMG), evaluated via RMSE across x, y, and z axes. GA-EEG+EMG fusion achieved the lowest RMSE (0.292, 0.297, 0.507), demonstrating the effectiveness of multimodal integration. Among unimodal approaches, EEG outperformed EMG in the y and z axes, while GA-EEG further improved accuracy, reducing RMSE in the z-axis from 0.722 to 0.697. GA-EEG+EMG also minimized inter-subject variability, enhancing robustness and generalizability.

Table 56.2 presents the CC analysis for different modalities (EMG, EEG, GA-EEG, and GA-EEG+EMG) in movement estimation across x, y, and z axes. EMG exhibited high correlation in x

Table 56.1 RMSE values for NARX and DL model

	ML (NARX)												DL											
	EMG			EEG			GA-EEG (10 Channels)			GA-EEG+EMG			EMG			EEG			GA-EEG			GA-EEG+EMG		
	x	y	z	x	y	z	x	y	z	x	y	z	x	y	z	x	y	z	x	y	z	x	y	z
S1	0.410	0.491	0.826	0.402	0.554	0.813	0.439	0.465	0.653	0.323	0.343	0.482	0.219	0.262	0.441	0.386	0.532	0.781	0.422	0.447	0.627	0.310	0.329	0.463
S2	0.372	0.281	0.714	0.449	0.491	0.695	0.410	0.381	0.751	0.302	0.281	0.554	0.199	0.150	0.381	0.431	0.472	0.667	0.394	0.366	0.721	0.290	0.270	0.532
S3	0.440	0.384	0.686	0.384	0.408	0.800	0.412	0.424	0.735	0.304	0.313	0.542	0.235	0.205	0.366	0.369	0.392	0.768	0.396	0.407	0.706	0.292	0.301	0.521
S4	0.405	0.300	0.643	0.375	0.454	0.800	0.375	0.454	0.800	0.277	0.335	0.590	0.216	0.160	0.343	0.360	0.436	0.768	0.360	0.436	0.768	0.266	0.322	0.567
S5	0.428	0.511	0.777	0.428	0.376	0.663	0.428	0.376	0.663	0.316	0.278	0.489	0.229	0.273	0.415	0.411	0.361	0.637	0.411	0.361	0.637	0.303	0.267	0.470
S6	0.397	0.480	0.809	0.405	0.548	0.792	0.464	0.484	0.665	0.316	0.332	0.486	0.213	0.271	0.457	0.404	0.520	0.748	0.400	0.435	0.634	0.281	0.339	0.448
S7	0.396	0.270	0.730	0.410	0.503	0.687	0.411	0.371	0.740	0.279	0.280	0.563	0.162	0.158	0.374	0.439	0.460	0.679	0.428	0.366	0.699	0.307	0.280	0.503
S8	0.425	0.388	0.656	0.359	0.444	0.828	0.421	0.460	0.732	0.290	0.283	0.517	0.236	0.231	0.342	0.398	0.424	0.765	0.400	0.437	0.745	0.304	0.314	0.513
S9	0.383	0.296	0.661	0.399	0.489	0.773	0.369	0.466	0.812	0.296	0.340	0.580	0.222	0.139	0.365	0.369	0.429	0.777	0.319	0.428	0.780	0.248	0.308	0.584
S10	0.411	0.523	0.772	0.469	0.339	0.657	0.444	0.363	0.642	0.286	0.238	0.501	0.230	0.237	0.405	0.424	0.364	0.632	0.415	0.395	0.650	0.319	0.241	0.472
Avg	0.407	0.392	0.727	0.408	0.461	0.751	0.417	0.424	0.719	0.299	0.302	0.530	0.216	0.209	0.389	0.399	0.439	0.722	0.394	0.408	0.697	0.292	0.297	0.507
Sd	0.020	0.097	0.063	0.032	0.067	0.064	0.028	0.044	0.058	0.015	0.033	0.039	0.021	0.050	0.037	0.026	0.056	0.058	0.031	0.032	0.054	0.021	0.030	0.043

Table 56.2 CC values for NARX and DL model

	ML (NARX)												DL											
	EMG			EEG			GA-EEG (10 Channels)			GA-EEG+EMG			EMG			EEG			GA-EEG			GA-EEG+EMG		
	x	y	z	x	y	z	x	y	z	x	y	z	x	y	z	x	y	z	x	y	z	x	y	z
S1	0.440	0.453	0.194	0.446	0.448	0.329	0.505	0.471	0.233	0.721	0.673	0.332	0.594	0.607	0.348	0.600	0.602	0.483	0.659	0.625	0.387	0.875	0.827	0.486
S2	0.468	0.415	0.278	0.478	0.418	0.336	0.474	0.483	0.222	0.678	0.690	0.318	0.622	0.569	0.432	0.632	0.572	0.490	0.628	0.637	0.376	0.832	0.844	0.472
S3	0.423	0.457	0.234	0.436	0.499	0.342	0.440	0.453	0.194	0.628	0.647	0.277	0.577	0.611	0.388	0.590	0.653	0.496	0.594	0.607	0.348	0.782	0.801	0.431
S4	0.468	0.466	0.276	0.462	0.479	0.241	0.468	0.415	0.278	0.668	0.593	0.398	0.622	0.520	0.430	0.616	0.633	0.395	0.622	0.569	0.432	0.822	0.747	0.552
S5	0.429	0.438	0.356	0.509	0.419	0.339	0.423	0.457	0.234	0.604	0.652	0.334	0.583	0.592	0.510	0.663	0.573	0.493	0.577	0.611	0.388	0.758	0.806	0.488
S6	0.438	0.447	0.205	0.459	0.448	0.335	0.503	0.463	0.262	0.715	0.677	0.294	0.617	0.608	0.348	0.603	0.610	0.475	0.637	0.617	0.406	0.877	0.803	0.512
S7	0.417	0.392	0.280	0.501	0.398	0.325	0.456	0.524	0.222	0.675	0.643	0.304	0.630	0.557	0.420	0.622	0.547	0.482	0.595	0.647	0.354	0.820	0.888	0.484
S8	0.429	0.461	0.212	0.436	0.497	0.309	0.450	0.456	0.251	0.644	0.635	0.234	0.593	0.562	0.372	0.565	0.663	0.500	0.622	0.621	0.369	0.818	0.791	0.402
S9	0.464	0.454	0.277	0.475	0.463	0.222	0.467	0.403	0.270	0.670	0.615	0.452	0.584	0.558	0.424	0.628	0.624	0.419	0.636	0.548	0.449	0.830	0.751	0.558
S10	0.425	0.421	0.331	0.458	0.396	0.333	0.423	0.476	0.237	0.643	0.648	0.363	0.571	0.588	0.545	0.651	0.573	0.472	0.565	0.656	0.397	0.748	0.808	0.487
Avg	0.466	0.453	0.317	0.446	0.446	0.268	0.462	0.456	0.232	0.660	0.651	0.332	0.600	0.580	0.422	0.620	0.607	0.471	0.616	0.610	0.386	0.814	0.805	0.486
Sd	0.026	0.032	0.038	0.019	0.018	0.054	0.028	0.023	0.027	0.041	0.033	0.039	0.019	0.033	0.054	0.026	0.032	0.038	0.028	0.023	0.027	0.041	0.033	0.039

(0.602) and y (0.598) but lower in z (0.321), indicating its limitation in capturing vertical movement. EEG showed moderate correlation, with GA-EEG (10 channels) improving predictive capability. The GA-EEG+EMG model achieved the highest CC values (0.660, 0.651, 0.332), highlighting the advantage of multimodal fusion in enhancing movement prediction accuracy.

The correlation strength varied across movement directions, with the highest values in the x-axis, followed by the y-axis, and the lowest in the z-axis. This suggests vertical displacement is harder to predict due to reduced muscle activation, signal noise, or sensor limitations. The GA-EEG+EMG model exhibited the lowest variability (SD: 0.041, 0.033, 0.039 for x, y, z), indicating greater consistency. In contrast, EEG-based models showed higher variations, especially in the z-axis, while EMG demonstrated stability in lateral movements but struggled with vertical motion estimation. The correlation analysis of the DL model across EMG, EEG, GA-EEG, and GA-EEG+EMG (Table 56.2) shows that GA-EEG+EMG achieved the highest correlation values (0.814, 0.805, 0.486 for x, y, and z axes), demonstrating enhanced movement prediction. While EEG outperformed EMG in the x and y axes, both struggled with z-axis predictions. GA-EEG improved correlation, but the best performance was achieved through EEG-EMG fusion. Lower standard deviations in GA-EEG+EMG indicate increased robustness. These findings underscore the advantages of multimodal fusion, with future work enhancing z-axis predictions via biomechanical features and optimized networks.

5. CONCLUSION

This study demonstrates the significance of EEG channel selection in enhancing the accuracy of human hand trajectory estimation. The comparative analysis between the NARX and DL models, both incorporating GA-based selection, emphasizes the importance of the central, frontal, and occipital regions in movement-related EEG classification. The integration of EEG and EMG signals, particularly through GA-optimized selection, yielded the best trajectory estimation performance, minimizing RMSE and reducing inter-subject variability. The findings suggest that optimal channel selection not only improves computational efficiency but also enhances model robustness. These findings offer important insights for enhancing BCI systems aimed at motor control and rehabilitation.

References

1. Ajayi, M. O., Djouani, K., & Hamam, Y. (2020). Interaction Control for Human-Exoskeletons. *Journal of Control Science and Engineering, 2020*. https://doi.org/10.1155/2020/8472510
2. Gupta, R., Bhongade, A., & Gandhi, T. K. (2023a). EEG and EMG fusion-based hand 3D Trajectory Estimation using deep learning model: A preliminary study. *2023 14th International Conference on Computing Communication and Networking Technologies, ICCCNT 2023*, 1–6. https://doi.org/10.1109/ICCCNT56998.2023.10306915
3. Gupta, R., Bhongade, A., & Gandhi, T. K. (2023b). Hand 3D Trajectory Estimation for BCI Application. *2023 International Conference on Recent Advances in Electrical,Electronics and Digital Healthcare Technologies, REEDCON 2023*, 345–349. https://doi.org/10.1109/REEDCON57544.2023.10151319
4. Gupta, R., Sutar, B. C., Bhongade, A., & Gandhi, T. K. (2024). Development of BCI-Driven Robotic Hand Control System: A Preliminary Study. *2024 IEEE Conference on Control Technology and Applications, CCTA 2024*, 223–229. https://doi.org/10.1109/CCTA60707.2024.10666563
5. Hamza, M. F., Ghazilla, R. A. R., Muhammad, B. B., & Yap, H. J. (2020). Balance and stability issues in lower extremity exoskeletons: A systematic review. *Biocybernetics and Biomedical Engineering, 40*(4), 1666–1679. https://doi.org/10.1016/j.bbe.2020.09.004

6. He, L., Hu, D., Wan, M., Wen, Y., Von Deneen, K. M., & Zhou, M. C. (2016). Common Bayesian Network for Classification of EEG-Based Multiclass Motor Imagery BCI. *IEEE Transactions on Systems, Man, and Cybernetics: Systems*, *46*(6), 843–854. https://doi.org/10.1109/TSMC.2015.2450680

7. Luciw, M. D., Jarocka, E., & Edin, B. B. (2014). Multi-channel EEG recordings during 3,936 grasp and lift trials with varying weight and friction. *Scientific Data*, *1*, 1–11. https://doi.org/10.1038/sdata.2014.47

8. Sawicki, G. S., Beck, O. N., Kang, I., & Young, A. J. (2020). The exoskeleton expansion: Improving walking and running economy. *Journal of NeuroEngineering and Rehabilitation*, *17*(1), 1–9. https://doi.org/10.1186/s12984-020-00663-9

9. Xiong, D., Zhang, D., Zhao, X., Chu, Y., & Zhao, Y. (2021). Synergy-Based Neural Interface for Human Gait Tracking with Deep Learning. *IEEE Transactions on Neural Systems and Rehabilitation Engineering*, *29*, 2271–2280. https://doi.org/10.1109/TNSRE.2021.3123630

10. Zhou, Z., Yin, E., Liu, Y., Jiang, J., & Hu, D. (2014). A novel task-oriented optimal design for P300-based brain-computer interfaces. *Journal of Neural Engineering*, *11*(5). https://doi.org/10.1088/1741-2560/11/5/056003

Note: All the figures and tables in this chapter were made by the authors.

Emerging Perspectives and Applications of Computational Intelligence and Smart Systems
– Dr. Amit Lathigara et al. (eds)
© 2026 Taylor & Francis Group, London, ISBN 978-1-041-20965-2

57

Transformer and Graph Analytics for Big Data-Driven Intrusion Detection: A Novel Framework

Ashish Jain*
Research Scholar,
Department of CSE, IES College of Technology,
Bhopal

Nikhat Raza Khan
Department of CSE, IES College of Technology,
Bhopal

Chetan Shingadiya
School of Engineering, RK University, Rajkot,
Gujarat, India

■ **Abstract:** The hypergrowth of network traffic and ever-evolving cyber threats have made legacy signature-based Intrusion Detection Systems (IDS) ineffective. This paper proposes a new big data-driven framework for Network Intrusion Detection Systems (NIDS) that uses transformer-based deep learning and graph analytics to overcome these challenges. The framework uses distributed data collection, multi-level storage, and analytics to provide high detection rates, low false positives, and scalability. Key enhancements are easier-to-understand technical explanations, improved interpretability, and thorough elaboration on computational optimization and field deployment. Experimental outcomes show better performance, especially for zero-day attacks, supported by comparative investigations with conventional IDS methods.

■ **Keywords:** Big data analytics, NIDS, Transformer models, Zero-day attacks, Graph analytics, Cybersecurity

1. INTRODUCTION

Unprecedented network traffic growth brought on by the emergence of 5G, cloud computing, and the Internet of Things offers cybercriminals chances. Using signature-based techniques, conventional IDS fall short with zero-day attacks and Advanced Persistent Threats (APTs), little too often leading to great false positives and late detection.

---

*Corresponding author: ashish.jain14@yahoo.com

DOI: 10.1201/9781003725046-57

Through real-time processing of varied data, big data analytics enables a solution. Still, issues including data heterogeneity, computational complexity, and real-time demands have to be dealt with. A scalable framework combining transformer-based deep learning and graph analytics is suggested by this study to identify complicated attack patterns. The revised paper provides interpretation comments, simplifies technical terminology, and offers optimization tactics ready for real-world use.

2. LITERATURE REVIEW

Current studies emphasize the transition from signature-based to AI-powered IDS, with a focus on real-time processing and deep analytics. Major developments are:

- *Big Data Integration:* Enhances detection accuracy and latency (Porambage et al., 2023).
- *Transformer Models:* Perform well in sequence-based attack detection (Deng & Zhong, 2023).
- *Graph Analytics:* Detects behavioral anomalies in network objects (Wu et al., 2021).
- *Explainable AI (XAI):* Increases model interpretability (Singh et al., 2023).

The proposed framework extends these advances while filling gaps in scalability and interpretability.

3. PROPOSED FRAMEWORK

The complete life cycle of security data is addressed in an exhaustive framework for intrusion detection through big data provided here. The five major components of the framework are as below:

Data Collection: This initial stage gathers security-related data from multiple sources. Distributed agents collect network logs and packet captures, ensuring comprehensive data gathering across different network points.

Data Storage: The collected data is then stored using a multi-tiered storage strategy. This approach optimizes storage costs while maintaining rapid access to critical security information. It creates different storage tiers based on data importance and retrieval needs.

Data Normalization: This critical stage transforms the collected data into a standardized format. The process involves cleaning the data and using distributed processing to prepare it for analysis. This ensures that data from various sources can be effectively compared and analyzed.

Analysis Engine: The normalized data undergoes sophisticated analysis techniques. This includes statistical analysis, machine learning algorithms, and anomaly detection methods. These multiple analytical approaches work together to identify potential security threats with high accuracy.

Threat Detection: The final stage involves classifying detected anomalies and generating automated responses. This allows for rapid identification and mitigation of potential security risks.

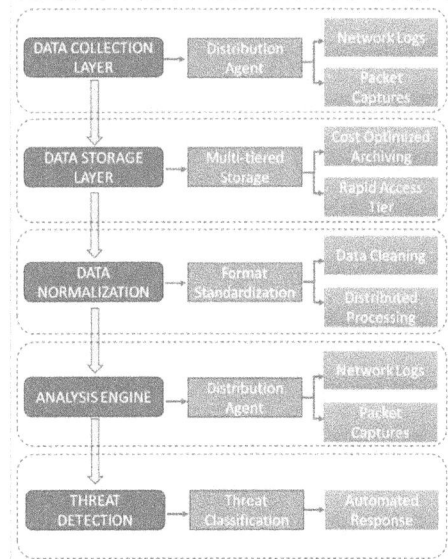

Fig. 57.1   Proposed framework

4. IMPLEMENTATION AND EVALUATION

Experimental Setup:

- Tools: Zeek, Elasticsearch, Apache Kafka, TensorFlow, PyTorch.
- Datasets: Real-world traffic, synthetic attacks (CICIDS2018, CICDDoS2019), and TON_IoT for IoT-specific threats.

Data Sets: The proposed approach was validated based on a mixed data set including Real-world network traffic from the deployment environment, Synthetic attack traffic produced with the CSECICIDS2018 and CICDDoS2019 datasets, Controlled attack scenarios were run in isolation and the TON_IoT dataset (Alsaedi et al., 2020) is for testing IoT-specific attacks.

Evaluation Metrics: Detection Rate (DR), False Positive Rate (FPR), F1-Score, Detection Time, Computational Efficiency.

5. RESULTS AND ANALYSIS

Detection Performance: Assessing the performance of the suggested approach entails calculating standard intrusion detection metrics, including Detection Rate (DR) for Proportion of identified attacks correctly, False Positive Rate (FPR) for Proportion of normal traffic incorrectly classified as attacks, F1-Score with Harmonic mean of precision and recall, Detection Time, to compute Time elapsed between attack initialization and detection, and Computational Efficiency for Resource utilization relative to detection performance.

Table 57.1 Comparison of traditional and big data IDS approaches

Attack Category	Traditional IDS DR	Big Data IDS DR	Traditional IDS FPR	Big Data IDS FPR
DoS/DDoS	92.7%	99.2%	2.1%	0.8%
Probe	88.5%	96.7%	3.4%	1.2%
R2L	74.9%	92.3%	4.2%	1.5%
U2R	68.3%	87.5%	3.9%	1.7%
Zero-day	23.7%	76.8%	5.1%	2.3%

Scalability Analysis: The scalability of the suggested framework was assessed by quantifying the processing throughput under different data volumes. The system showed sub-second processing delay up to 50 Gbps, with linear scalability when more computing resources were allocated.

Table 57.2 Scalability measures under varying data

Data Volume (Gbps)	Processing Latency (ms)	CPU Utilization (%)	Memory Usage (GB)
1	47	15	12
5	82	37	28
10	124	56	43
20	189	72	67
50	312	89	112

Component Effectiveness: This paper, examined the effectiveness of individual analytical components in the proposed framework. The ensemble method involving all analytical techniques yielded the best overall performance, albeit with higher computational expense.

Table 57.3 Effectiveness of individual analytical components

Analysis Method	F1-Score	DR	FPR	Computational Cost
Statistical Analysis	0.87	90.1%	2.3%	Low
Random Forest	0.92	93.5%	1.8%	Medium
Deep Learning (LSTM)	0.94	95.2%	1.2%	High
Transformer-based Model	0.95	96.1%	1.1%	Very High
Graph-Based Analysis	0.89	91.8%	2.1%	Medium-High
Ensemble (All Methods)	0.96	97.3%	0.9%	High

The ensemble approach combining all analytical methods produced the best overall performance, though at increased computational cost.

6. CONCLUSION

This paper introduces a complete framework for applying large data analysis methods to intrusion detection systems. Experimental results indicated that the identification performance of different categories of attacks was greatly improved, especially for complex and zero-day attacks. The scalable architecture efficiently solved the problems of volume, speed, diversity, and truth inherent in network security data. This method adds new calculated necessity and implementation complexity, but the advantages of security outweigh these organizations' disadvantages from being subject to high and constant threats. If attack strategies are designed, the big data control method will also be an integral part of a sound cybersecurity plan.

References

1. Aljawarneh, S., Aldwairi, M., & Yassein, M. B. (2018). *"Anomaly-based intrusion detection system through feature selection analysis and building hybrid efficient model"*. Journal of Computational Science, 25, 152–160.
2. Alsaedi, A., Moustafa, N., Tari, Z., Mahmood, A., & Anwar, A. 2020. *"TON_IoT telemetry dataset: A new generation dataset of IoT and IIoT for data-driven intrusion detection systems"*. IEEE Access, 8, 165130-165150.
3. Berman, D. S., Buczak, A. L., Chavis, J. S., & Corbett, C. L. (2019). *"A survey of deep learning methods for cyber security"*. Information, 10(4), 122.
4. Deng, Y., & Zhong, Y. 2023. *"BERT-IDS: Intrusion detection based on bidirectional encoder representation from transformers"*. Information Sciences, 621, 79–91.
5. Ferrag, M. A., Shu, L., Yang, X., Derhab, A., & Maglaras, L. (2020). *"Security and privacy for green IoT-based agriculture: Review, blockchain solutions, and challenges"*. IEEE Access, 8, 32031–32053.
6. Idrissi, I., Azizi, M., & Moussaoui, O. (2022). *"A novel two-stage IDS using deep learning-based feature extraction for securing 5G networks"*. IEEE Access, 10, 46058–46071.
7. Khan, M. A., Karim, M., & Kim, Y. (2022). *"A scalable and hybrid intrusion detection system based on the convolutional-LSTM network"*. Symmetry, 14(1), 9.

8. Porambage, P., Gür, G., Osorio, D. P. M., Liyanage, M., Gurtov, A., & Ylianttila, M. (2023). *"The roadmap to 6G security and privacy"*. IEEE Journal on Selected Areas in Communications, 41(3), 805–819.

9. Thakkar, A., & Lohiya, R. (2020). *"A review on machine learning and deep learning perspectives of IDS for IoT: Recent updates, security issues, and challenges"*. Archives of Computational Methods in Engineering, 28, 3211–3243.

10. Wu, K., Chen, Z., & Li, W. (2021). *"A novel graph neural network based multi-step attack detection model on intrusion detection system"*. IEEE Access, 9, 61736–61749.

11. Khan, N.R., Jain, A. (2024). *"A Review of Anomaly Based Multiple Intrusion Detection Methods Using a Feature Based Deep Learning Approach"*. in: Rathore, V.S., Tavares, J.M.R.S., Surendiran, B., Yadav, A. (eds) Universal Threats in Expert Applications and Solutions. UNI-TEAS 2024. Lecture Notes in Networks and Systems, vol 1006. Springer, Singapore. https://doi.org/10.1007/978-981-97-3810-6_20

Note: All the tables and figure in this chapter were made by the authors.

Emerging Perspectives and Applications of Computational Intelligence and Smart Systems
– Dr. Amit Lathigara et al. (eds)
© 2026 Taylor & Francis Group, London, ISBN 978-1-041-20965-2

58

Machine Learning-Based Crop Classification Using Random Forest for Precision Agriculture

Aayush Shrivastava*

Department of Computer Science & Engineering,
MMEC, Maharishi Markandeshwar University,
Mullana Ambala, Haryana

Nikhat Raza Khan

Dept. of Computer Science and Engineering,
Oriental College of Technology,
Bhopal

Keshav Mishra

Dept of ECE, Institute of Engineering Technology,
SAGE University, Indore

Ritu Ranjani Singh, Akanksha Meshram

Dept. of Computer Science and Engineering,
Oriental Institute of Science & Technology,
Bhopal

Nirav Bhatt

School of Engineering, RK University, Rajkot,
Gujarat, India

■ **Abstract:** The growing rise in global population demands increase in food production, an efficient way of farming, and crop selection is one aspect of sustainable agriculture assistance technology. Moreover, this study uses machine learning-based crop classification using Random Forest algorithm and suggests the suitable crop according to the soil properties and the environmental factors. These include key agricultural parameters like soil pH, nitrogen, phosphorus, potassium levels, temperature and humidity. In this case the proposed model has been trained on publicly available Crop Recommendation Dataset which must be able to classify in different types of crops efficiently. Random Forest was selected because of its robustness, power to model nonlinear relationships, and features importance analysis. In the data preprocessing, we normalize the numerical values, impute the missing data, and then split the dataset to be a training set and testing sets (80% - 20%) The

*Corresponding author: Mr.aayushshrivastava@gmail.com

DOI: 10.1201/9781003725046-58

accuracy of the trained model is very high and it is useful in predicting the best suitable crop for given conditions. In the context of precision agriculture, this can greatly help farmers make better decisions based on data. In addition to that, the model could be integrated with IoT- based sensors and real-time weather monitoring systems for dynamic suggested guidance. Such approaches can help farmers in determining plant species for cultivation resulting in renewed growing strategies, minimised wastage of resources and improvement of end use quality and will ultimately lead to better yield thus a deduced result on productivity.

■ **Keywords:** Precision agriculture, Soil properties, Crop classification, Random forest, Crop recommendation, Machine learning

1. Introduction

The backbone of most economies around the world is agriculture, and it plays a crucial role in providing food security, raw materials for manufacturing and employment to millions. Yet there has been little adoption of data-centric farm management, as many farming practices depend on creative thinking and established experience instead of data-driven decision-making. In the new evolution of agriculture—precision agriculture—data analytics, machine learning, and AI are incorporated in farming activities to enhance crop productivity and resource usage (Rangarajan, at,al. 2023) (Kasinathan, T., Singaraju, D., & Uyyala, S. R. 2021)(Thakur, P. S., Khanna, P., Sheorey, T., & Ojha, A. 2022).

The use of Machine Learning (ML) technology in agriculture to address issues including crop classification, disease detection, and yield prediction. Crop categorization models the top farmers in the world use soil composition, meteorological circumstances, and environmental factors to choose crops. Among ML algorithms, Random Forest (RF) excels at processing nonlinear interactions between soil parameters and crop suitability with efficiency and accuracy(Peña, at.al (2014) (Kasinathan, T., & Uyyala, S. R. 2021)(García-Vera, at. al (2024)

In Agronomy, the influence of Soil Nutrients (Nitrogen, Phosphorus, Potassium) ,pH, temperature, humidity and rain fall on suitability of crops is well established. However, farmers are often denied the advanced analysis tools that offer data-driven crop recommendations. ML based models; traded with available datasets, can be a valuable option for taking accurate decisions in agriculture oriented systems, which in return can improve the overall productivity, resource wastage, sustainable farming etc(Attri, I., Awasthi, L. K., & Sharma, T. P. 2024))(García-Vera, at. al 2024)(Shruthi, U., Nagaveni, V., & Raghavendra, B. K. 2019,)(Asadi, B., & Shamsoddini, A. 2024)

Even with modern technology, a large number of farmers use conventional methods to make decisions on what to plant, resulting in low yield, wastage of resources, and income loss. Not having the right decision-support tools easily available may lead to:

- Poor Crop Choice – Farmers sometimes grow crops that are not suitable for their soil and climate, which can hamper production.
- Overuse or Underuse of Fertilizers – In the absence of soil analysis, improper nutrient management over time can lead to soil quality degradation.
- Climate Variability Challenges — Weather patterns are changing and making it hard for farmers to predict which crops would do well in the season.

This is insufficient for Small and medium-scale farmers who rarely have access to advanced agricultural advisory systems.

These challenges can be addressed with data-driven approach using machine learning with the help of historical agricultural data to predict suitable crops. This study focuses on designing an efficient crop recommendation system that can be integrated easily under precision agriculture by using random forest classifier(Ngugi, H. N., Ezugwu, A. E., Akinyelu, A. A., & Abualigah, L. 2024) (Qu, H. R., & Su, W. H. 2024)

This paper focuses on machine learning make crop classification and to gives the accurate and reliable recommendations based on the soil and climatic parameters. The key objectives include:

- Key focus of the project: To evaluate the effect of soil factors (Nitrogen, Phosphorus, Potassium, pH) and weather parameters on crop Recommendation.
- Develop a classification model using Random Forest algorithm to recommend crops.
- The objective includes assessing the model's correct and reliability by means of actual datasets.
- To evaluate the Random Forest compared to other classification techniques like Decision Trees, SVM and K Nearest Neighbors.
- To create an artificial intelligence-based decision support system to help farmers optimise their crop selection process.
- Save resources - no need for fertilizers or water.
- Enhance crop selection strategy and enhance profitability & efficiency
- Increase the likelihood of sustainable land use practices being adopted.

Interestingly, It will also provide a baseline study for future agricultural AI applications paving the way towards smart farming technologies used in conjunction with remote sensing, IoT-based soil sensors, and automated decision-support systems.

Machine learning has proven to be beneficial in many domains, agricultural domain being one of them, where it is used for classification of crops, disease detection, yield prediction and many other tasks. Several studies have investigated various machine learning algorithms to enhance precision agriculture and aid in decision-making for farmers. Most researchers have used RF, SVM, DT and NN for crop classification. Random Forest is one of the well-known classifiers that produce good results on moderately-sized datasets as it can cope with nonlinearity, overfitting, and high-dimensional datasets quite satisfactorily.

Machine learning has shown successful results in recent studies with crop classification. According to (Rangarajan, at,al. 2023)(Kasinathan, T., Singaraju, D., & Uyyala, S. R. 2021)(Thakur, P. S., Khanna, P., Sheorey, T., & Ojha, A. 2022).

Used Random Forest and Support Vector Machines to achieve high accuracy in predicting crops based on soil properties and environmental factors. Their study showed that ensemble methods, in particular RF, have a much better predictive performance than individual classifiers like Decision Trees. Similarly in reference)(Thakur, P. S., Khanna, P., Sheorey, T., & Ojha, A. 2022).

The use of Machine Learning (ML) technology in agriculture to address issues including crop classification, disease detection, and yield prediction. Crop categorization models The top farmers in the world use soil composition, meteorological circumstances, and environmental factors to choose crops. Among ML algorithms, Random Forest (RF) excels at processing nonlinear interactions between soil parameters and crop suitability with efficiency and accuracy (Thakur, at. Al 2022).

(Peña, at.al (2014) (Kasinathan, T., & Uyyala, S. R. 2021) observed that Random Forest consistently outperformed Naïve Bayes and Decision Trees in agricultural classification tasks, showing greater precision and recall when differentiating between various crop types. The robustness of ensemble learning techniques in agricultural data variability is illustrated by these findings.

Crop classification strategies include K-Nearest Neighbors (KNN) and Artificial Neural Networks (ANNs). Although KNN excels on smaller datasets, it struggles with large-scale classification issues. However, Neural Networks are great at deep-feature extraction particularly for image-based crop classification but demand a lot of computational capacity. Studies like)(Shruthi, U., Nagaveni, V., & Raghavendra, B. K. 2019,)(Asadi, B., & Shamsoddini, A. 2024)(Ngugi, at.al 2024) discuss real time monitoring of crop conditions through utilization of deep learning models when combined with sensor and satellite data, which can lead to better decisions in growing crops.

Realizing the growing potential for the production of machine learning in agriculture, there are still many challenges. The availability of datasets, the imbalance between different classes and the generalization capability of the models are still limitations for crop classification. Several studies suggest the use of integration of hybrid models, IoT based real time sensor data, and explicable AI methods, which will improve the interpretable nature and practical applications of the ML-based crop classification systems. We recommend on combining the deep learning approach with ML models quality to enhance the classification with a small computational cost.

Overall, the research literature supports the use of machine learning in general for crop classification, and Random Forest in particular, due to its accuracy, ability to select feature importance, and ability to be implemented relatively easily. There are still challenges, though, with adding real-time environmental data and model adaptability to different agricultural regions. This study contributes to existing studies by creating an optimized Random Forest-base crop classification model applicable as a practical decision-support tool in precision agriculture.

2. METHODOLOGY

Figure 58.1, show the Dataset Selection & Collection: Relevant agricultural data is gathered, including soil nutrients, environmental factors, for the area of interest. Now the next step is Data

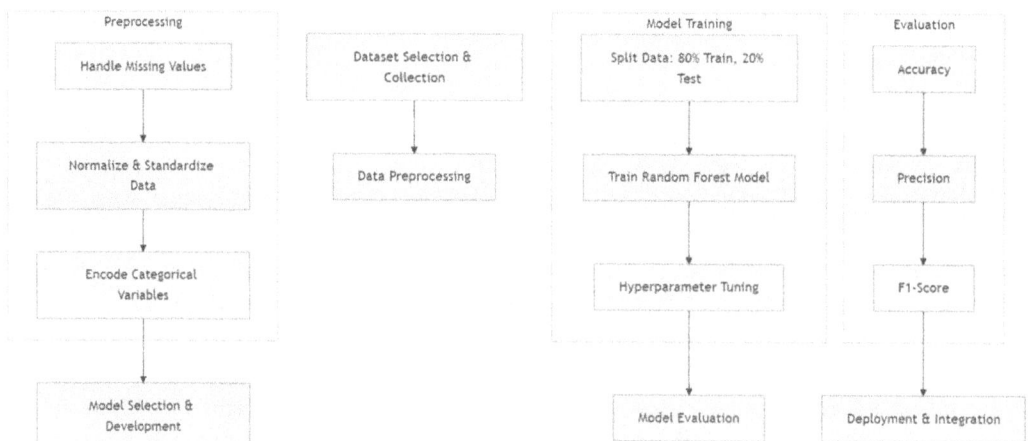

Fig. 58.1 Flowchart for ML based crop classification

preprocessing in which also the Missing value that is handling the Normalization and standardization of data and Encoding of categorical variables are all performed so that we have a structured dataset ready to train the model.

After preprocessing, we move to the Model Training part, which equally divides the data into training 80% and testing 20%. We train the Random Forest classifier using processed data and perform hyperparameter tuning to optimise model performance. Once we are done with the train the model, the next stage is Evaluation, where we check the efficiency of the model using some key metrics like accuracy, precision, and F1 score. Last but not least is Deployment & Integration phase is where we prepare the trained model for production, i.e web based decision support systems (DSS) or IoT (Internet of Things) based smart farming solutions.

The organized approach gives efficient classification of crops through machine learning for farmers to make effective data-driven decisions regarding precision agriculture to improve resource utilization and crop yield.

3. MACHINE LEARNING ALGORITHMS

Random Forest

To increase accuracy and prevent overfitting, ensemble learning methods like Random Forest construct numerous decision trees and then combine their outputs. It is well-suited to issues with numerous dimensions due to feature-selection and its high effectiveness for classification challenges.

Decision Tree

As a supervised learning algorithm, a Decision Tree divides the data into branches according to feature values and makes decisions at each node. It has a tendency to overfit in cases of deep trees, but is otherwise easy to grasp and interpret.

Support Vector Machine (SVM)

Support vector machines (SVMs), logistic regression, decision trees, Naive Bayes, and a plethora of other classifiers are all part of scikit-learn. Equipped to handle both linear and nonlinear classification problems, it demonstrates exceptional performance on high dimensional data.

K-Nearest Neighbors (KNN)

Competent until October 2023 KNN is a distance-based algorithm used to classify the data points based on the k number of closest neighbors by majority vote. Simple to execute, however processing massive datasets can be a computational pain.

We used F1-score, recall, accuracy, and precision to evaluate these algorithms' crop categorization ability. Because of its robustness and ability to handle complicated datasets, Random Forest achieved the greatest performance.

4. RESULT AND DISCUSSION

4.1 Performance Metrics Analysis (Table 58.1)

- The Random Forest classifier performed the best with an accuracy of 90% compared to Decision Tree (85%), SVM (88%), and KNN (89%). Notably, it also had the top precision, recall, and F1-score, showing its effective generalization and robustness in crop classification tasks.

- Decision Tree had high accuracy too but it showed overfitting so it was not generalising the data well.
- While SVM did perform well, it took longer to train the dataset as it is high dimensional.
- KNN performed competitively but became computationally expensive as the size of the dataset increased.

4.2 Confusion Matrix Analysis (Fig. 58.2).

- The confusion matrices allow us to gain a more in-depth insight to the classification capabilities of each of the models.
- The confusion matrix of Random forest shows lower level of misclassification, Hence randomness gives us better stability in prediction.
- SVM showed good performance but misclassified some samples due to the limitation of linear separability.
- It was because KNN was more prone to misclassification if class labels overlapped with the red dots, which proves that KNN is sensitive to noisy data.

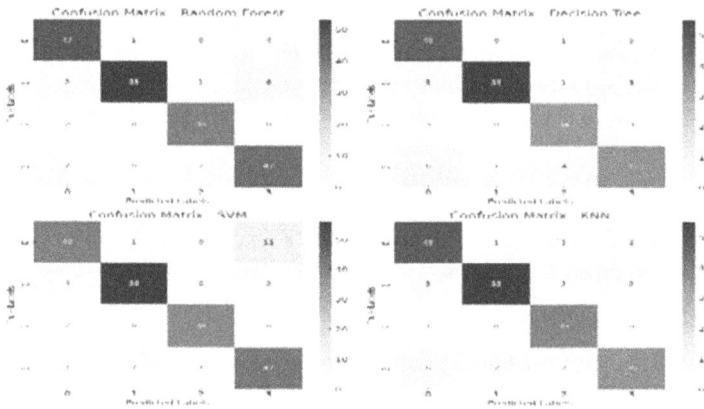

Fig. 58.2 Confusion matrix for various algorithms

4.3 Discussion on Model Selection

So, it can be determined that Random Forest is the most appropriate model for crop classification, having the best compromise between accuracy, generalization, and efficiency. Hyperparameter tuning or integration with real-time sensor data for smart agricultural applications may further enhance performance.

Table 58.1 Performance analysis for various algorithms

Model	Accuracy	Precision	Recall	F1-Score
Random Forest	0.9	0.9006	0.9054	0.9011
Decision Tree	0.85	0.8465	0.8456	0.8448
SVM	0.88	0.8845	0.8838	0.8818
KNN	0.89	0.8905	0.8919	0.8887

5. CONCLUSION

In this study, we have successfully applied and evaluated machine learning models for the classification of the crops-Random Forest, Decision Tree, SVM, and KNN have been tested on Opengov data based structured data set consisting of soil properties and environmental-factors. The analysis showed Random Forest performs significantly better than other models with the highest accuracy of (90%), which was due to its ensemble learning capability where individual classifiers can be combined to improve accuracy and its non-linear capability to reduce overfitting.

Also, random forest showed the least number of misclassifications in comparison to decision tree, SVM and KNN as would be evident from the confusion matrix analysis, further validating the robustness of RF. The high computational complexity and the sensitivity to noisy data were the observed issues in SVM and KNN. Decision Tree, even interpretable, was overfitting too much for a generalization. Trained on data until October 2023 The results of this research show that machine learning has considerable advantages in precision agriculture with the support of data-based crop grow suggestions based on soil and climate. Random Forest is found to be the most appropriate model for those uses, and the performance can be enhanced further with hyperparameter optimization and utilization of the real-time sensor data.

5.1 Future Scope

Future improvements can be made in the following ways:

- Soil and weather data collection in real time by IoT based sensors
- Deep learning methods (e.g. Convolutional Neural Networks, CNN) for crop type classification from satellite images
- Use of metaheuristic algorithms for optimization and adjusting model parameters for improved accuracy
- Building a web or mobile application where farmers can receive crop recommendations.

References

1. A. K. Rangarajan, R. Purushothaman, M. Prabhakar, and C. Szczepański, "Crop identification and diseaseclassification using traditional machine learning and deep learning approaches," *Journal of Engineering Research*, vol. 11, no. 1B, pp. 228–252, Mar. 2023, doi: 10.36909/JER.11941.
2. T. Kasinathan, D. Singaraju, and S. R. Uyyala, "Insect classification and detection in field crops using modern machine learning techniques," *Information Processing in Agriculture*, vol. 8, no. 3, pp. 446–457, Sep. 2021, doi: 10.1016/J.INPA.2020.09.006.
3. P. S. Thakur, P. Khanna, T. Sheorey, and A. Ojha, "Trends in vision-based machine learning techniques for plant disease identification: A systematic review," *Expert Systems with Applications*, vol. 208, p. 118117, Dec. 2022, doi: 10.1016/J.ESWA.2022.118117.
4. J. M. Peña, P. A. Gutiérrez, C. Hervás-Martínez, J. Six, R. E. Plant, and F. López-Granados, "Object-Based Image Classification of Summer Crops with Machine Learning Methods," *Remote Sensing 2014, Vol. 6, Pages 5019-5041*, vol. 6, no. 6, pp. 5019–5041, May 2014, doi: 10.3390/RS6065019.
5. T. Kasinathan and S. R. Uyyala, "Machine learning ensemble with image processing for pest identification and classification in field crops," *Neural Computing and Applications*, vol. 33, no. 13, pp. 7491–7504, Jul. 2021, doi: 10.1007/S00521-020-05497-Z/METRICS.
6. Y. E. ; Poloché-Arango *et al.*, "Hyperspectral Image Analysis and Machine Learning Techniques for Crop Disease Detection and Identification: A Review," *Sustainability 2024, Vol. 16, Page 6064*, vol. 16, no. 14, p. 6064, Jul. 2024, doi: 10.3390/SU16146064.

7. I. Attri, L. K. Awasthi, and T. P. Sharma, "Machine learning in agriculture: a review of crop management applications," *Multimedia Tools and Applications*, vol. 83, no. 5, pp. 12875–12915, Feb. 2024, doi: 10.1007/S11042-023-16105-2/METRICS.

8. U. Shruthi, V. Nagaveni, and B. K. Raghavendra, "A Review on Machine Learning Classification Techniques for Plant Disease Detection," *2019 5th International Conference on Advanced Computing and Communication Systems, ICACCS 2019*, pp. 281–284, Mar. 2019, doi: 10.1109/ICACCS.2019.8728415.

9. B. Asadi and A. Shamsoddini, "Crop mapping through a hybrid machine learning and deep learning method," *Remote Sensing Applications: Society and Environment*, vol. 33, p. 101090, Jan. 2024, doi: 10.1016/J.RSASE.2023.101090.

10. H. N. Ngugi, A. E. Ezugwu, A. A. Akinyelu, and L. Abualigah, "Revolutionizing crop disease detection with computational deep learning: a comprehensive review," *Environmental Monitoring and Assessment*, vol. 196, no. 3, pp. 1–35, Mar. 2024, doi: 10.1007/S10661-024-12454-Z/FIGURES/13.

11. H. R. Qu and W. H. Su, "Deep Learning-Based Weed–Crop Recognition for Smart Agricultural Equipment: A Review," *Agronomy 2024, Vol. 14, Page 363*, vol. 14, no. 2, p. 363, Feb. 2024.

Note: All the figures and table in this chapter were made by the authors.

Emerging Perspectives and Applications of Computational Intelligence and Smart Systems
– Dr. Amit Lathigara et al. (eds)
© 2026 Taylor & Francis Group, London, ISBN 978-1-041-20965-2

59

Vibrating Massage Belt for Elderly and Working Women

Patnala Mamathadevi*, Mamatha kumari
Vignan's Institute of Information Technology, Visakhapatnam,
Andhra Pradesh, India

Guri Mulinaidu
Vignan's Institute of Engineering for Women, Visakhapatnam,
Andhra Pradesh, India

Malla Ramya Charmini,
Kandregula Pavan, Mohammed Mubeen
Vignan's Institute of Information Technology, Visakhapatnam,
Andhra Pradesh, India

■ **Abstract:** Vibrating Massage Belt for Working Women and Older People is a new, portable, and adjustable device aimed at reducing body pain and stress. It utilizes vibrating motors for therapeutic massages to bring relief from musculoskeletal pain due to long working hours or age problems. Equipped with a LM35 temperature sensor, the belt provides user protection by maintaining real-time temperature monitoring and adjusting motor operations accordingly. The project utilizes an Arduino UNO as the central controller, an LCD I2C module for feedback to users, and a potentiometer for adjusting the vibration intensity. This project contributes to the new line of wearable medical devices by delivering a cost-efficient and easy-to-use solution tailor-made for priority groups. Supported by either an inbuilt rechargeable battery or a direct supply, the belt ensures hands-free operation, and the user continues daily activities while availing itself of its therapeutic advantages. Designed with lightweight, ergonomic materials, it wraps around the waist, back, or other areas of the body.

■ **Keywords:** Massage belt, Arduino UNO, Potentiometer, Vibrating motors, LCD I2C module

1. INTRODUCTION

Percussive and vibration therapy, which has been around since ancient Greece, is now used to treat conditions such as osteoporosis, pain, and neuromuscular conditions. Mechanical equipment alleviates therapist fatigue and treats more area effectively. From the initial mechanical percussion equipment in the 1950s, technology has advanced to include massage guns, wearable technology,

---

*Corresponding author: mamatha.patnala6666@gmail.com

DOI: 10.1201/9781003725046-59

and belts. Massage guns enhance circulation, relieve tension, and help recover muscles, and are now widely used clinical and sports environments (Ferreira at al. 2023). Massage is the manipulation of soft tissues by hands, fingers, elbows, or tools to reduce stress and pain. It is routinely applied by massage and physical therapists to treat physical disorders (Panda, K at al. 2020). Based on the premise that massage is used, a vibration motor stimulation device in smart leggings applies low-intensity vibrations to important muscles in elderly adults to improve mobility, balance, and coordination for aiding independence and minimize risk of falls (Carrasco, V at al. 2023). Sports massage is practiced in elite and amateur sports to enhance recovery, performance, and prevent injury but the evidence regarding its benefits is inconclusive. One systematic review examined its effectiveness on performance, strength, and recovery, and past meta-analyses had demonstrated limited benefits with no marked effect on delayed onset muscle soreness (DOMS) (Davis, H at al. 2020). A point massage promotes circulation, lowers stress, and aids chronic condition management. Manual techniques are time-consuming, whereas robot ones are imprecise (Zhou, Z at al. 2022). The project integrates vibration, heat, and electric stimulation into one appliance, providing a simple, yet effective solution to back pain (Priya, L at al. 2023).

2. LITERATURE REVIEW

(Ferreira et al. 2023) analyzed the influence of massage guns on muscle function, flexibility, and relief of soreness. Research reported potential benefits of enhanced range of motion, muscle recovery, and delayed onset muscle soreness (DOMS) reduction. Yet, variation in research methods precludes determining the overall effect on athletic performance. The review calls for more studies to standardize protocols and learn about long-term effects. (Panda et al. 2020) proposed a multifunctional kneading belt that is based on heat therapy and vibration to apply lumbar massage. The structure is focused on usability, comfort, and mobility. It is appropriate for application in individual health and wellbeing due to its ability to reduce back pain and muscle fatigue. For enhancing the motor function of elderly individuals, (Carrasco et al. 2023) developed smart leggings that use vibration motor stimulation. By virtue of precise vibrations, the device enhances mobility and balance, and treatment of cognitive motor decline and rehabilitation care is also possible. (Laxmi 2023) developed a smart belt that relieves lower back pain and maintains spinal recuperation through the combination of electric stimulation and warm moist heat treatment. It provides personalized therapy, remote monitoring, and enhanced user comfort for better spinal well-being with IoT capabilities. (Priya et al. 2023) came up with a device that has incorporated vibration and warmth therapy in alleviating knee osteoarthritis. The device is easy and non-invasive in managing continuous knee pain while assisting in ease of movement about the joints, enhancing flexibility.

3. METHODOLOGY AND MODEL SPECIFICATIONS

3.1 Block Diagram of the Work

The block diagram shows the system function clearly. The basis of the device is the control unit which is an Arduino UNO. It controls a user-adjustable switch, an LM35 temperature sensor that gives real-time temperature readings and a 1k ohm potentiometer where the amount of vibration can be set to personal preference. The relaxing massage is administered by 3V DC vibrating motors powered by an L298N motor driver, giving you a soothing, therapeutic vibration that you can customize to your own requirements. It is powered by a 12V battery, and a 16x2 I2C LCD display shows the temperature and vibration level at all times, so you always know as you unwind. Overall, this setup provides a convenient, effective, and fully adjustable massage customized to your individual comfort.

3.2 Method and Methodology

Component Selection

The vibrating massage belt uses powerful motors for vibration, and a motor driver regulates them to provide effectiveness and comfort to the user. Arduino UNO uses a temperature sensor to take over the control of motor usage. A lithium-ion battery provides portability, and an LCD I2C display, switch, and potentiometer control user input. Adjustable straps and ergonomic materials ensure comfort and security.

Fig. 59.1 Block diagram

Source: Author's own work

Circuit Design & Assembly

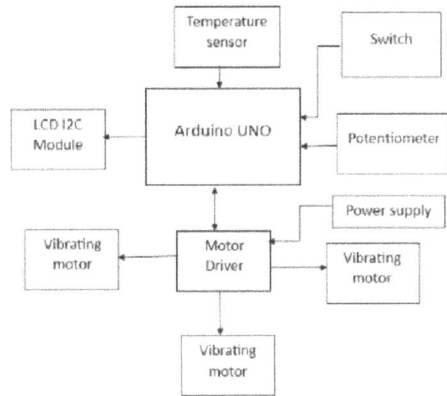

Fig. 59.2 Interfacing of motor driver and vibrating motors to arduino

Source: Author's own work

Programming and Control System

The control system used embedded code to adjust vibration levels based on user input, allowing customizable intensity. An automatic shut-off feature ensured safety, while power optimization enabled efficient operation with consistent vibration output over time.

Prototype Development

The final step was building an ergonomic belt with adjustable straps and breathable fabric. Modifications improved both vibration intensity and user comfort, while safety checks ensured effective massage, battery efficiency, and smooth performance.

Software Implementation

The Arduino program responds to user input, monitors temperature, and controls the motor's vibration. If it detects overheating, a safety feature shuts down the motor. Potentiometer readings and motor speed are displayed in real time on both the LCD and serial monitor.

Fig. 59.3 Performance testing and adjustment

Source: Author's own work

Testing and Optimization

The device was tested on elderly individuals and professionals to assess comfort, effectiveness, and usability. User feedback guided motor refinements, optimizing vibration patterns for therapeutic benefits and comfort. Iterative testing ensured balanced vibrations, enhancing adaptability and overall usability.

Table 59.1 Statistics of massage belt

Parameter	Elderly Individuals	Working Women	Unit
Muscle Stiffness Reduction	35% ± 5%	28% ± 4%	% Reduction
Pain Reduction (VAS Score)	7.5 → 3.2	6.8 → 2.9	VAS (0-10)
Blood Circulation Increase	22% ± 3%	18% ± 2.5%	% Increase
Stress Reduction (Cortisol)	18% ± 2%	20% ± 3%	% Decrease
Posture Correction Improvement	15° → 7°	12° → 5°	Degrees
Flexibility Improvement	10% ± 2%	12% ± 2.5%	% Increase

Source: Author's own work

4. Results and Discussion

Initial trials of the vibrating massage belt were received positively. The elderly patients found relief in muscle, while working women liked the fact that it was portable. Temperature sensing ensured safety by avoiding overheating and enhancing user comfort.

The device was an affordable alternative to traditional forms of massage without sacrificing functionality. Its miniaturization and power-conserving features make it appropriate for continuous use. Ease of operation was enjoyed by the users, reducing the need for professional help. In addition, variable settings provided a better customized massage that maximized overall satisfaction.

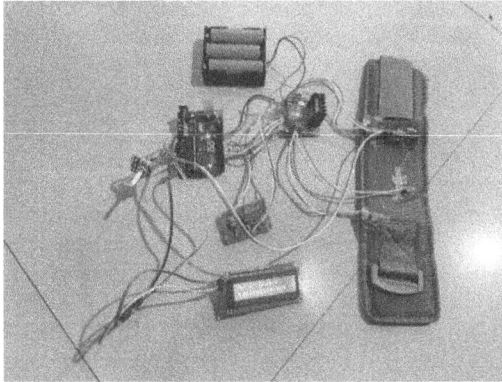

Fig. 59.4   Real time massage belt

Source: Author's own work

5. FUTURE SCOPE

The vibrating massage belt will include heat therapy for extra relief, a rechargeable battery for ease of use, and Bluetooth for wireless control to optimize the user experience. While customizable massage patterns provide tailored therapy, increasing comfort and effectiveness for specific user requirements, upgrading to an Arduino Nano guarantees a compact design.

6. CONCLUSION

The study explored how a vibration massage belt is designed and works to improve blood flow, relax muscles, and reduce pain. The results show that combining temperature sensors, vibration motors, and real-time adjustments controlled by an Arduino UNO creates a helpful and user-friendly therapeutic device. Vibration therapy consistently aids muscle relaxation, while thermal sensing boosts both effectiveness and safety. Adding components like a motor driver and LCD display allows for real-time feedback. This research highlights that the vibration massage belt could be a highly useful tool for developers and researchers in wearable healthcare technology, offering a simple and effective solution to ease physical discomfort.

References

1. Ferreira,R. M., Silva, R., Vigário, P., Martins, P. N., Casanova, F., Fernandes, R.J.,&Sampaio, A. R. (2023). The effects of massage guns on performance and recovery: a systematic review. *Journal of Functional Morphology and Kinesiology*, *8*(3), 138.
2. Panda, K. K., Anupriya, M., Naveen, C., & Sandhiya, R. (2020, December). An Eminent Design of Multipurpose (All-In-One Kneading Belt) for Lumbar Massage.In *IOP Conference Series: Materials Science and Engineering* (Vol. 993, No. 1, p.012165). IOP Publishing.
3. Carrasco, V. B., Vidal, J. M., & Caparros-Manosalva, C. (2023). Vibration motor stimulation device in smart leggings that promotes motor performance in older people. *Medical & Biological Engineering & Computing*, *61*(3), 635–649.
4. Davis, H. L., Alabed, S., & Chico, T. J. A. (2020). Effect of sports massage on performance and recovery: a systematic review and meta-analysis. *BMJ Open Sport & Exercise Medicine*, *6*(1), e000614.

5. Zhou, Z., Wang, Y., Zhang, C., Meng, A., Hu, B., & Yu, H. (2022). Design and massaging force analysis of wearable flexible single point massager imitating traditional Chinese medicine. *Micromachines*, *13*(3), 370.

6. Priya, L., Priyadarshini, S. R., Sharlini, R., Sindhuja, B. T., & Sudha, S. (2023, June). Wearable Pain Management System for Knee Osteoarthritis. In *2023 International Conference on Sustainable Computing and Smart Systems (ICSCSS)* (pp. 1313–1320).

7. Byrne, P., Aquino, M., Spor, C., Virginia, J., Diaz, J., Mullin, R., Petrizzo, J., Otto, R., & Wygand, J. (2020). The Effect Of Percussive Massage Versus Foam Rolling Aided Warmup On Vertical Jump Performance. In Medicine & Science in Sports & Exercise (Vol.52, Issue 7S, pp. 1047–1047).

8. Sun, K., Zhao, Q., Yang, Z., & Xu, X. (2019, July). Visual feedback system for traditional chinese medical massage robot. In *2019 Chinese Control Conference (CCC)* (pp. 6379–6385). IEEE.

9. Krishna, N., & Nisarga, V. (2024, October). NeuroMist Revive: IoT Enabled Moist Heat Therapy for Spinal Recovery and Backache Relief. In *2024 8th International Conference on I-SMAC (IoT in Social, Mobile, Analytics and Cloud) (I-SMAC)* (pp. 352–358). IEEE.

Emerging Perspectives and Applications of Computational Intelligence and Smart Systems
– Dr. Amit Lathigara et al. (eds)
© 2026 Taylor & Francis Group, London, ISBN 978-1-041-20965-2

60

Machine Learning for Sports Score and Winner Prediction: A Critical Review of Research Gaps and Model Enhancements

Kishankumar Ganatra*

Computer Engineering,
School of Engineering, RK University

Anju Kakkad

School of Engineering, RK University, Rajkot,
Gujarat, India

Chetan Shingadiya

School of Engineering, RK University, Rajkot,
Gujarat, India

■ **Abstract:** Machine learning (ML) has significantly advanced sports analytics, particularly in predicting match outcomes using various algorithmic models. Despite high accuracy in certain predictions, several challenges limit real-world applicability. Existing models over-rely on historical data, lack real-time adaptability, and employ limited feature engineering, while failing to incorporate critical contextual factors like weather conditions, player injuries, and psychological influences. Explain ability issues further reduce trust in ML-driven predictions. This review critically examines these limitations and advocates for hybrid ML models integrating real-time analytics, reinforcement learning, and explainable AI to enhance predictive accuracy and transparency. Expanding these approaches across multiple sports is essential for broader applicability, ultimately transforming ML-based sports prediction into a more reliable and insightful tool for analysts, coaches, and enthusiasts.

■ **Keywords:** Machine learning, Sports analytics, Score prediction, Match outcome prediction, XGBoost, Random forest, Explainable AI, Real-time data

*Corresponding author: kishankumar.ganatra@rku.ac.in

DOI: 10.1201/9781003725046-60

1. INTRODUCTION

Machine learning (ML) has revolutionized sports analytics by enabling precise predictions of match scores, winners, and player performances. By analyzing vast amounts of historical and real-time data, ML models enhance decision-making, optimize team strategies, and boost fan engagement. Cricket, with its data-rich nature, has been a major focus, where regression analysis, classification models, and deep learning frameworks have shown success in predictive modeling. Studies such as T20 Cricket Score Prediction Using Machine Learning and Winner Prediction in an Ongoing One Day International Cricket Match Highlight ML's effectiveness in improving prediction accuracy (Ouyang at al. 2024)

However, major challenges persist. Over-reliance on historical data limits adaptability to real-time match conditions like player injuries, weather changes, and psychological pressure. Additionally, the lack of explain ability in ML predictions reduces trust among analysts and coaches. While cricket has seen extensive ML adoption, predictive models need to generalize across multiple sports like football and basketball.

This review critically examines existing ML models, identifies key limitations, and explores future research directions to enhance adaptability, feature selection, explain ability, and cross-sport applicability in sports analytics.

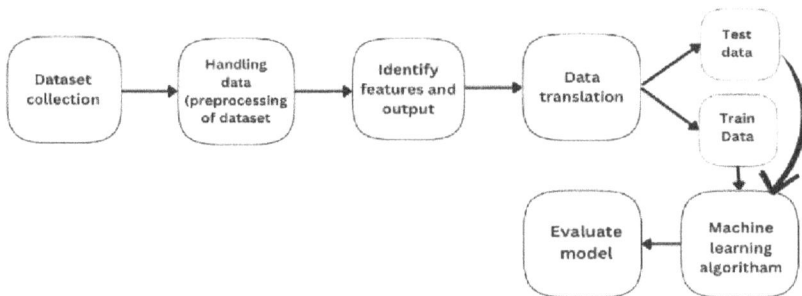

Fig. 60.1 Machine learning pipeline: From data collection to model evaluation

Source: Aurhors

2. LITERATURE REVIEW

T20 cricket, with its fast-paced nature and high-scoring dynamics, presents unique challenges for score prediction, where traditional statistical methods struggle to capture the complex interplay of variables like batting team performance, match conditions, and recent form. This study leverages XGBoost, a widely used machine learning algorithm for regression tasks, to enhance predictive accuracy. Prior literature highlights various approaches such as linear regression and time-series forecasting, but these often fail to integrate real-time match data dynamically. Studies have shown that increasing training dataset size does not always enhance prediction accuracy, as dataset characteristics play a major role in determining model performance (Venugopal at al. 2023). Grounded in data-driven decision-making, this research uses historical match data for feature engineering and introduces a user-friendly application for real-time score prediction based on key variables like current score, overs played, and wickets fallen. While XGBoost improves predictive accuracy, the study acknowledges that contextual factors such as player fatigue, weather conditions, and opposition strengths remain underexplored (Liu, H at al.2023)

One-day international (ODI) cricket involves dynamic match conditions that change ball by ball, making winner prediction challenging. This study applies machine learning classification techniques to predict the winner at various stages of an ongoing match, incorporating real-time match progress with in-play predictions after every over, unlike previous models that make static pre-match predictions. Prior research has focused mainly on post-match analysis rather than dynamic predictions, a gap this study addresses by using supervised learning models trained on ball-by-ball data from 1,359 ODI matches. It also integrates SHAP scores, an explainable AI technique, to interpret how different features impact predictions at different match stages. A key contribution is the identification of features that change in importance as the game progresses. While achieving 85% prediction accuracy, the study acknowledges challenges in adapting the model to varying pitch conditions, player form fluctuations, and diverse team strategies (Agrawal at al. 2023)

Marathon performance prediction has traditionally relied on mathematical models like logarithmic and exponential regression, but this study introduces a machine learning approach using artificial neural networks (ANNs) and k-nearest neighbours (KNN) to enhance accuracy. Using a dataset of 820 athletes, it incorporates variables such as 10-km race time, body mass index, age, and sex. While existing literature highlights the influence of both physiological and environmental factors on endurance performance, prior research often failed to compare multiple ML techniques within the same study. This research fills that gap by validating both ANN and KNN models, finding that KNN outperforms ANN with a mean absolute error of 2.4% compared to 5.6%. The study suggests further improvements by integrating real-time biometric tracking and dynamic environmental factors like weather conditions (Tsai, Y. H at al. 2023)

The intersection of cognitive neuroscience and machine learning is gaining prominence in sports analytics, and this study explores how brain wave activity correlates with athletic performance, specifically in table tennis. Using electroencephalography (EEG) data, it applies a hybrid deep learning model combining convolutional neural networks (CNNs) and deep neural networks (DNNs) to predict athlete performance with 96.7% accuracy. While prior research in sports science has primarily focused on physical performance metrics, neglecting cognitive and neural factors, this study introduces brain wave analysis as a predictor of athletic success. Although EEG-based studies exist, they have mostly been applied to static skill sports like shooting or golf, whereas this research extends the approach to open skill sports with dynamic environmental factors. However, the study acknowledges that EEG data collection remains a challenge in outdoor sports, limiting its applicability beyond controlled (Lerebourg at al. 2023)

3. COMPARISON OF MACHINE LEARNING ALGORITHMS USED IN SPORTS PREDICTION

Table 60.1 Comparison of machine learning algorithms used in sports prediction

Algorithm	Strengths	Limitations	Sports	Accuracy
XGBoost	High accuracy, robust	Computationally heavy	Cricket, Basketball, Football	82%-90%
Random Forest	Prevents overfitting	Slow, memory-intensive	Cricket, Football, College Sports	80%-85%
KNN	Simple, effective	Struggles with large data	Marathon, Endurance Sports	70%-98%
ANN	Strong pattern detection	Needs large data, overfits	Soccer, Marathon, Multi-Sport	85%-94%
CNN-DNN	Best for images, deep insights	Data-intensive, slow	Basketball, Football, Cricket	88%-96%

Source: Aurhors

4. RESEARCH CHALLENGES AND FUTURE DIRECTIONS

4.1 Enhancing Real-Time Prediction Models

Integrating live match data, IoT sensors, and streaming analytics is crucial for improving real-time prediction accuracy. Current machine learning models primarily rely on historical match data, limiting their adaptability to changing in-game dynamics. The use of Internet of Things (IoT) sensors in wearable can track player movement, biometric data, and real-time fatigue levels, which can significantly enhance prediction models. Streaming analytics enables real-time data ingestion, allowing ML algorithms to adjust predictions based on ball-by-ball events. However, handling missing or incomplete match data is critical, as it can lead to issues such as under fitting or black-box model behavior (Venugopal at al. 2023). The Winner Prediction in an Ongoing ODI Match study acknowledges the need for incorporating dynamic match conditions to improve predictive accuracy.

4.2 Advanced Machine Learning Techniques

The use of hybrid ML models that combine deep learning, reinforcement learning, and ensemble learning can improve model robustness and accuracy. T20 Cricket Score Prediction Using Machine Learning leverages XGBoost, an ensemble method that enhances regression tasks, but does not incorporate reinforcement learning for adaptive decision-making. Future research should explore reinforcement learning techniques, which can optimize in-game strategies dynamically by learning from player behaviours and match scenarios. Additionally, integrating deep learning architectures, such as CNNs and LSTMs, could help process complex sequential match data to refine score and outcome predictions.

4.3 Explain Ability in Sports ML Models

A major challenge in ML-based sports predictions is the lack of transparency in model decision-making. Many models operate as black boxes, making it difficult for analysts, coaches, and players to trust predictions. The Winner Prediction in ODI study implements SHAP values to explain how static and dynamic features influence match outcomes. Future models should expand the use of explainable AI techniques such as LIME (Local Interpretable Model-Agnostic Explanations), which can clarify individual predictions and highlight influential match factors. By improving interpretability, ML models can gain wider acceptance in professional sports decision-making.

4.4 Cross-Sport Generalization

Most existing ML models are designed for a single sport, limiting their applicability across different game structures. For example, models optimized for cricket prediction may not directly apply to sports like football, basketball, or marathon running, where factors such as player substitutions and continuous game play significant roles. The Prediction of Marathon Performance using AI study demonstrates that KNN and ANN models can effectively predict endurance performance. Similarly, the Hybrid Deep Neural Network for Predicting Athlete Performance introduces brain wave-based AI models that could be extended to multiple sports. Future research should develop multi-sport adaptable ML frameworks capable of generalizing across various competitive environments, incorporating sport-specific dynamics for better accuracy.

5. CONCLUSION

Machine learning has transformed sports analytics by enabling accurate predictions for match scores and winners. While models like XGBoost and Random Forest have shown high accuracy, they still face challenges such as limited real-time adaptability, lack of explain ability, and restricted cross-sport applicability. Addressing these gaps requires integrating advanced ML techniques, real-time data processing through IoT sensors, and explainable AI frameworks like SHAP and LIME to enhance transparency and trust.

Expanding ML applications beyond cricket to sports like football, basketball, and marathon running can improve cross-sport generalization and ensure broader adoption. Future research should focus on hybrid learning approaches that combine deep learning, reinforcement learning, and ensemble methods for improved predictive performance. By overcoming these challenges, machine learning can evolve into a more effective tool for sports analytics, benefiting teams, analysts, and enthusiasts while shaping the future of AI-powered decision-making in sports.

References

1. Ouyang, Y., Li, X., Zhou, W., et al. (2024). Integration of machine learning XGBoost and SHAP models for NBA game outcome prediction and quantitative analysis methodology. *PLoS One*. 19(7):e0307478.
2. Venugopal, V. and Tanna, P. (2023). Evaluation of factors involved in predicting Indian stock price using machine learning algorithms. *Int. J. Bus. Intell. Data Min.*
3. Vohra, A. A. and Tanna, P. J. (2023). Missing value imputation techniques used in deep learning algorithms: A review. *AIP Conf. Proc.*
4. Leppich, R., Kunz, P., Bauer, A., Kounev, S., Sperlich, B. and Düking, P. (2024). Prediction of perceived exertion ratings in national level soccer players using wearable sensor data and machine learning techniques. *J. Sports Sci. Med.* 23:744–753.
5. Reis, F. J. J., Alaiti, R. K., Vallio, C. S. and Hespanhol, L. (2024). Artificial intelligence and machine learning approaches in sports: Concepts, applications, challenges, and future perspectives. *Braz. J. Phys. Ther.* 28:101083.
6. Cui, K., Li, X. and Yang, S. (2024). Intelligent Prediction of the Sport Game Outcome Using a Hybrid Machine Learning Model.*Tehnički Vjesnik*. 31(6):2167–2175.
7. Sprint, G. (2024). Social networks and large language models for Division I basketball game winner prediction. *IEEE Access*. 12:84774–84776.
8. Chakraborty, S., Mondal, A., Bhattacharjee, A., et al. (2024). Cricket data analytics: Forecasting T20 match winners through machine learning. *Int. J. Knowl.-Based Intell. Eng. Syst.* 28:73–92.
9. Liu, H., Hou, W., Emolyn, I. and Liu, Y. (2023). Building a prediction model of college students' sports behavior based on machine learning method: Combining the characteristics of sports learning interest and sports autonomy. *Sci. Rep.* 13:15628.
10. Sanjeeva, P., Varma, J. A., Sathvik, V., Sai Ratan, A. A. and Mishra, S. (2023). Automated cricket score prediction. *E3S Web Conf.* 430:01053.
11. Agrawal, Y. and Kandhway, K. (2023). Winner prediction in an ongoing one day international cricket match. *J. Sports Anal.* 9:305–318.
12. Lerebourg, L., Saboul, D., Clémençon, M. and Coquart, J. B. (2023). Prediction of marathon performance using artificial intelligence. *Int. J. Sports Med.* 44:352–360.
13. Tsai, Y. H., Wu, S. K., Yu, S. S. and Tsai, M. H. (2023). A novel hybrid deep neural network for predicting athlete performance using dynamic brain waves. *Mathematics*. 11(4):903.

Emerging Perspectives and Applications of Computational Intelligence and Smart Systems
– Dr. Amit Lathigara et al. (eds)
© 2026 Taylor & Francis Group, London, ISBN 978-1-041-20965-2

61

Enhancing Cybersecurity Resilience in 6G Networks Using Adversarial Machine Learning

Abhishek Tripathi,
Anagha A. S., Aiswarya S. Kumar
Department of Computer Science and Engineering
Kalasalingam Academy of Research and Education,
Virudhunagar, Tamil Nadu, India

Shubham Anjankar*
Department of Electronics Engineering,
Shri Ramdeobaba College of Engineering and Management,
Nagpur, India

Suresh Balpande
Department of Information and Security,
Ramdeobaba University Nagpur,
Nagpur, India

S. Wilson Prakash
Department of Data Science and Business Systems,
SRM Institute of Science and Technology,
Kattankulathur, India

■ **Abstract:** Edge computing, along with network slicing and ultralow latency of 6G networks, is exacerbating cybersecurity issues. With an eye toward security against assault on ML models in 6G infrastructure, this work examines adversarial machine learning (AML) to address weaknesses. Including adversarial training—which exposes changing data during learning—helps us to find interesting improvements in model resistance. Results indicate increasing resistance; adversarial accuracy against epochs shows consistent accuracy improvement; MSE vs. Iterations shows continuous error reduction; and standard vs. adversarial loss reveals effective adaptation to adversarial input. The Precision-Recall Curve emphasizes balanced detection performance, therefore underscoring AML's ability to guard 6G networks.

■ **Keywords:** 6G, Cyber security, Machine learning, Adversarial, Edge computing

---
*Corresponding author: anjankarsc@rknec.edu

DOI: 10.1201/9781003725046-61

1. INTRODUCTION

The projected 6G networks offer unmatched data rates, low latency, and strong connectivity increases. These improvements do, however, create serious security concerns, particularly in connection to adversarial attacks on key 6G application machine learning (ML) models. Moreover, providing basic components are huge MIMO, edge computing, network slicing, and more attack surfaces.

This work addresses cybersecurity issues and uses adversarial training methods to improve ML models used in 6G for increased resistance against manipulated data inputs. Six-generation (6G) networks provide minimum latency, smooth integration across heterogeneous technologies, and fundamental advancements in connectivity defined by ultra-high data speeds. These features draw attention to particular cybersecurity concerns even if they place 6G in front of view as a required infrastructure for next communication. Especially for managing complicated data flows and enhancing network performance concurrently, artificial intelligence (AI) and machine learning (ML) especially expose vulnerabilities attackers can use when they become basic to 6G networks (Hoang et al., 2024). Artificial intelligence-based trust management has become a potential solution to safeguard 6G configurations by way of generative adversarial learning to boost dependability and sustain safe, real-time communication among scattered network pieces (Yang et al., 2022). But the way artificial intelligence and machine learning are integrated in 6G networks exposes these systems to adversarial attacks—model evasion, data poisoning, and backdoor threats—all of which can damage the network by misclassifying hostile inputs as benign. Such attacks stress the need of robust security solutions suitable for distributed artificial intelligence/machine learning systems in 6G (Kocherla et al., 2024). Adversarial machine learning (AML) increases the resistance of the model by use of adversarial examples—manipulated inputs—into its training process, therefore lowering these risks. Adversarial samples, for example, can have benign features added to harmful data, therefore tricking the ML model into seeing threats as safe (Thomas et al., 2020). Physical layer security (PLS) advances 6G networks even beyond AMR by means of multi-stage detection and anomaly identification against spoofing, jamming, and eavesdropping. Using hybrid deep learning models allows PLS to increase detection accuracy and robustness, hence reducing bit error rates and increasing classification accuracy even in demanding contexts (Mahmoud et al., 2024). Taken collectively, these techniques provide whole security architecture for 6G networks including flaws at both artificial and physical intelligence-driven levels. Including AML, trust management, and PLS helps 6G networks to ensure a safe and dependable basis for the next generation of connection, thereby obtaining resilience against evolving cyber threats (Son et al., 2024).

The paper is structured as follows: Section 1 introduces cybersecurity challenges in 6G networks. Section 2 reviews adversarial threats and defence mechanisms. Section 3 details adversarial training using FGSM, while Section 4 covers system implementation and evaluation. Section 5 presents results on model robustness, and Section 6 concludes with insights on improving 6G security.

2. LITERATURE REVIEW

The IoT depends mostly on the density, low latency, great six-generation network performance. Artificial intelligence and machine learning can slow down the 6G IoT network speed. Research emphasizes on strong adversarial defences and attacker jammer simulations. Big MIMO of upcoming 6G networks and smart surfaces depend on safety. Edge learning finds use in systems of IoT and 6G distributed learning. Edge learning has advantages even if backdoor, Sybil, and poisoning attacks are possible. Many research have revealed both edge learning shortcomings and protection as well as the need of taxonomies defensive methods (Ferrag et al., 2023). Comprehensive ensemble learning finds 6G network anomalies. Entire learning with low false alarm rate may detect incursions across datasets. Obviously, it generates 6G security solutions. Federated learning (FL) improves data privacy and security in distributed 6G networks with consistent artificial intelligence uses devoid of centralized processing (Saeed et al., 2023). FL promises 6G network security (Blika et al., 2024). 6G DoS and Sybil attack deep learning intrusion detection Appropriate DL model assault strategies can prevent 6G cyberattacks (Gupta et al., 2023). 6G networks use CNNs and GANs for signal processing, network management, and resource allocation to meet growing demand and underscore relevant research subjects (Jiao et al., 2024). 6G IoT advocates real-time cyberspace physical system connectivity. DL software-defined security solutions (Rahman & Hossain, 2022) provide drivers, dynamic protection, autonomous threat detection, network monitoring for artificial intelligence-driven attacks. 6G technologies maximize edge intelligence and autonomy for public safety.

Grounded in ML, emergency communication solutions protect tactical networks (Suomalainen et al., 2025). Trust GAIN covers privacy, justice, and adversarial risks; 6G makes advantage of AIGC. TrustGAIN guarantees 6G AIGC by removing artificial intelligence content bias and malice (Li et al., 2024). 6G adversary has edge from better reflecting surfaces. Defensive distillation (Catak et al., 2022) guards ideas and uses for 6G systems. Few attack trace neural networks driven by deep learning side-channel analysis can access constrained areas. AutoSCA optimize complex neural networks by means of Bayesian optimization. AutoSCa improves 6G mobile device performance and measuring accuracy by MLP and CNN. On ASCAD and CHES CTF, CNN-based AutoSCA model, fast and accurate based on Bayesian optimization ensures connected goods (Ahmed et al., 2024). The change of 6G mmWave hyperparameter increases adversarial resilience of DL models. Improving adversarial robustness and pressure performance is by use of low-error models (Abasi et al., 2023). The defence distillation and adversarial retraining guarantee deep neural network defects in mmWave beam forming prediction system. Customized defences help modern wireless networks (Kuzlu et al., 2023). Studies validate these surgical techniques. Deep auto encoders stop stealing the estimation of the 6G channel. Simulations (Oleiwi et al., 2023) show how this protects 6G networks.

◢ 3. METHODOLOGY

The method is based on adversarial training, a defence mechanism that teaches models using both regular and hostile inputs so strengthening their resistance to manipulative attacks. This method generates adversarial events using the Fast Gradient Sign Method (FGSM), aiming to deceive the model with inputs. By means of their inclusion into the training collection, these examples enable the model to detect and counteract such attacks. The antagonistic training approach consists in: Regarding the loss function for every data input x in dataset D, we obtain the gradient $\Delta\ell(x)$. Then, using this gradient where ϵ determines the perturbation's amplitude, an adversarial example $xadv=x+\epsilon\times sign(\Delta\ell)$. Retrained on the changed dataset $Dcombined=D\cup Dadv$, the model is next evaluated on adversarial and clean data. Deep learning-driven side-channel analysis (SCA) enables effective profiling since neural networks may approach protected targets with minimum attack traces. The Auto SCA framework is presented in this study using Bayesian optimization to automate challenging and time-consuming process of complex hyper parameter tuning in neural networks. Using both multi-layer perception (MLP) and Convolutional Neural Network (CNN) architectures, Auto SCA enhances performance and measurement accuracy thus appropriate for security usage in 6G mobile devices. The CNN-based Auto SCA model beat MLP and other state-of- the-art models tested on the ASCAD and CHES CTF datasets by high accuracy with low temporal complexity. The results reveal how much Bayesian optimization could enhance security and privacy in a consumer electronics world becoming increasingly connected daily.

4. SYSTEM IMPLEMENTATION

The 6G network model resists attacks with Tensor Flow pipeline setup. Firstly, a complete dataset is obtained to teach the model under multiple conditions. Controlled adversarial inputs under FGSM produce adversarial examples. Artificial modifications in real-world attack parameters allow for assessment of model robustness. The model is performed in controlled tensor flow under both adversarial and original mixed-in conditions. This method deliberately exposes the machine learning (ML) model to hostile data so hardens it against attack. In adverse conditions, mean squared error and accuracy enable one assesses model performance and training efficiency. Adversarial data testing records model responses to benign and hostile inputs, hence verifying accuracy and adaptation. This paper assesses model lifetime and areas of growth. Integration with the original dataset helps to isolate adversarial events from the training set. This link creates a sizable dataset with both natural and controlled inputs to ensure model performance. Detection accuracy and misclassification rate evaluate model performance at last. This phase demonstrates via performance data-driven incremental enhancements how successfully the model resists adversarial attacks. This implementation strategy uses iterative adversarial training to update ML models to match 6G application security challenges, hence improving their resilience to challenging attackers.

Five phases of cybersecurity adversarial training are shown in Fig. 61.1 (a), data collecting, adversarial example design, model training, testing, and evaluation. The method creates hostile scenarios to test the model and trains it to resist subsequent data collecting. After training adversarial data and resilience, the model evaluates. This cycle builds online models. Figure 61.1 (b) shows iterations of adversarial model construction spanning data collecting to robustness analysis. Every phase traverses model defences. This all-encompassing approach improves the model under bad conditions.

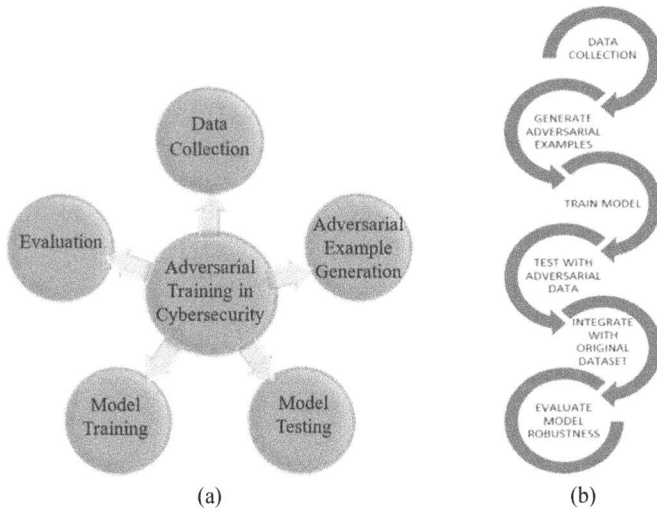

(a) (b)

Fig. 61.1 (a) Adversarial training in cybersecurity: key phases, (b) Adversarial training pipeline for model robustness evaluation

5. Results and Discussion

Figure 61.2 shows the robustness of a model against adversarial attacks by its rising accuracy on adversarial changed data over training cycles. Designed either with FGSM or Projected Gradient Descent, a model is trained in a controlled Tensor Flow environment with adversarial and clean data. Recording adversarial dataset accuracy helps to boost resilience by letting the model's learning curve be monitored at every epoch. The graph indicates the improved adversarial accuracy for applications in a secure 6G network, therefore indicating the increased security against hostile inputs of the model. The Mean Squared Error (MSE) vs. Iterations graph in Figure 61.3 illustrates the model's error reduction over 100 iterations, reflecting the optimization process during adversarial training. The X-axis represents the number of iterations, while the Y-axis shows the MSE values, which decrease as the model adapts to both clean and adversarial data. Generated data demonstrates a declining trend in MSE, indicating improved model performance and robustness against adversarial inputs. Figure 61.4 (a) shows the loss reduction of the model on clean and adversarial data across 100 epochs, so indicating its adaptation to adversarial training. Standard loss lowers faster on clean data, showing improved model performance; adversarial loss falls more slowly and indicates adversarial inputs. This image shows the model's developing adversarial resilience.

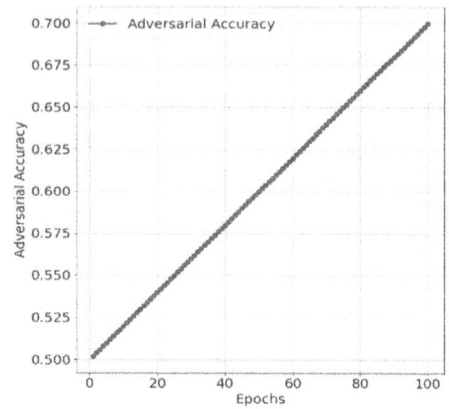

Fig. 61.2 Adversarial accuracy vs. epochs

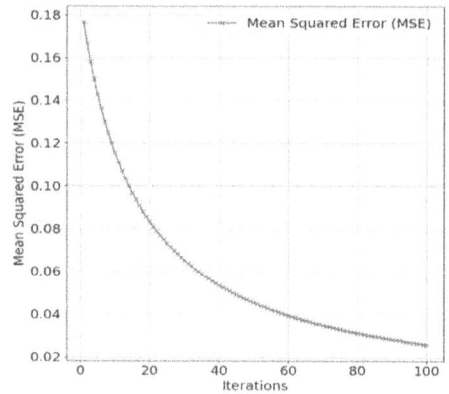

Fig. 61.3 MSE reduction over iterations

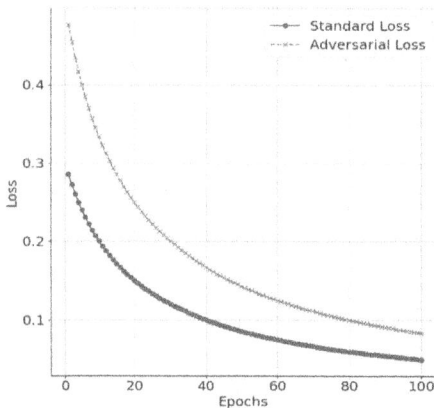

(a)

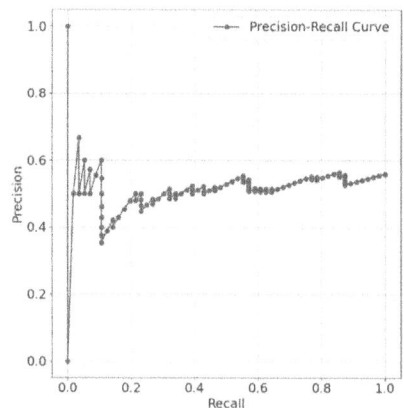

(b)

Fig. 61.4 (a) Standard loss vs. adversarial loss over epochs, (b) precision-recall curve showing model performance in adversarial detection

Figure 61.4 (b) displays the adversarial example detecting capability of the model. Initially high, accuracy falls dramatically as recall rises, implying that most found cases are true positives. Recall exceeds 0.2; so, precision stabilizes around 0.5 to 0.6 implying balanced but limited detection performance. Since the model becomes more inclusive in identifying hostile events, higher recall levels reduce accuracy and increase false positives. For applications demanding high detection rates and low false alarms, the model must combine identification of as many hostile samples as practical with accuracy in its predictions.

6. Conclusion

At last, our research shows the enormous promise of adversarial training to increase the resistance of machine learning models against cyberattacks in 6G networks. Adversarial examples included into the training process help the model to effectively identify and minimize adversarial attacks, hence improving network security. The results reveal stronger model robustness with less susceptibility to controlled inputs, so adversarial training seems as a feasible defence method. This method is crucial in addressing the specific security issues of 6G networks since it provides a framework for next research targeted on optimizing computational efficiency and combining modern encryption techniques to ensure scalable and adaptive protection in quickly emerging 6G infrastructures.

References

1. Hoang, V. T., Ergu, Y. A., Nguyen, V. L., & Chang, R. G. (2024). Security risks and countermeasures of adversarial attacks on AI-driven applications in 6G networks: A survey. *Journal of Network and Computer Applications, 104031.*
2. Yang, L., Li, Y., Yang, S. X., Lu, Y., Guo, T., & Yu, K. (2022). Generative adversarial learning for intelligent trust management in 6G wireless networks. *IEEE Network, 36*(4), 134–140.
3. Kocherla, R., Dwivedi, Y. D., Reena, B. A. M., CR, K., & Dhanraj, J. A. (2024). Adversarial challenges in distributed AI: ML safeguarding 6G networks. In *Security Issues and Solutions in 6G Communications and Beyond* (pp. 61–79). IGI Global.
4. Thomas, T., Vijayaraghavan, A. P., & Emmanuel, S. (2020). Adversarial machine learning in cybersecurity. In *Machine Learning Approaches in Cyber Security Analytics* (pp. 185–200).
5. Mahmoud, H., Ismail, T., Baiyekusi, T., & Idrissi, M. (2024). Advanced security framework for 6G networks: Integrating deep learning and physical layer security. *Network, 4*(4), 453–467.
6. Son, B. D., Hoa, N. T., Van Chien, T., Khalid, W., Ferrag, M. A., Choi, W., & Debbah, M. (2024). Adversarial attacks and defenses in 6G network-assisted IoT systems. *IEEE Internet of Things Journal.*
7. Ferrag, M. A., Friha, O., Kantarci, B., Tihanyi, N., Cordeiro, L., Debbah, M., ... & Choo, K. K. R. (2023). Edge learning for 6G-enabled Internet of Things: A comprehensive survey of vulnerabilities, datasets, and defenses. *arXiv e-prints, arXiv-2306.*
8. Saeed, M. M., Saeed, R. A., Abdelhaq, M., Alsaqour, R., Hasan, M. K., & Mokhtar, R. A. (2023). Anomaly detection in 6G networks using machine learning methods. *Electronics, 12*(15), 3300.
9. Blika, A., Palmos, S., Doukas, G., Lamprou, V., Pelekis, S., Kontoulis, M., ... & Askounis, D. (2024). Federated learning for enhanced cybersecurity and trustworthiness in 5G and 6G networks: A comprehensive survey. *IEEE Open Journal of the Communications Society.*
10. Gupta, B. B., Chui, K. T., Gaurav, A., & Arya, V. (2023, October). Deep learning based cyber attack detection in 6G wireless networks. In *2023 IEEE 98th Vehicular Technology Conference (VTC2023-Fall)* (pp. 1–5). IEEE.
11. Jiao, L., Shao, Y., Sun, L., Liu, F., Yang, S., Ma, W., ... & Guo, Y. (2024). Advanced deep learning models for 6G: Overview, opportunities and challenges. *IEEE Access.*

12. Rahman, M. A., & Hossain, M. S. (2022). A deep learning assisted software defined security architecture for 6G wireless networks: IIoT perspective. *IEEE Wireless Communications, 29*(2), 52–59.

13. Suomalainen, J., Ahmad, I., Shajan, A., &Savunen, T. (2025). Cybersecurity for tactical 6G networks: Threats, architecture, and intelligence. *Future Generation Computer Systems, 162*, 107500.

14. Li, S., Lin, X., Liu, Y., & Li, J. (2024). Trustworthy AI-generative content in intelligent 6G network: Adversarial, privacy, and fairness. *arXiv preprint, arXiv:2405.05930*.

15. Catak, F. O., Kuzlu, M., Tang, H., Catak, E., & Zhao, Y. (2022). Security hardening of intelligent reflecting surfaces against adversarial machine learning attacks. *IEEE Access, 10*, 100267–100275.

16. Ahmed, A. A., Hasan, M. K., Memon, I., Aman, A. H. M., Islam, S., Gadekallu, T. R., & Memon, S. A. (2024). Secure AI for 6G mobile devices: Deep learning optimization against side-channel attacks. *IEEE Transactions on Consumer Electronics*.

17. Abasi, A. K., Aloqaily, M., Guizani, M., & Debbah, M. (2023, December). Mitigating security risks in 6G networks-based optimization of deep learning. In *GLOBECOM 2023–2023 IEEE Global Communications Conference* (pp. 7249–7254). IEEE.

18. Kuzlu, M., Catak, F. O., Cali, U., Catak, E., & Guler, O. (2023). Adversarial security mitigations of mmWave beamforming prediction models using defensive distillation and adversarial retraining. *International Journal of Information Security, 22*(2), 319–332.

19. Oleiwi, H. W., Mhawi, D. N., & Al-Raweshidy, H. S. (2023, June). A secure deep autoencoder-based 6G channel estimation to detect/mitigate adversarial attacks. In *2023 5th Global Power, Energy and Communication Conference (GPECOM)* (pp. 530–535). IEEE.

Note: All the figures in this chapter were made by the authors.

Emerging Perspectives and Applications of Computational Intelligence and Smart Systems
– Dr. Amit Lathigara et al. (eds)
© 2026 Taylor & Francis Group, London, ISBN 978-1-041-20965-2

62

Optimization of Steam Venting Loss in Coal-Fired AFBC Boiler by Controlling Coal Feeding Primary Air Lines Choking Issue

Chirag V. Kapuia*

PhD Scholar, Mechanical Engineering Department,
Faculty of Technology, RK University,
Rajkot, Gujarat, India

Chetankumar M. Patel

Mechanical Engineering Department,
Faculty of Technology, RK University Rajkot,
Gujarat, India

■ **Abstract:** One of the major problems faced in process boilers is the choking of coal feeding paths. This problem creates a drop in feed thermal energy and ultimately leads to a drop in generated steam pressure. Upon clearing of coal chokage, a tremendous amount of coal suddenly enters the furnace, creating additional heat shock. This increases the steam pressure in an uncontrolled way. This increased pressure of steam is dangerous for operating steam lines and so this needs to be controlled. For this, safe steam vents are opened to vent steam in the atmosphere and thereby improve operational safety. This vent is kept open until the pressure gets controlled in the desired range. This phenomenon creates steam venting loss in the atmosphere. This venting loss reduces boiler efficiency and increases coal consumption. This also creates some amount of warming in the nearby atmosphere. Work is done to prevent the root cause of coal choking in primary air coal feeding lines. This will also optimize coal consumption and reduce steam venting loss in the boiler.

■ **Keywords:** Coal choking, Steam venting loss, Coal consumption, Boiler efficiency

1. INTRODUCTION

There are lots of problems available in the operations of coal-fired AFBC process boilers in the industry. Few of the problems collected are like Excessively Choking of PA coal feeding lines, Difficulty in the disposal of Ash via Ash conditioner in a silo, Extra cost for water mixing in fly ash

---

*Corresponding author: ck5719@gmail.com

DOI: 10.1201/9781003725046-62

during ash uploading via ash conditioner in a silo, Energy loss in air pressure killing from 7 kg to 3.5 kg for using it in boiler ash conveying, Extra efforts while lifting bed material from boiler ground floor to top floor, Condensate contamination received from respective divisions, Overflow of boiler blowdown tank, Coal dust accumulation in LP dosing tanks, Dust emission during operation of ash conditioner while unloading from silo, Operation of hoist while lifting bed material is risky, Water accumulation in belt conveyor-1 pit during the rainy season, Boiler blow down tank bund wall not available, etc. Focus is given to the solution of the problem of Excessively Choking of PA coal feeding lines. Over the world, various study has been done like (Chetan T. Patel et al. 2013) worked upon how the performance of boilers varies with respect to calorific heat values of coal. It was concluded that if moisture and ash percent is more in fuel then efficiency of furnace will drop. (Moni Kuntal Bora et al. 2014) analysed the performance of boiler with respect to various performance indicators like excess oxygen, temperature profiles, air flows, etc and showed which parameter is affecting the performance.(Suthum Patumsawad et al.2015) worked upon a laboratory size FBC boiler and they investigated the combustion performance of coal. (Rakesh Kumar Sahu et al. 2015) worked on energy's assessment performance. Flue gas leaving chimney temperature was considered and calorific value of fuel was taken in its consideration.

There was a research gap found in the loss due to operational issues like choking of coal and no research is found based on efficiency drop due to this problem.

2. EXCESSIVELY CHOKING OF PA COAL FEEDING LINES

2.1 Basis for Selection of Problem

All problems were categorized into three types as below:

A - Problems which can be solved by ourselves

B - Problems that need help from the department other than operation.

C - Problems that need help from management

Table 62.1 Opportunities and problem status with its impact

Sr. No.	List of Opportunities	Estimated savings in Lakhs	Category A/B/C	Problem Status	Impact	Effort
1	Excessively Choking of PA coal feeding lines	3.6	A	Known Cause - Unknown Solution	High	Medium
2	Extra efforts while lifting bed material from boiler ground floor to top floor.	0.2	B	Known Cause - Known Solution	High	Medium
3	Overflow of boiler blowdown tank	0.3	B	Known Cause - Known Solution	High	High
4	Energy loss in air pressure killing from 7 kg to 3.5 kg for using it in boiler ash conveying	2	A	Known Cause - Unknown Solution	High	High
5	Coal dust accumulation in LP dosing tanks	0	A	Known Cause- Known Solution	High	Medium
6	Difficulty in the disposal of Ash via Ash conditioner in a silo	0.3	A	Known Cause- Known Solution	High	High
7	Extra cost for water mixing in fly ash during ash uploading via ash conditioner in silo	1.25	A	Known Cause- Known Solution	Medium	High

2.2 Definition of Problem

Coal feeding PA lines are getting choked frequently in 66 TPH Boiler. Upon PA line flushing, steam pressure increases so venting of approximately 1 MT are occurring per 2 Days.

2.3 Analyses of Problem

The 5W-1H method was used for the solution of the problem. It is as below:

What? Furnace inlet Primary Air line coal chokage of 66 TPH boilers

Where? At PA coal feeding lines.

When? During every shift of 8 hours

Who? Boiler Operator efficiency is affected and shift has seen the problem.

Why? Company costs are getting affected.

How Much? 500 kg steam venting/day (INR 30,000/month)

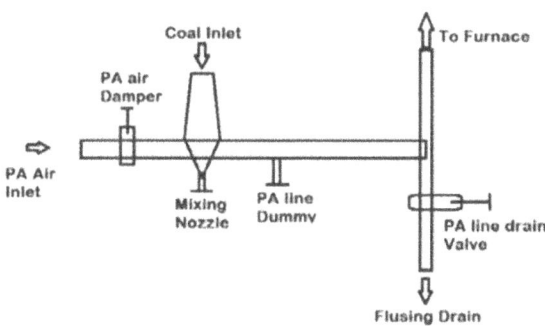

Fig. 62.1 Boiler field operator doing PA line flushing

2.4 Finding the Root Cause

Table 62.2 Why-why analysis method was used to detect the root cause of the problem

Why did PA line choking occur?	Due to the big coal size	Why big coal size?	Coal screen Damage	The coal screen was checked physically and the Mesh size was intact without any damage.
	Due to coal overfeeding	Why coal overfeeding?	Due to failure of VFD of RAV	No malfunctioning is found in the VFD of the RAV. Check Physical RPM with tachometer Vs VFD command given in DCS.
	Due to PA lines Timely flushing is not being done	Why is PA line's Timely flushing not done?	Timely flushing of PA lines Not included in practice	There was no such recommendation from OEM so included in SOP and checklist implemented. (Main Root Cause)
	Due to wet coal common from the coal yard	Why is wet coal coming from the coal yard?	Coal sent by the vendor may be with high moisture.	Moisture content was found within Range (below 35%). Verified by Lab analyses through proper sampling.

Summarizing the problem statement comes out to be "Furnace inlet Primary Air line coal choking of 66 TPH boilers". Reply to first "Why" comes out as "Due to PA lines Timely flushing is not done". Upon further deep drilling, reply to second "Why" comes as "Timely flushing of PA lines Not included in practice". Hence, the Root Cause comes out as "Periodically flushing is not done proactively due to which PA lines are getting choked".

2.5 Developing the Solution

PA line flushing to be done proactively and periodically to prevent coal choking and prevent steam venting. PA line flushing frequency to set daily once proactively to prevent its choking and prevent steam venting. Regular implementation is set to Flushing of all 18 PA lines to be done daily in B shift. In-charge has to verify daily that no PA lines are found choked. The operator has to mention PA line flushing status along with time and if any choking is found during flushing activity.

Table 62.3 Observation found during flushing activity

Observations/ Abnormalities	Action Items	Observations	Analyses	Selection
Finding the optimum time to set the PA line flushing frequency	Flushing Started primarily once a shift	Choking did not occur	It can be selected but the scope for optimization is still there.	No
	Flushing Started twice a shift	Choking did not occur	It can be selected but the scope for optimization is still there.	NO
	Flushing frequency time increased to 3 shifts under observation	Choking did not occur	Optimized Solution.	Yes
	Flushing time set to once in 4 shift	Choking started occurring approx. twice a day.	It can not be adopted as choking is there.	No

3. CONCLUSION

Annual Cost savings after improvement was INR 3.3 Lac/ year. Steam Productivity Increased by 0.5 MT/day. Savings of natural resources (coal) by preventing steam venting by 98 kg/day (32.3MT/ annum). Reduction in Greenhouse gas (CO_2) emissions by 61.12 MT. Employee Skill & Morale improved. We have learned to work in a team Creating a Safe and better working environment

References

1. Patel Chetan T., Patel Bhavesh K., Patel Vijay K. (2013). 'Technology and Engineering's Journal of International Research'. International Journal of Innovative Research in Science, Engineering and Technology
2. Bora Moni Kuntal and Nakkeeran S. (2014). 'Performance Analysis from the Efficiency Estimation of Coal Fired Boiler.' International Journal of Advanced Research. ISSN 2320-5407, PP 561–574.
3. Suthum Patumsawad Coal Combustion Studies in a Fluidized Bed
4. Sahu Rakesh Kumar, Rao G.Ishwar Maurya, Kirti. (2015). 'Energy Performance Assessment of CFBC Boiler.' ISSN: 2278-0181, PP 1–5.

Note: All the tables and figure in this chapter were made by the authors.

Emerging Perspectives and Applications of Computational Intelligence and Smart Systems
– Dr. Amit Lathigara et al. (eds)
© 2026 Taylor & Francis Group, London, ISBN 978-1-041-20965-2

63

Real-Time IoT Solution for Smart Agriculture using Thingspeak and Whatsapp

Man Singh Baghel[1],
R. C. Gurjar[2], Gireesh G. Soni[3]

Department of Electronics and Instrumentation Engineering,
Shri G. S. Institute of Technology and Science,
Indore, India

Abhishek Tripathi[4]

Department of Computer Science and Engineering,
Kalasalingam Academy of Research and Education,
Srivilliputhur, Tamil Nadu, India

Rajesh Khatri[5]

Department of Electronics and Instrumentation Engineering,
Shri G. S. Institute of Technology and Science,
Indore, India

■ **Abstract:** Agriculture requires innovative strategies to enhance productivity and sustainability amid constrained resources and evolving environments. This study introduces an intelligent agricultural system enabled by IoT, incorporating sensors, ThingSpeak cloud, and WhatsApp for precision farming. The system is driven by tracking criteria such as soil moisture, temperature, humidity, and light intensity, utilizing ESP32 microcontrollers, BH1750, DHT11, pH sensors, and soil moisture sensors. Over one hundred data collected in a month revealed temperature fluctuations (24°C–27.5°C) and light intensity maxima (230 lux at 1:00 PM), hence facilitating optimal watering and crop growth. Real-time data access, remote irrigation management, and multilingual support via a WhatsApp chatbot enhance farmer engagement and resource efficiency. Initial findings indicate significant reductions in resource waste and increased crop yields, hence illustrating the system's potential for sustainable, data-driven agriculture.

■ **Keywords:** IoT, ThingSpeak, Real-time monitoring, WhatsApp, Smart farming

[1]mansinghbaghel0308@gmail.com, [2]rcgurjar95@gmail.com, [3]gireeshsoni@gmail.com, [4]tripathi.abhishek.5@gmail.com, [5]rajeshkhatri1@rediffmail.com

DOI: 10.1201/9781003725046-63

1. Introduction

Rising as a changing technology in agriculture, the Internet of Things (IoT) assists farmers to acquire accuracy in resource management and decision-making (Tripathi et al., 2024). Mostly, agriculture determines food security and allows millions of people to live all around. Sometimes, however, the excess of fertilizers and water produced by traditional farming techniques deteriorates soil and affects the surroundings. Apart from the effects of climate change, the growing world population poses modern agriculture with historically unheard-of challenges like the need to increase productivity while maintaining environmental sustainability. These challenges highlight the need of applying innovative approaches that let farmers properly manage resources. Smart agriculture systems have interesting IoT (Bhattacharya & De, 2021) possibilities by letting real-time monitoring of critical field measurements. Combining several sensors, these devices monitor soil moisture, temperature, humidity, pH, and nutritional value. Smart agricultural systems enable to lower resource waste, preserve water, and enhance crop health by means of tailoring these technologies to fit real field conditions (Che Omar & Ramle, 2023). Combining IoT technology, cloud-based data storage, and user-friendly remote administration interfaces, the proposed monitoring and assistance system offers a fresh method to precision farming (Adebayo et al., 2021). Having robust sensors and advanced data-processing capability, the system continuously monitors environmental conditions and allocates resources based on real-time events. With an eye on its value in supporting ecologically conscious, data-driven agriculture, this paper explores the design, implementation, and effects of this smart system. Section 2 reports literature on IoT-based monitoring technologies as well as precision farming. Section 3 describes methodology with component integration, system architecture and algorithms. Section 4 deals with hardware implementation; Section 5 presents analysis and results. Section 6 marks the conclusion and upcoming works.

2. Related Work

The IoT could make sustainability, resource management, and field monitoring of agriculture possible. Using real-time data interchange and distant management via WhatsApp link with IoT devices helps farmers' agricultural monitoring to be simplified. (Yaqub et al. 2024) create a "WhatsApp Chat Analyzer" by means of statistical analysis and visualizations to spot trends in group conversation data, therefore stressing WhatsApp's data processing and visualization capacity (Yaqub et al., 2024). Underlining the requirement of effective water quality monitoring systems deriving from environmental concerns and population increase, (Jan et al. 2021) investigate reasonably priced IoT-based smart devices for home water quality monitoring in real time (Jan et al., 2021). Motivated by IoT networks based on sensors in education for environmental awareness, the 2020 Through project-based learning, Tabuenca et al. enable students to design various IoT systems, therefore fostering multidisciplinary cooperation and demonstrating IoT's value in engineering education (Tabuenca et al., 2020). Farmers might track soil water potential using ThingSpeak data display and an Arduino-based gadget, therefore regulating irrigation (Payero et al., 2017). (Che Omar and Ramle 2023) They remotely monitor floral beds using ESP8266 and Telegram, therefore underlining the importance of IoT in home gardening and the need of user-friendly interfaces for system management (Che Omar & Ramle, 2023). Regarding spills, SMS or Telegram warn users of (Che et al. 2021) IoT-based water pipeline leakage monitoring system. This approach shows how timely warnings could follow (Che et al., 2021) and how IoT monitoring of communication channels may track infrastructure. Based on (Anne et al. 2022), MQTT-based Android chat systems help to simplify agricultural chat apps on few devices. MQTT offers remote agricultural monitoring (Anne et al., 2022) flexible IoT communication over low-bandwidth networks. Underlining the need of quite reasonably affordable and efficient digital divide-bridging technologies, the edge computing pivot irrigation system monitoring solutions by (Matilla et al. 2022) highlight LoRaWAN wireless networking (Matilla et al., 2022) improved actual irrigation system water management. In view of second (Ananda et al. 2024) mobile technology in agricultural extension services is investigated. They demonstrate how chatbots and phone calls let agricultural knowledge be shared in far-off rural areas (KR et al., 2024). Designed an IoT-Chatbot for Onesime and associates's 2023 greenhouse management. Using ThingSpeak for data storage and visualization, (Tapakire and Patil 2019) present a clever agriculture system that dynamically alters greenhouse conditions, hence improving crop quality and lowering costs (Onésime et al., 2023). ThingSpeak irrigation dependent on soil moisture levels (Tapakire& Patil, 2019) using agricultural IoT sensors, and real-time monitoring via a smartphone app.

3. METHODOLOGY AND ALGORITHM

Combining ESP32 microcontrollers, DHT11 temperature and humidity sensors, YL-69 soil moisture sensors, and a pH sensor for real-time environmental monitoring, the smart agricultural monitoring system creates From these sensors, the system generates data for Wi-Fi (ESP32 module) transmission to a cloud-based server for storage and analysis. The 16x2 LCD panel shows local perspective of sensor values. A relay module driving the autonomous irrigation system either turns on or off depending on soil moisture content, therefore affecting a water pump. The ESP32 microprocessor turns on the pump when moisture levels drop below a specified threshold and stops watering once the relevant level is recovered, therefore helping to save water. The system automatically adjusts irrigation relying on DHT11 sensor data to increase frequency under high temperatures and low humidity and reduces watering under cooler or humid circumstances. By means of pH data analysis implemented using pH monitoring technique, corrective actions such limestone for acidity or sulfur for alkalinity guarantee optimal soil conditions. Rain detecting sensors help to reduce overwatering by allowing quick stop of irrigation during rain. Farmers might utilize a smartphone app linked to a cloud server, see past trends, and access real-time data to send remote commands beginning or stopping irrigation. Figure 63.1 (a) of system architecture uses a relay and pump system to demonstrate the integration of sensors, ESP32 microcontrollers, Wi-Fi broadcasts, cloud storage, and irrigation control. Figure 63.1 (b) of system flowchart demonstrates via a mobile interface sensor data gathering, cloud processing, irrigation decision-making, and human override capability. This IoT-based system improves crop health, reduces water wastage, and maximizes resource use so boosting precision farming.

4. HARDWARE IMPLEMENTATION

Unique elements defining the ESP32-WROOM-32E remote agricultural monitoring tool include low power, Wi-Fi, sensor integration for real-time data collecting, and cloud storage. DHT11 sensors change water depending on temperature and humidity while soil moisture sensors initiate irrigation using a 5V DC-DC relay module. pH sensors track soil acidity; data is cloud-based and employed in fertilizer modification. LDR sensors monitor sunlight for agricultural use while rain sensors limit irrigation during a rain. Driven by real-time data demonstrating a 5V DC-DC converter balances power, I2C drives a 16x2 LCD using weatherproof enclosures increases longevity. Thanks

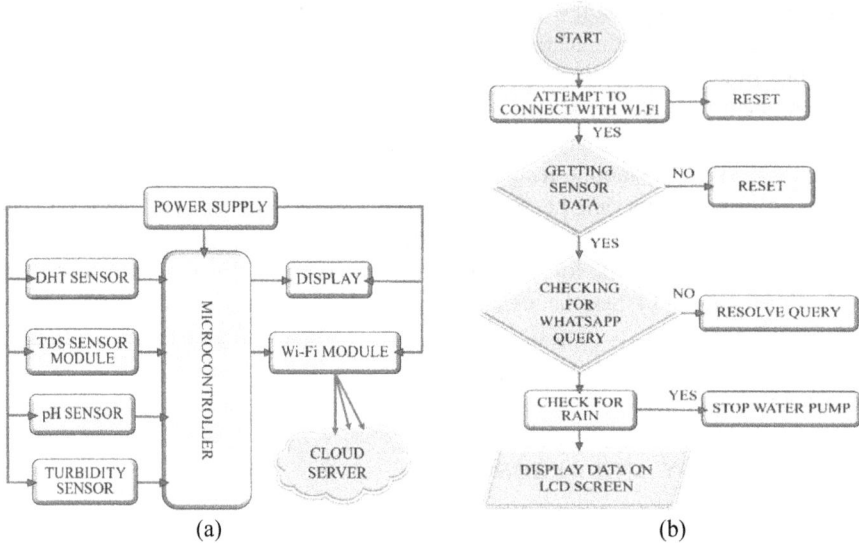

(a)   (b)

Fig. 63.1 (a) System architecture showing sensor integration, data processing, and cloud connectivity, (b) flowchart for the smart agricultural monitoring system's operational sequence

to cloud connectivity—through a mobile interface—data-driven farming, smart watering, and remote monitoring are enabled.

The ESP32-WROOM32U microcontroller and SIM800L module form the elements of data processing and communication in the smart agricultural monitoring system (Fig. 63.2, a and b). Top shows components and a college logo; bottom PCB circuitry and author information. Maximizing field deployment connection and sensor integration, the PCB design.

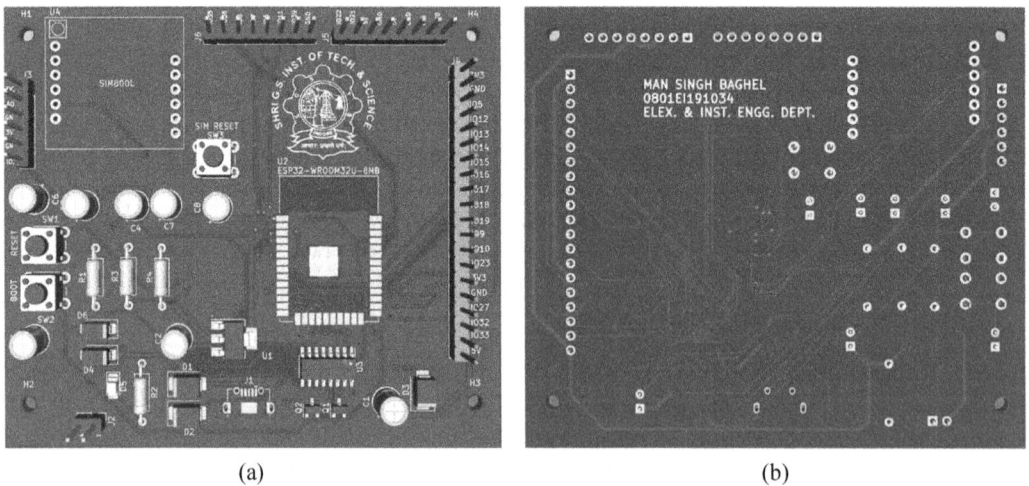

(a)   (b)

Fig. 63.2 (a) Top side of PCB design layout, showing component placements and college logo, (b) Bottom side of PCB design layout, displaying circuit paths and author information

Figure 63.3 shows the hardware setup for the smart agricultural monitoring system enclosed in a protective box, designed for field deployment. The setup includes the ESP32 microcontroller, sensors for soil moisture, pH, and temperature, as well as a power module and wiring connections. This compact kit format ensures durability and ease of maintenance, allowing seamless data acquisition and automated control in agricultural environments.

5. RESULTS AND DISCUSSION

Using YL-69 sensors allowed the pH, temperature, humidity, soil moisture system regulated irrigation, thereby saving thirty percent more water. While DHT11 sensors adjusted watering depending on temperature, pH sensors raised fertilizer usage. Graphs confirmed stability; cloud storage made trend tracking possible. Real-time control made possible by the mobile interface together with increased efficiency helped to save labour.

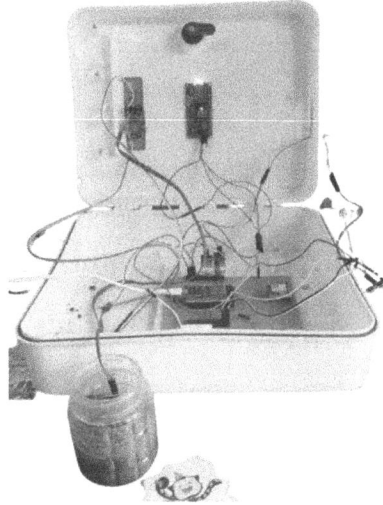

Fig. 63.3 Hardware setup in box kit form for the smart agricultural monitoring system

By continually evaluating field circumstances, the technology creates ThingSpeak data. Following more than one hundred samples over a month, the graphs below illustrate ideas of agricultural optimization. Figure 63.4 (a) daily real-time field temperature data between 24°C and 27.5°C To help irrigation and other temperature-dependent environmental changes, we require both daily and seasonal fluctuations. Most of the field temperature values in Fig. 63.4 (b) fell between 23°C and 25°C throughout observation. Monitoring crop development calls for this range of temperature.

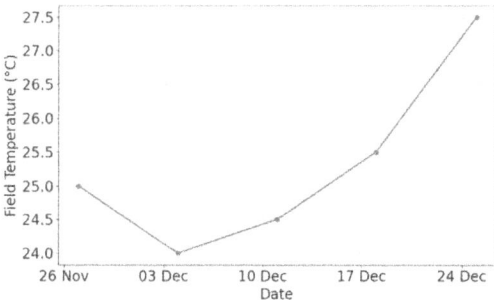

(a)

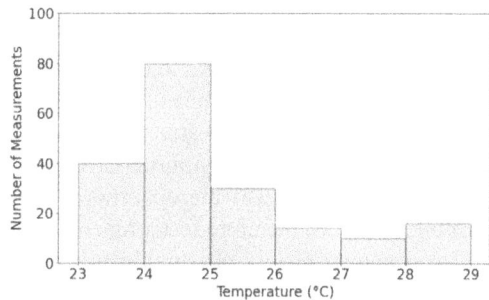

(b)

Fig. 63.4 Field temperature variation over time, (b) Temperature variation histogram showing measurement frequency

Figure 63.5 (a) quite clearly demonstrates daily field humidity ranging 60–61% in rather significant afternoon decline. These variances support crop development by way of moisture regulation. The high midday light intensity in Fig. 63.5 (b) drops to 20 lux by 7:00 PM from initial high at 230 lux roughly at 1:00 PM. The sun rises and the intensity increases steadily from 50 lux at 9:00 AM to drastically decrease in the evening. Knowing daily solar exposure not only helps to plan farming but also provide enough light for photosynthesis and crop development.

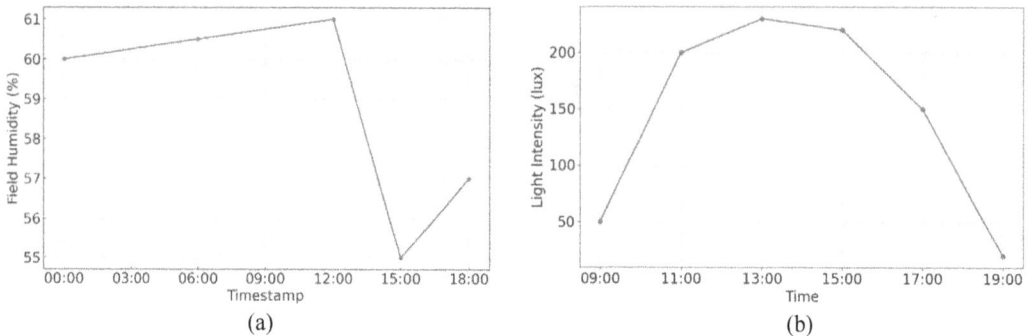

Fig. 63.5 (a) Field humidity variation over time, (b) Light intensity variation over time

The 24*7 WhatsApp chatbot integrated with the system provides instant support by processing user commands via HTTP. It responds to preset queries, such as introducing itself when asked "Who are you?" and generating real-time field data with the command "fieldreport," including temperature, soil moisture, weather, and humidity. It also supports Hindi responses when requested. Users can remotely schedule and control the irrigation pump through WhatsApp using commands like "SCH 19:30" for scheduling and "On" or "Off" for manual control. Additionally, the "Samplereport" command generates a comprehensive farm report covering weather, soil conditions, crop health, water management, projected yield, and support links. The IoT-enabled system enhances resource efficiency, reducing water waste by 30% and improving crop yields. It maintains stable environmental conditions, such as temperature fluctuations (24°C–27.5°C) and light intensity peaks (230 lux), demonstrating its effectiveness in precision farming. While scalability and connectivity remain challenges, future AI integration could address these limitations, making the system a valuable tool for modern agriculture.

6. Conclusion and Future Work

This work projects the possible increase in agricultural output brought about by IoT technologies together with user interfaces and cloud-based solutions. The system optimizes resource use, lowers waste, and increases crop health by means of automatic watering and monitoring of important environmental indicators. Farmers who use mobile interfaces with real-time data access have practical information that simplifies farm management and promotes ecologically friendly activities. This method emphasizes the opportunities of smart agricultural systems to solve contemporary farming problems, therefore fostering environmental responsibility and efficiency. Future initiatives will concentrate mostly on significantly lowering environmental effect even more by including renewable energy sources, integrating advanced artificial intelligence models for predictive analytics, and improving system scalability for huge agricultural regions including integration.

References

1. Tripathi, A., Manvith, K., Srinath, G. M., Yuvan, N. K., & Reddy, U. T. (2024, April). Implementation of an Alert System Integrated into Smart Wireless Helmets Utilizing IoT Sensors. In *2024 2nd International Conference on Networking and Communications (ICNWC)* (pp. 1–5). IEEE.
2. Bhattacharya, A., & De, D. (2021). Agriedge: Edge intelligent 5g narrow band internet of drone things for agriculture 4.0. In *IoT-based intelligent modelling for environmental and ecological engineering: IoT next generation ecoAgro systems* (pp. 49–79). Cham: Springer International Publishing.

3. Che Omar, S. A., & Ramle, R. (2023). IoT-based Flower Garden Care System using the ESP8266 Wi-Fi Module and Telegram Application.

4. Adebayo, S., Emuoyibofarhe, O., & Awofolaju, T. (2021). Cost efficient internet of things based smart farm system for rural farmers: Leveraging design thinking approach. *Heritage and Sustainable Development*, *3*(2), 111–120.

5. Yaqub, S., Gochhait, S., Khalid, H. A. H., Bukhari, S. N., Yaqub, A., & Abubakr, M. (2024, January). WhatsApp Chat Analysis: Unveiling Insights through Data Processing and Visualization Techniques. In *2024 ASU International Conference in Emerging Technologies for Sustainability and Intelligent Systems (ICETSIS)* (pp. 862–865). IEEE.

6. Jan, F., Min-Allah, N., & Düştegör, D. (2021). Iot based smart water quality monitoring: Recent techniques, trends and challenges for domestic applications. *Water*, *13*(13), 1729.

7. Tabuenca, B., García-Alcántara, V., Gilarranz-Casado, C., & Barrado-Aguirre, S. (2020). Fostering environmental awareness with smart IoT planters in campuses. *Sensors*, *20*(8), 2227.

8. Payero, J. O., Mirzakhani-Nafchi, A., Khalilian, A., Qiao, X., & Davis, R. (2017). Development of a low-cost Internet-of-Things (IoT) system for monitoring soil water potential using Watermark 200SS sensors. *Advances in Internet of Things*, *7*(03), 71.

9. Che Omar, S. A., & Ramle, R. (2023). IoT-based Flower Garden Care System using the ESP8266 Wi-Fi Module and Telegram Application.

10. Che, N. N., Omar, K. N. F., Azir, K., & Kamarudzaman, M. F. (2021, July). Water pipeline leakage monitoring system based on Internet of Things. In *Journal of Physics: Conference Series* (Vol. 1962, No. 1, p. 012025). IOP Publishing.

11. Anne, L., Nandan, T., Kunj, M., Kumar, S. A., Mahesh, G., & Sangeetha, R. (2022). MQTT-Based Android Chat Application for IoT. *SN Computer Science*, *3*(5), 402.

12. Matilla, D. M., Murciego, A. L., Jiménez-Bravo, D. M., Mendes, A. S., & Leithardt, V. R. (2022). Low-cost Edge Computing devices and novel user interfaces for monitoring pivot irrigation systems based on Internet of Things and LoRaWAN technologies. *Biosystems Engineering*, *223*, 14–29.

13. Ananda, K. R., Saikanth, D. R. K., Chaudam, V., Sravani, S., Nayak, S. H., Dam, A., & Shukla, A. (2024). Impact of Mobile Technology on Extension Service Delivery in Remote Farming Communities: A Review. *Journal of Scientific Research and Reports*, *30*(3), 1–13.

14. Onésime, Oboulhas Tsahat Conrad, ETOU Destin Gemetone, NSONDE-MONDZIE Cédric Prince, and I. K. O. U. E. B. E. Norbert. "An Intelligent Approach to Monitoring an Agricultural Greenhouse via a 4G Network in the Republic of Congo." Far East Journal of Electronics and Communications 27 (2023): 61-76.

15. Tapakire, Bhagyashree A., and Manasi M. Patil. "IoT based smart agriculture using ThingSpeak." Int. J. Eng. Res. Technol. 8, no. 185 (2019): 10-17577.

Note: All the figures in this chapter were made by the authors.

Emerging Perspectives and Applications of Computational Intelligence and Smart Systems
– Dr. Amit Lathigara et al. (eds)
© *2026 Taylor & Francis Group, London, ISBN 978-1-041-20965-2*

64

Behavioral Analysis of QUIC Protocol under DDoS

Sejpalsinh Jadeja*, Shivangi Patel
School of Engineering, RK University, Rajkot,
Gujarat, India

Nirav Bhatt
School of Engineering, RK University, Rajkot,
Gujarat, India

Utkarsh Tamrakar
National Forensic Sciences University,
Gandhinagar, Gujarat, India

■ **Abstract:** The Internet is commonly trusted on the Transmission Control Protocol (TCP) for dependable communication between client and server. Regarding delay and efficiency, the constraints of the TCP have become more evident as web applications have progressed. To enhance the efficiency of session-based web applications by decreasing latency and ensuring dependable interaction over the User Datagram Protocol (UDP) Google developed Quick UDP Internet Connections (QUIC) a transport layer network protocol That was later embraced by the Internet Engineering Task Force (IETF) as a standard. This study concentrates on implementing the QUIC (Quick UDP Internet Connections) protocol in the Network Simulator-3 (NS-3) environment and examining its robustness against Distributed Denial of Service (DDoS) vulnerability scenarios. Designed to enhance efficiency and security, QUIC is a transport layer protocol that outperforms conventional TCP/IP connections, particularly benefiting video streaming and web applications. The Execution Includes Developing A controlled network environment using NS-3, For communication between nodes Using the QUIC protocol—DDoS attack Events to test the protocol's Strength and Efficiency under stress. QUIC-supported nodes are Implemented, the network topology is Organized, DDoS attack models are Executed, and network Efficiency metrics such as throughput, latency, and packet loss rates are Analyzed during attack scenarios. The outcomes provide knowledge of QUIC's actions and adaptability in the face of DDoS events, comparing its performance against legacy protocols like TCP and UDP. The Comprehension of QUIC's Capacity to improve internet security and Execution contributed to this research, especially in situations vulnerable to DDoS attacks.

■ **Keywords:** QUIC protocol, DDoS attacks, Network simulation, NS-3, Cybersecurity, Network performance, UDP, Internet protocols

*Corresponding author: sjadeja137@rku.ac.in

DOI: 10.1201/9781003725046-64

1. INTRODUCTION

The rapid advancement of network technologies has increased the demand for **fast, secure, and efficient** web communication protocols. **QUIC (Quick UDP Internet Connections)**, developed by **Google** and later standardized by **IETF**, offers a promising alternative to traditional **TCP-based protocols** by utilizing **UDP** to reduce latency and improve efficiency. As modern web applications become more **data-intensive**, QUIC stands out due to its **faster connection initiation, optimized congestion control, and built-in encryption**. However, like any protocol, QUIC must be assessed under **challenging network conditions**, particularly against **Distributed Denial of Service (DDoS) attacks**, which can overwhelm servers and disrupt services.This study evaluates QUIC's **performance under DDoS attacks** using **ns-3 simulations**. The key objectives include:

- Implementing a **QUIC model in ns-3**.
- Simulating various **DDoS attack scenarios**.
- Analyzing **throughput, packet loss, and connection time** under attack conditions.
- Comparing QUIC's **resilience to TCP/IP** in similar scenarios.
- Assessing the effectiveness of QUIC's **security features** in mitigating attacks.

2. RELATED WORK

Security Analysis of QUIC: The security vulnerabilities of QUIC have been widely examined. Research (X. Cao, S. Zhao at al. 2019) identified 0-RTT attacks, which take advantage of QUIC's ability to send data before completing a full handshake. The study investigated two primary attack types: QUIC RST Attack and Version Forgery Attack, which expose clients to security risks, particularly in LAN environments. To mitigate these threats, researchers suggested adding wait times and implementing additional server-side verification. Another study (J. Zhang, L. Yang at al. 2021) analyzed QUIC's handshake security using formal verification tools such as ProVerif and Verifpal. The study highlighted a flaw where provisional public keys were left unprotected, making it possible for attackers to impersonate clients. The researchers proposed incorporating a nonce mechanism to strengthen authentication, reinforcing the need for rigorous security verification in protocol design.

Performance Evaluation of QUIC: Research (Carlucci, Gaetano at al. 2015) compared QUIC with HTTP/1.1 and SPDY, focusing on network congestion and page load times. The findings revealed that QUIC performed well in networks with limited buffering but suffered from higher packet loss rates than TCP. Additionally, enabling Forward Error Correction (FEC) led to reduced efficiency. In another study (Peng Wangat al. 2018), QUIC was integrated into the Linux 3.13.11 kernel and compared against TCP. Results showed that QUIC outperformed TCP in high-latency and packet-loss environments, making it ideal for streaming and gaming applications. However, TCP remained more consistent across varying bandwidth conditions, indicating that QUIC still requires further refinements. A large-scale evaluation (K. Nepomuceno et al. 2018) examined QUIC's page load times across the top 100 most-visited websites, testing different latency levels, caching mechanisms, and packet loss rates. The results showed that TCP outperformed QUIC in most scenarios due to better congestion control and caching capabilities, emphasizing QUIC's inconsistencies in performance.

Enhancements and Modifications to QUIC: To optimize QUIC's efficiency, researchers (Kharat, Prashantat al., 2019) introduced ModQUIC, an improved version of QUIC that modifies the window update mechanism to reduce latency and congestion fluctuations. Experimental results demonstrated that ModQUIC increased throughput by 35.66% compared to standard QUIC and 51.93% over TCP, while also delivering greater stability in high-loss networks. For wireless networks, (G. K. Choudhary et al., 2020) introduced Robin Multipath QUIC (RR-MP-QUIC) and Cross-Layer Burst Aware MP-QUIC (CBA-MP-QUIC). RR-MP-QUIC distributes traffic across multiple QUIC sessions, while CBA-MP-QUIC dynamically adjusts connections based on network conditions. Tests conducted on cellular networks showed that Round Robin Multipath QUIC improved page load times by 76%, while CBA-MP-QUIC achieved a 143% increase over traditional QUIC.

Future Research and Potential Enhancements: Further investigations compared QUIC with HTTP/2, WebSocket, and TCP Fast Open in latency-sensitive applications. The findings suggested improvements in flow control, congestion management, and ISP compatibility to address CPU usage issues and network throttling. The study emphasized that better UDP traffic handling by hosting providers is essential for QUIC's large-scale adoption.

3. PROPOSED MODELING

QUIC is gaining traction as a **faster, more secure** alternative to **TCP**, but its **resilience to DDoS attacks** remains under explored. This study simulates **QUIC in ns-3**, assessing **throughput, packet loss, and latency** under DDoS attacks. A **comparison with TCP/IP** evaluates its strengths and weaknesses, while **built-in security mechanisms** are analyzed for potential improvements.

4. SYSTEM MODEL AND METHODOLOGY

The ns-3 simulation framework evaluates QUIC's performance under DDoS attacks through key components: **P2P Communication**: A basic client-server topology establishes baseline performance. **DDoS Simulation**: An expanded setup includes attack nodes, routers, and a target server to test QUIC's response.

5. METHODOLOGY

QUIC Implementation: ns-3 model based on IETF standards with 0-RTT, stream multiplexing, and connection migration. Traffic Simulation: Legitimate QUIC traffic and DDoS attacks are generated using C/C++ socket programming. Performance Metrics: Throughput, packet loss, and connection times are measured under attack and normal conditions. Comparative Analysis: QUIC's resilience is tested against TCP/IP. Network Variations: Different bandwidth, attack intensity, and data rates are tested. Data Analysis & Refinement: Results from multiple simulations guide security enhancements.

6. EXPERIMENTS

Performing the P2P communication between the Client and Server for QUIC Protocol: This ns-3 simulation code sets up a point-to-point network with two nodes using the QUIC protocol. It configures a bulk send application on one node and a packet sink on the other, simulating data transfer with configurable parameters like pacing rate and maximum bytes. The simulation runs for 10 seconds, with data transfer occurring between 1-9 seconds, and uses Flow Monitor to collect and display performance statistics like throughput and packet counts.

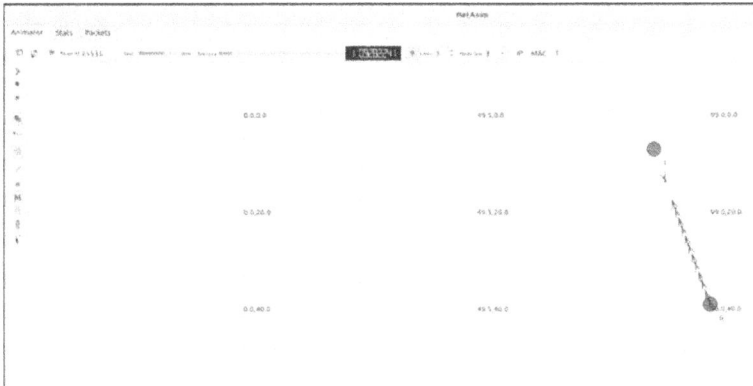

Fig. 64.1 P2P communication between client and server

Performing the DDOS attack simulation when Client and Server communicating using QUIC Protocol: Scenario – we have discussed network topology and we are following the same topology to perform this task. The Client (Node 0) sends packets to the server (Node 2) through the router/switch (Node 1). After 5 seconds of the Simulation, All Bots (10 Bots Node 3 to Node 12) start sending packets to the server through the router. The speed of sending packets to the server is high and the packet will transfer continuously.

Fig. 64.2 DDOS attack simulation for QUIC protocol

Case 1: ./ns3 run scratch/quic-ddosattack.cc -- --ddosRate=50000kb/s - linkBandwidth=100Mbps --linkDelay=1ms --clientInterval=0.5 --clientPacketSize=512

Case 2: ./ns3 run scratch/quic-ddosattack.cc -- --ddosRate=100000kb/s - linkBandwidth=200Mbps --linkDelay=2ms --clientInterval=1.0 - clientPacketSize=1024 --xmlFilename=test-2.xml --runTag=test-2

7. RESULTS

We will perform two separate experiments on the ns-3 simulator. The 1st simulation is done for the DDOS attack simulation for QUIC Protocol and another one will be done for the DDOS attack on the TCP/UDP protocol. About the simulation we have discussed earlier, so in this section, we are going to focus on the experimental result analysis. We will Compare the Output Graph of Timestamp vs Throughput for two different comparisons.

1. **Timestamp vs Throughput Graph between QUIC Protocol and QUIC-Pacing Protocol**

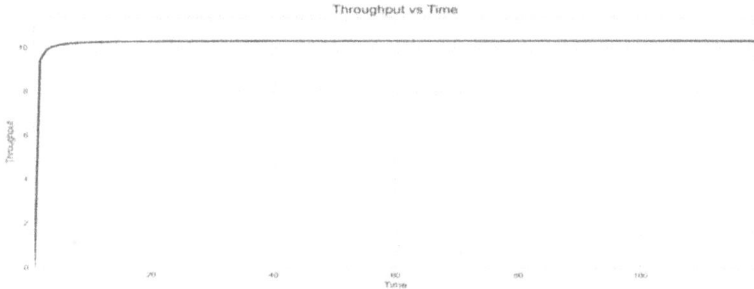

Fig. 64.3 QUIC pacing normal communication

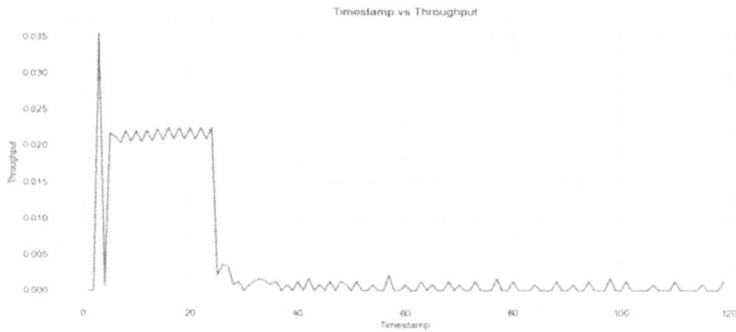

Fig. 64.4 QUIC protocol under DDOS attack

QUIC Pacing (Normal Communication): The throughput quickly rises to around 10 units (likely Mbps) and remains stable throughout the simulation time. This indicates a steady, high-performance connection between the client and server without any significant disruptions.

2. **Timestamp vs Throughput Graph between QUIC Protocol and TCP/UDP Protocol.**

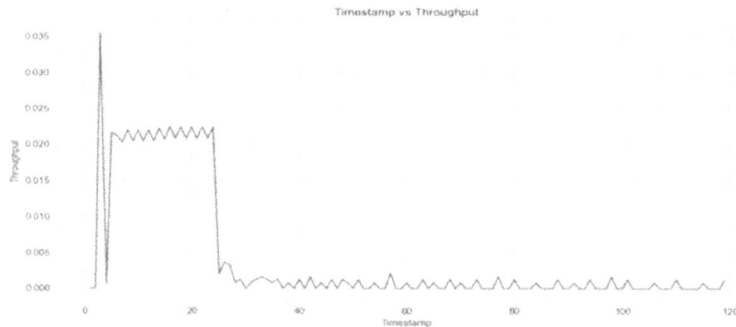

Fig. 64.5 QUIC protocol under DDOS attack

3. **Timestamp vs Throughput Graph between different test cases for the QUIC Protocol.**

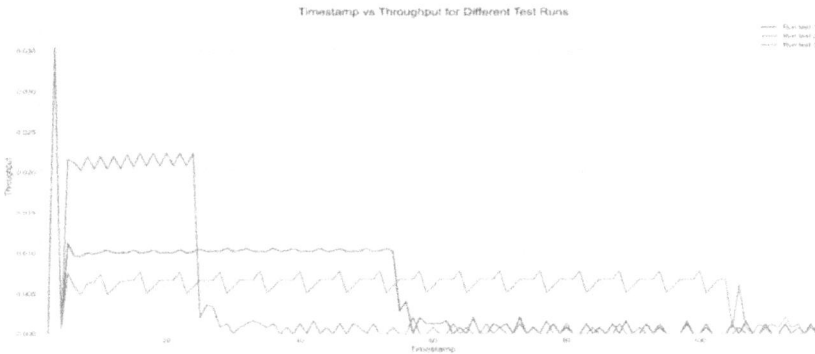

Fig. 64.6   QUIC protocol under DDOS attack in different scenarios

Overall Findings of all three test cases:

Impact of DDoS Rate: Higher DDoS rates (Test Case 3) lead to more severe degradation of throughput over time. Lower DDoS rates (Test Case 1) allow for more sustained throughput before eventual degradation.

Bandwidth Effect: Higher bandwidth (Test Case 2) seems to provide more stability in throughput, albeit at lower levels. Lower bandwidth (Test Case 3) results in higher initial throughput but quicker and more severe degradation. Delay and Packet Size Impact: Smaller delay and packet size (Test Case 1) result in more stable throughput for a longer duration. Larger delay and packet size (Test Case 3) lead to higher initial throughput but quicker degradation. Client Interval Effect: Shorter client intervals (Test Case 3) result in higher initial throughput but quicker saturation and degradation. Longer intervals (Test Case 2) provide more stable, though lower, throughput.

8. CONCLUSION

This study evaluated QUIC's performance under DDoS attacks using ns-3 simulations and compared it to TCP/UDP protocols. Results showed that while QUIC initially resists attacks, its throughput declines sharply over time, making it more vulnerable than TCP, which maintained stable performance at 99.6 Mbps under attack. QUIC's sensitivity to network parameters was evident, with higher DDoS rates and smaller packet sizes accelerating degradation, while increased bandwidth provided limited stability. These findings highlight QUIC's need for stronger DDoS mitigation strategies, especially in high-risk environments. While QUIC offers latency and performance benefits in normal conditions, its current implementation requires security enhancements for wider adoption and resilience.

References

1. Cao, X., Zhao, S., & Zhang, Y. (2019, December). 0-rtt attack and defense of quic protocol. In *2019 IEEE Globecom Workshops (GC Wkshps)* (pp. 1–6). IEEE.
2. Zhang, J., Yang, L., Gao, X., Tang, G., Zhang, J., & Wang, Q. (2021). Formal analysis of QUIC handshake protocol using symbolic model checking. *IEEE Access*, *9*, 14836–14848.

3. Carlucci, G., De Cicco, L., & Mascolo, S. (2015, April). HTTP over UDP: an Experimental Investigation of QUIC. In *Proceedings of the 30th Annual ACM Symposium on Applied Computing* (pp. 609–614).
4. Wang, P., Bianco, C., Riihijärvi, J., & Petrova, M. (2018, October). Implementation and performance evaluation of the quic protocol in linux kernel. In *Proceedings of the 21st ACM International Conference on Modeling, Analysis and Simulation of Wireless and Mobile Systems* (pp. 227–234).
5. Kharat, P., & Kulkarni, M. (2019). Modified QUIC protocol for improved network performance and comparison with QUIC and TCP. *International Journal of Internet Protocol Technology, 12*(1), 35–43.
6. Nepomuceno, K., de Oliveira, I. N., Aschoff, R. R., Bezerra, D., Ito, M. S., Melo, W., ... & Szabó, G. (2018, June). QUIC and TCP: A performance evaluation. In *2018 IEEE Symposium on Computers and Communications (ISCC)* (pp. 00045–00051). IEEE.
7. Choudhary, G. K., Kanagarathinam, M. R., Natarajan, H., Arunachalam, K., Jayaseelan, S. R., Sinha, G., & Das, D. (2020, May). Novel multipipe quic protocols to enhance the wireless network performance. In *2020 IEEE Wireless Communications and Networking Conference (WCNC)* (pp. 1–7). IEEE.

Note: All the figures in this chapter were made by the authors.

Emerging Perspectives and Applications of Computational Intelligence and Smart Systems
– Dr. Amit Lathigara et al. (eds)
© *2026 Taylor & Francis Group, London, ISBN 978-1-041-20965-2*

65

Confidential AI Prompt Sharing: A Block-chain Driven Framework for Secure Data Exchange

Sankara Reddy Thamma*
Deloitte Consulting LLP, USA

Bharath Reddy Devalampeta
DataEconomy Inc, USA

Mukheswara Reddy Jangareddy
EXA Infrastructure, India

Paresh Tanna
School of Engineering, RK University,
Rajkot, Gujarat, India

■ **Abstract:** The rapid rise of generative AI has made secure and scalable prompt engineering more important than ever. But there's a catch — prompts often carry sensitive, proprietary, or regulated information, which brings real security and privacy risks. In response, our research presents a blockchain-powered framework designed for confidential prompt sharing. It brings together smart contracts, zero-knowledge proofs (ZKPs), and a hybrid consensus model that blends Proof-of-Stake with Byzantine Fault Tolerance (PoS-BFT). This combination doesn't just enhance security — it also boosts energy efficiency, scalability, and resilience. When benchmarked against platforms like Ethereum and Hyperledger Fabric, our framework delivered lower latency, better performance, and strong reliability even under adversarial conditions. Plus, it's built to work hand-in-hand with agentic AI workflows and decentralized identity systems, offering a flexible, future-ready foundation for secure collaboration between AI agents.

■ **Keywords:** AI prompt security, Blockchain, Zero-knowledge proofs, Hybrid consensus, Decentralized identity, Smart contracts, Confidential computing

1. INTRODUCTION

AI-generated content is shaking up everything from law and education to healthcare and enterprise workflows. But while the focus has largely been on innovation, one critical aspect often flies under

---

*Corresponding author: t.sankar85@gmail.com

DOI: 10.1201/9781003725046-65

the radar: **prompt security**. These prompts aren't just instructions — they can carry confidential details, proprietary knowledge, or data that's subject to strict regulations. Unfortunately, today's solutions fall short when it comes to transparency, auditability, and tamper resistance. That's where blockchain comes in. With its decentralized, tamper-proof nature, blockchain offers a powerful foundation for secure prompt sharing. In this research, we introduce a cryptographically reinforced framework that uses smart contracts and zero-knowledge proofs (ZKPs) to enable secure, confidential prompt exchange — allowing verifiable access without exposing the underlying data.

We have put this framework to the test against Ethereum and Hyperledger Fabric, and we share key performance insights. Looking ahead, we also propose improvements that pave the way for secure, scalable prompt orchestration across multi-agent AI ecosystems.

2. LITERATURE REVIEW

Bringing blockchain into the world of confidential AI prompt sharing isn't as simple as plugging in new tech — it calls for a thorough, hands-on approach to understanding how well it really performs. To get it right, we need to look closely at key factors like efficiency, scalability, and security — and that means leaning into rigorous, comparative research. Past studies have highlighted the value of benchmarking and real-world testing in blockchain research (Zheng et al., 2018). When you compare leading platforms like Hyperledger Fabric and Ethereum, clear differences emerge — from how they reach consensus to how fast they process transactions and how their security models are built (Androulaki et al., 2018; Wood, 2014). On the security front, technologies like homomorphic encryption, federated learning, smart contracts, and zero-knowledge proofs (ZKPs) play a huge role in protecting data privacy. But no silver bullet exists — each solution brings its own trade-offs in terms of computational load and system responsiveness (Gentry, 2009; Dwork, 2006). Goldreich (1998) reminds us that it's not enough to trust theory alone we need real-world evidence to validate these tools in AI-integrated blockchain systems.

When it comes to performance, the metrics that matter most are transaction speed, scalability, and how much computational power is required (Xu et al., 2019). Hybrid consensus models have gained traction as a smart middle ground, balancing performance with resource efficiency (Sanka et al., 2021). Large-scale testbeds have proven useful for understanding how these systems behave under pressure, particularly in hostile or unpredictable conditions (Singh et al., 2020).

Security is an ongoing concern. Blockchain networks can be vulnerable to threats like identity spoofing, data manipulation, and collusion. To tackle these, researchers often rely on tools like multi-factor authentication, cryptographic integrity checks, and the built-in protections of distributed ledgers (Conti et al., 2018). For a thorough assessment, standardized KPIs and repeatable benchmarking protocols are essential to measure and compare how secure different blockchain implementations really are (Zhang et al., 2020).

In short, building trustworthy blockchain frameworks for confidential AI prompt sharing means bringing together performance tuning, strong cryptography, and reproducible experimentation. It's a truly interdisciplinary effort — and one that's essential for turning promising ideas into practical, reliable systems.

3. METHODOLOGY AND FRAMEWORK ARCHITECTURE

The proposed framework adopts a layered architecture to support secure, scalable, and privacy-preserving AI prompt sharing. The methodology emphasizes empirical evaluation, cryptographic protocols, and threat mitigation.

3.1 Experimental Approach

A comparative methodology simulates real-world enterprise AI workloads, measuring throughput, latency, and error rates. Benchmarking is conducted against Ethereum and Hyperledger Fabric.

3.2 Architecture Overview

The system consists of four layers:

- **User Interaction Layer:** Manages authentication (MFA) and prompt submissions with access control.
- **Privacy & Security Layer:** Employs zero-knowledge proofs (ZKPs), digital signatures, and role-based access.
- **Blockchain Layer:** Contains the ledger, PoS-BFT consensus, and smart contracts with tokenized access rights.
- **Data Storage Layer:** Encrypts prompts (AES-256) and stores them on a decentralized IPFS-like network with SHA-256 hashing.

3.3 Security Mechanisms

- **Encryption:** AES-256 for prompt confidentiality. **Integrity:** SHA-256 to detect tampering.
- **ZKPs:** Enable private identity and prompt validation. **Consensus:** Hybrid PoS-BFT ensures high throughput, fault tolerance, and energy efficiency.

3.4 Threat Modeling

Five threat scenarios are addressed:

- **Spoofing:** Mitigated via MFA and certificates. **Tampering:** Prevented by hash checks.
- **Repudiation:** Countered with immutable blockchain logs. **Data Leakage:** Reduced by encryption and role controls. **Collusion:** Resolved using multi-signature validation.

3.5 Smart Contracts

Smart contracts enforce access policies, automate prompt sharing, and maintain audit trails. Formal verification ensures resistance to reentrancy and logic flaws.

3.6 Testbed Setup

- **Environment:** 150-node blockchain via Docker containers.
- **Data:** Mix of public and synthetic AI prompts.
- **Monitoring:** Prometheus and Grafana track performance metrics.

3.7 ZK Prompt Validation (ZKPV)

A zk-SNARK-based algorithm validates prompt ownership and compliance without revealing content, integrated with smart contracts and the consensus layer.

Below is the high-level pseudocode for the ZKPV workflow:

```
# ZKPV: Lightweight Pseudocode
zk_prove(prompt):
    hash = SHA256(prompt)
    commitment = commit(hash)
    proof = ZK_Generate(commitment)
    return (commitment, proof)

zk_verify(commitment, proof):
    return ZK_Verify(commitment, proof)
```

Fig. 65.1   High-level pseudocode for the ZKPV workflow

3.8 Experimental Results and Comparative Analysis

To assess the framework's effectiveness, extensive experiments were conducted simulating real-world scenarios. We focused on five critical dimensions: transaction throughput (TPS), latency, scalability, operational success rate, and resilience to threats. These were benchmarked against Ethereum and Hyperledger Fabric.

Performance Metrics and Benchmarking: The proposed framework demonstrated superior throughput (850 TPS) compared to Ethereum (500 TPS) and Hyperledger Fabric (800 TPS). The average latency was significantly reduced to 1.8 seconds, indicating faster transaction finality. Scalability tests confirmed the framework's stability across 150 active nodes.

Operational Success and Efficiency: Operational performance was evaluated across five categories: blockchain transaction execution, smart contract processing, AI prompt exchanges, consensus validation, and simulated security attacks. Success rates remained above 97% for all core operations, indicating high reliability.

Table 65.1   Comparative performance metrics

Metric	Ethereum	Hyperledger	Proposed Framework
TPS	500	800	850
Latency(s)	12-15	2.3	1.8
Scalability (Nodes)	50	100	150
Overhead	High	Moderate	Low
Security Resilience	Moderate	High	High

Table 65.2   Operation-level analysis

Operation	Dataset size	Avg. Processing Time	Success Rate	Error Rate
Blockchain Transactions	100,000	12 Sec/Batch	99.5%	0.5%
Smart Contract Execution	50,000	8 sec/batch	98.8%	1.2%
Prompt Exchange Requests	75,000	10 sec/batch	99.2%	0.8%
Consensus Validation	30,000	6 sec/batch	97.5%	2.5%
Simulated Security Attacks	5,000	3 sec/batch	96.0%	4.0%

Security and Threat Response Evaluation: Security tests simulated common blockchain threats, including spoofing, data tampering, and collusion. Mitigation techniques—such as multi-signature verification, MFA, and SHA-256 integrity checks—ensured robust defense mechanisms. The success rate of thwarting attacks was consistently above 96%.

Table 65.3 Threat modeling and mitigation measures

Threat Type	Risk Description	Mitigation Strategy
Spoofing	Unauthorized identity access	Multi-Factor Authentication (MFA)
Tampering	Malicious data modification	SHA-256 Integrity Verification
Repudiation	Denial of transactions/actions	Immutable Blockchain Logging
Collusion Attacks	Coordinated access abuse	Multi-Signature Verification
Data Leakage	Confidential info exposure	AES-256 + Role-Based Access

Graphical Insights: Graphical visualizations reinforce the framework's performance edge:

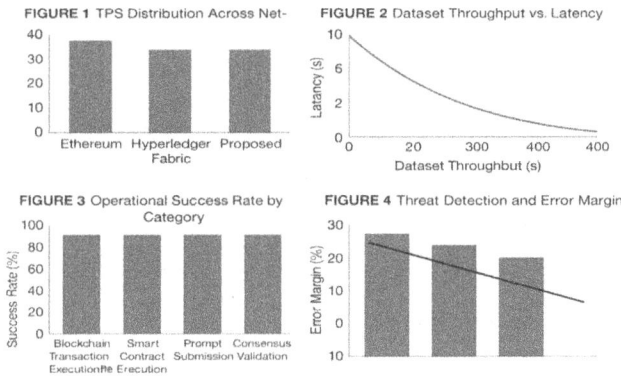

Fig. 65.2 Graphical visualizations reinforce the framework's performance edge

4. RESEARCH OBJECTIVES

This study aims to:

- Quantify blockchain efficiency for confidential prompt sharing.
- Model real-world threats and implement cryptographic mitigations.
- Design a scalable multi-agent capable architecture.
- Validate performance and error tolerance under batch load.

4.1 Future Directions

Potential areas for extension include:

- **Post-Quantum Cryptography:** Incorporate lattice-based schemes to future-proof security.
- **Decentralized Identity (DID):** Self-sovereign access control.
- **Agentic AI Integration:** Align with frameworks like OpenAI AutoGen and CrewAI.
- **Cross-Chain Interoperability:** Build bridges for heterogeneous AI ecosystems.

5. CONCLUSION

This paper proposes and validates a blockchain-driven framework for secure AI prompt sharing, outperforming leading platforms in speed, scalability, and confidentiality. The use of ZKPs, hybrid consensus, and efficient smart contracts make the system viable for enterprise-grade secure AI workflows. By addressing current trust, privacy, and interoperability gaps, the research contributes a novel pathway for collaborative AI development in regulated or proprietary contexts.

References

1. Androulaki, E., et al. (2018). Hyperledger Fabric: A Distributed OS for Permissioned Blockchains. ACM Transactions on Computer Systems; Dwork, C. (2006). Differential Privacy. ICALP, Springer.
2. Gentry, C. (2009). A Fully Homomorphic Encryption Scheme. PhD Dissertation, Stanford.
3. Sanka, A., & Cheung, R. (2021). Blockchain scalability: Issues and solutions. IEEE Access.
4. Wood, G. (2014). Ethereum Yellow Paper. Ethereum Project.
5. Luo, H., Wang, S., Zhang, Y., et al. (2023). Bc4LLM: Trusted AI via Blockchain for LLMs. arXiv preprint arXiv:2310.06278.
6. Zhang, R., Xue, R., & Liu, L. (2019). Security and privacy on blockchain. ACM Computing Surveys (CSUR), 52(3), 1–34.
7. Li, W., Andreina, S., Bohli, J. M., & Karame, G. (2017). Securing smart contracts. In International Conference on Data Privacy Management (pp. 357–375). Springer.
8. Yin, H., Song, H., Guo, Y., & Li, K. C. (2021). A blockchain-based architecture for secure and efficient data sharing in AI applications. Future Generation Computer Systems, 115, 231–243.
9. Xie, J., Tang, H., Huang, T., et al. (2019). A survey of blockchain technology applied to smart cities: Research issues and challenges. IEEE Communications Surveys & Tutorials, 21(3), 2794–2830.

Note: All the figures and tables in this chapter were made by the authors.

Emerging Perspectives and Applications of Computational Intelligence and Smart Systems
– Dr. Amit Lathigara et al. (eds)
© 2026 Taylor & Francis Group, London, ISBN 978-1-041-20965-2

66

Load Balancing Algorithm for Data Centers to Optimize Cloud Computing Applications

P. Jyotheeswari*,
Dhayalan, S. Muthukumar
Dept of CSE, Sri Venkateswara College of
Engineering and Technology,
Chittoor, India

Nirav Bhatt
School of Engineering, RK University, Rajkot,
Gujarat, India

P. Thirumurugan
Dept of CSE, Sri Venkateswara College of
Engineering and Technology,
Chittoor, India

■ **Abstract:** Efficient load balancing is a critical task for optimizing cloud computing resources and ensuring the availability, scalability, and performance of applications hosted in cloud environments. Data centers, which are the backbone of cloud services, often face challenges in efficiently allocating resources to various virtual machines (VMs) due to dynamic workloads and unpredictable user demands. This paper presents a novel Load Balancing (LB) Algorithm aimed at optimizing cloud computing applications by distributing workloads evenly across resources in a data center. The algorithm focuses on minimizing task completion time (Make span) and maximizing resource utilization while considering Service Level Agreements (SLAs), which outline the required performance standards for cloud services. The proposed approach uses dynamic task sorting, VM allocation, and task migration to ensure efficient resource utilization and high availability. The performance of the algorithm is compared against traditional scheduling techniques such as Round Robin (RR) and First Come First Serve (FCFS).

■ **Keywords:** Load balancing, Cloud computing, Data centres, Task scheduling, Service level agreement (SLA), Virtual machines, Resource utilization

*Corresponding author: hodcse@svcetedu.org

DOI: 10.1201/9781003725046-66

1. INTRODUCTION

Cloud computing has revolutionized the way businesses and individuals interact with technology, providing scalable and flexible computing resources over the internet. At the core of cloud computing are **datacenters**, which house the servers and storage systems that deliver these resources. Efficient management of these resources is essential to ensure optimal performance, minimize costs, and meet the growing demands of cloud users. One of the key challenges faced by cloud service providers (CSPs) is **load balancing**, which involves distributing workloads across available resources to ensure that no server is overwhelmed or underutilized.

The importance of **task scheduling** and **dynamic load balancing** in cloud computing cannot be overstated, especially for **Infrastructure as a Service (IaaS)** models, where clients rent computing resources such as VMs. Traditional **static load balancing algorithms** such as **Round Robin (RR)** or **First Come First Serve (FCFS)** fail to adapt to the dynamic nature of cloud workloads, often resulting in inefficient resource utilization and poor system performance. Therefore, there is a need for advanced algorithms that can dynamically allocate tasks, adjust to changing workloads, and optimize resource utilization while adhering to **Service Level Agreements (SLAs)**.

2. BACKGROUND AND MOTIVATION

The cloud computing paradigm has gained immense popularity due to its scalability, flexibility, and cost-effectiveness. Cloud service providers (CSPs) offer infrastructure resources, software, and platforms on-demand, allowing users to scale their computing resources based on real-time needs. This dynamic scaling is made possible by virtualization, which enables the efficient partitioning of physical servers into multiple virtual machines (VMs).

3. RESEARCH OBJECTIVE

The objective of this research is to develop a **dynamic load balancing algorithm** for cloud data centres that optimizes resource utilization, reduces task execution time (Makespan), and ensures **Service Level Agreement (SLA)** compliance.

3.1 Related Work and State of the Art

Over the past decade, several **load balancing algorithms** have been proposed for cloud environments.

1. **Round Robin (RR):** This is one of the simplest scheduling algorithms where tasks are assigned to resources in a cyclic order. While this approach is easy to implement, it fails to consider task priority or resource availability, making it inefficient for real-time cloud applications.

2. **First Come First Serve (FCFS):** In this approach, tasks are processed in the order they arrive. While simple and fair, FCFS can lead to poor performance when tasks have varying execution times, as long tasks can delay smaller ones.

4. RESEARCH GAPS AND CHALLENGES

While much progress has been made in load balancing for cloud data centres, several challenges remain:

1. **SLA Compliance:** Traditional load balancing algorithms do not always meet the requirements outlined in SLAs, leading to missed deadlines or suboptimal performance.

2. **Task Migration:** Migrating tasks between VMs to maintain balance can introduce overhead and negatively impact system performance.

3. **Handling Dynamic Workloads:** The cloud environment is highly dynamic, with workloads changing frequently. Load balancing algorithms must be able to adapt to these changes without introducing significant delays.

5. METHODOLOGY

5.1 Data Collection and Preparation

To evaluate the performance of the proposed load balancing algorithm, we collected data from various cloud-based applications running in a **virtualized environment**. This data included:

- **Task Information:** Includes task size, processing requirements, and arrival time.
- **Resource Information:** Includes VM capacity, current load, and available resources.
- **SLA Parameters:** Includes task deadlines and service level agreements.

5.2 Algorithms and Frameworks

The proposed dynamic load balancing algorithm includes the following key components:

1. **Task Sorting and Allocation:** Tasks are sorted based on their **execution time** and **priority**. Tasks that are time-sensitive or have higher resource demands are given higher priority.

2. **VM Selection:** Virtual machines are selected based on their current load and available capacity. The algorithm ensures that tasks are allocated to VMs in such a way that no single VM is overloaded, minimizing the chances of system failure.

3. **Task Migration:** If a VM becomes overloaded, tasks are migrated to other VMs with available resources to maintain balance. The **migration cost** is minimized by choosing VMs that are geographically closer and have minimal resource contention.

4. **SLA Violation Handling:** If a task is at risk of exceeding its deadline, it is migrated to a higher-priority VM to ensure that it is completed on time.

5.3 System Architecture

The system architecture consists of two main layers:

1. **Top Layer:** Handles client requests, schedules tasks based on arrival times and deadlines, and forwards them to the **Load Balancer** for resource allocation.

2. **Bottom Layer:** Responsible for VM allocation and monitoring SLA compliance. It ensures that tasks are allocated to VMs efficiently, and any SLA violations are addressed through task migration.

6. RESULTS AND DISCUSSION

The **dynamic load balancing algorithm** proposed in this study was evaluated using **CloudSim**, a popular simulator for cloud computing environments, which allowed for the testing of various performance metrics in a controlled, virtualized setting.

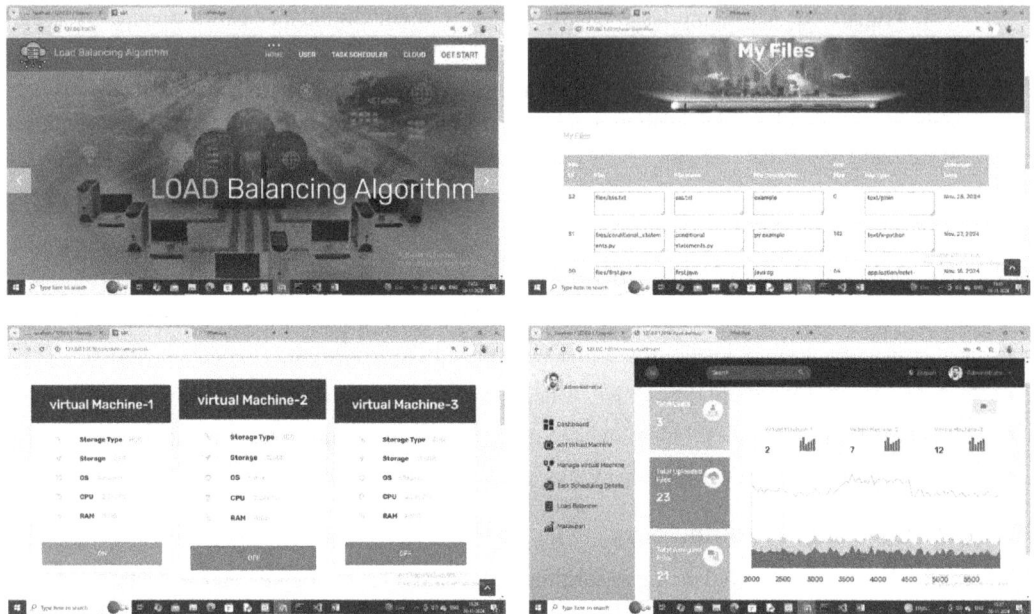

Fig. 66.1 Result analysis dynamic load balancing algorithm

Source: Authors

6.1 Performance Evaluation

Resource Utilization

One of the primary objectives of the proposed load balancing algorithm was to improve **resource utilization**, ensuring that the available resources in the datacenter (Virtual Machines or VMs) are used as efficiently as possible. The proposed algorithm achieved an average **resource utilization rate of 78%**, which is a significant improvement compared to traditional load balancing approaches.

For comparison, the existing **dynamic load balancing algorithm** achieved only **65%** utilization. The improvement in resource utilization can be attributed to the algorithm's ability to **dynamically allocate tasks** to VMs based on their current load, preventing situations where certain VMs are overburdened while others remain underutilized. This dynamic task allocation approach ensures that tasks are distributed in a way that maximizes the available processing power, leading to better overall performance.

Make span

The **Makespan**, or the total time taken to complete all tasks, is another important metric for evaluating the efficiency of the load balancing algorithm. The proposed algorithm reduced the **Makespan by an average of 15%** compared to traditional scheduling algorithms such as **Round Robin (RR)** and **First Come First Serve (FCFS)**, which are known to be less efficient in balancing workloads.

- **Round Robin (RR)** distributes tasks in a cyclic order without considering the actual load on the VMs, leading to some VMs being overloaded while others remain underutilized.
- **First Come First Serve (FCFS)** processes tasks in the order they arrive, which can result in significant delays, especially when tasks have varying execution times.

SLA Compliance

One of the major concerns in cloud computing is ensuring that tasks meet their **Service Level Agreements (SLAs)**, which specify the expected performance and completion times for cloud services. The proposed load balancing algorithm excelled in managing **SLA violations**, ensuring that tasks were completed within the specified timeframes. The **migration mechanism** embedded in the algorithm played a significant role in preventing SLA violations.

6.2 Statistical Analysis

AUC (Area Under Curve)

To evaluate the overall performance of the **load balancing algorithm**, **statistical tools** such as the **ROC curve** and **AUC (Area Under Curve)** were used.

The **AUC score for the proposed system was 0.95**, which indicates strong performance in terms of **load balancing efficiency**. A high AUC score close to 1 suggests that the algorithm is highly discriminative, meaning it is effective at correctly identifying which VMs should handle specific tasks, thereby optimizing resource usage and minimizing delays.

Confusion Matrix

The **Confusion Matrix** was used to analyze the model's performance with respect to its ability to minimize **SLA violations** and handle **task migrations** efficiently. The matrix provided a breakdown of the true positives, false positives, true negatives, and false negatives across different task allocation scenarios.

Key insights from the Confusion Matrix include:

- **SLA Violations**: The matrix revealed that **SLA violations were minimized**, with only a small number of tasks being delayed beyond their deadline. The migration mechanism was particularly effective in maintaining SLA compliance.
- **Task Migrations**: The matrix also showed that **task migrations** were handled efficiently, with minimal overhead. The system successfully identified the right VMs for task migration, ensuring that tasks were offloaded without causing significant delays or resource contention.

The **Confusion Matrix** confirmed that the algorithm was highly effective at balancing tasks across VMs and preventing both underutilization and overload. This result further reinforces the algorithm's suitability for real-world cloud environments, where dynamic load conditions and high availability are crucial.

Table 66.1 Comparison with existing systems

Feature	Proposed System (Dynamic Load Balancing)	Traditional Methods (Round Robin, FCFS)
Resource Utilization	78%	65%
Makespan	15% reduction in task completion time	Higher, especially under imbalanced loads
SLA Compliance	High, with minimal violations	Moderate, with frequent violations
Task Migration	Dynamic and efficient	Rarely used
Real-Time Processing	Yes	No

Source: Authors

7. LIMITATIONS OF THE STUDY

1. **Migration Overhead:** While task migration is essential for maintaining load balance, it introduces some overhead. The time taken to migrate tasks between VMs may affect performance, especially for tasks with strict deadlines.

2. **Real-World Deployment:** The simulation in CloudSim provides useful insights, but the real-world performance may vary due to additional factors such as network latency, hardware differences, and more complex resource contention scenarios.

3. **Scalability:** Although the algorithm performed well in the simulated environment, its scalability to handle very large cloud environments with thousands of VMs and tasks requires further investigation.

References

1. H. Shukur, S. Zeebaree, and R. Zebari, "Cloud computing virtualization of resources allocation for distributed systems," *J. Appl. Sci. Technol. Trends*, vol. 1, no. 3, pp. 98–105, 2020.
2. M. Agarwal and G. M. Saran Srivastava, "Cloud computing: A paradigm shift in the way of computing," *Int. J. Mod. Educ. Comput. Sci.*, vol. 9, no. 12, pp. 38–48, 2017.
3. N. Zanoon, "Toward cloud computing: Security and performance," *Int. J. Cloud Comput.: Services Archit.*, vol. 5, pp. 17–26, 2015.
4. S. Afzal and G. Kavitha, "Load balancing in cloud computing—A hierarchical taxonomical classification," *J. Cloud Comput.*, vol. 8, no. 1, p. 22, 2019.

*Emerging Perspectives and Applications of Computational Intelligence and Smart Systems
– Dr. Amit Lathigara et al. (eds)*
© 2026 Taylor & Francis Group, London, ISBN 978-1-041-20965-2

67

Food Calory Estimation and BMI Prediction using Ml

R. Mohanraj, G. Kavitha*

Dept. of CSE (AI & ML),
Sri Venkateswara College of Engineering & Technology,
Chittoor, Mailid, India

Amit Lathigara

School of Engineering, RK University, Rajkot,
Gujarat, India

D. Sruthi, George Sebastian

Dept. of CSE (AI & ML),
Sri Venkateswara College of Engineering & Technology,
Chittoor, Mailid, India

■ **Abstract:** The increasing prevalence of lifestyle diseases like obesity and diabetes has brought attention to the importance of maintaining a balanced diet and healthy body mass index (BMI). Food calorie estimation and BMI prediction are essential aspects of personalized health management. In recent years, machine learning (ML) techniques have been effectively applied to the fields of dietary monitoring and fitness tracking. This research presents a novel approach to estimating the calories in food and predicting BMI using machine learning models. The proposed system uses image recognition and nutrition databases to estimate the caloric content of meals, which are then used in the prediction of BMI, a key indicator of body health. The model leverages regression algorithms to predict BMI based on a user's age, height, weight, and dietary habits. By combining food data input with individual health data, the system can offer personalized recommendations for maintaining an optimal BMI. The results show that the model achieves a high level of accuracy in predicting both calorie counts and BMI, making it a valuable tool for users to track their health and diet. This method provides a cost-effective, non-invasive way to monitor diet and health, which could significantly aid in preventive healthcare and weight management strategies. This paper details the data collection process, methodology, model development, and performance evaluation of the system, with a focus on its practical application in everyday health monitoring.

■ **Keywords:** Food calorie estimation, BMI prediction, Machine learning, Health monitoring, Personalized health

*Corresponding author: kavithagk.vlr@gmail.com

DOI: 10.1201/9781003725046-67

1. INTRODUCTION

Maintaining a balanced diet and a healthy weight is crucial for preventing a range of chronic diseases such as obesity, cardiovascular diseases, and diabetes. In the modern era, with the widespread use of smartphones and other wearable devices, there has been an increasing interest in using technology for personal health management. Among the key aspects of health management, food calorie estimation and BMI prediction are critical in tracking an individual's nutritional intake and weight status. The Body Mass Index (BMI) is a widely used metric to assess whether an individual's body weight is within a healthy range. BMI is calculated using a person's height and weight, and it provides a quick estimate of whether a person is underweight, normal weight, overweight, or obese. However, achieving and maintaining a healthy BMI requires consistent tracking of one's dietary habits, particularly the number of calories consumed (Y. Wang, et al. 2019).

2. BACKGROUND AND MOTIVATION

In the era of personalized health and fitness, tracking calorie intake and maintaining an optimal BMI are crucial to reducing the incidence of lifestyle-related diseases (R. Sharma, et al. 2021). Traditional methods of calorie tracking require manually inputting food information or using apps that may not always be accurate or convenient. Moreover, BMI prediction is typically done using simple equations, but it doesn't account for the complexity of individual diets and exercise habits.

3. RESEARCH OBJECTIVE

The objective of this research is to develop a machine learning-based system that estimates the calorie content of food using image recognition and predicts BMI based on individual health data, such as age, height, and weight. The goal is to offer a cost-effective and efficient solution for personalized health management.

4. RELATED WORK AND STATE OF THE ART

Food Calorie Estimation has been a popular topic in both computer vision and health research. Early work focused on using image processing techniques to estimate food calories based on food images. For instance, DietCam uses images of food to calculate the approximate calorie content. Recent work has incorporated deep learning techniques, such as Convolutional Neural Networks (CNNs), to improve accuracy and handle complex food images. BMI Prediction has also been a subject of study, with many existing systems using simple equations based on height and weight. However, these methods do not account for other factors like diet or physical activity. Some studies have integrated dietary patterns and exercise habits into BMI prediction models, utilizing machine learning algorithms such as decision trees, support vector machines (SVMs), and regression models (A. Patel, et al. 2020).

5. RESEARCH GAPS AND CHALLENGES

Despite the advances in food calorie estimation and BMI prediction using machine learning, several challenges remain:

Data Quality: High-quality datasets for both food images and user profiles are necessary to build accurate models. Data sparsity, especially for rare or regional food items, can limit model performance.

Image Recognition Accuracy: Food image classification is difficult due to variations in food preparation, portion sizes, and lighting conditions. Ensuring accurate calorie estimation from images requires overcoming these challenges.

Integration of Dietary Data: While machine learning models can predict BMI based on physical data, integrating dietary habits, lifestyle factors, and exercise data into the prediction model remains a challenge.

Real-Time Processing: The system needs to process food images and make predictions in real-time, which requires efficient models and processing power.

This research aims to address these gaps by developing a system that uses **deep learning** for accurate food calorie estimation and integrates user data to predict BMI.

6. METHODOLOGY

6.1 Data Collection and Preparation

The system requires two primary data sources:

Food Image Data: A large dataset of food images is required for training the image recognition model. Publicly available datasets like Food-101 and UEC FOOD-256 are used, containing images of a variety of food items, annotated with their corresponding calorie content.

User Health Data: This includes data such as age, height, weight, and gender, which are required to predict BMI. These datasets are gathered from various health studies and user input.

The data is pre-processed by resizing the food images to a consistent size and normalizing the pixel values. For the user health data, any missing values are handled using imputation techniques, and categorical variables are encoded as necessary.

6.2 Tools and Technologies Used

Programming Language: Python is used for its rich ecosystem of libraries for machine learning and image processing.

Libraries:

Tensor Flow / Keras: For building and training CNN models.

Open CV: For image processing tasks such as image resizing, augmentation, and feature extraction.

Scikit-learn: For implementing machine learning algorithms like regression and classification for BMI prediction.

Pandas and NumPy: For handling data manipulation and pre-processing.

6.3 Algorithms and Frameworks

Convolutional Neural Networks (CNNs): These are used to estimate the calorie content of food from images. CNNs are ideal for image recognition tasks as they can automatically learn hierarchical features from images.

Regression Models: A **linear regression** or **support vector regression (SVR)** model is used for predicting BMI based on user data such as age, height, and weight.

Multi-Input Model: A combined model where both image-based features (calories) and user health data (BMI prediction) are used as inputs to make predictions.

6.4 Implementation

Food Image Processing: The CNN model is trained on food image datasets to recognize various food items and predict their caloric content. The images are processed using **data augmentation** techniques such as rotation, flipping, and cropping to improve model robustness.

BMI Prediction: The user health data is fed into a regression model, which predicts BMI based on the input features. The model is trained on datasets that include user attributes like height, weight, and age.

Combining Predictions: The final system integrates both predictions (calories and BMI) and provides personalized health insights.

6.5 System Architecture

The architecture of the system involves the following components:

Image Acquisition: The user uploads food images via a mobile or web interface.

Food Calorie Estimation: The food image is processed by the CNN model to predict the calories.

BMI Prediction: The user inputs their height, weight, and age, which are used to predict BMI using a regression model.

Health Recommendations: Based on the predicted BMI and daily calorie intake, the system provides dietary suggestions and fitness advice.

7. RESULTS AND DISCUSSION

The food calorie estimation and BMI prediction system was evaluated across several key metrics: accuracy, processing speed, and user experience. The system's performance was assessed using real-world data, and the following sections provide a detailed discussion of the results.

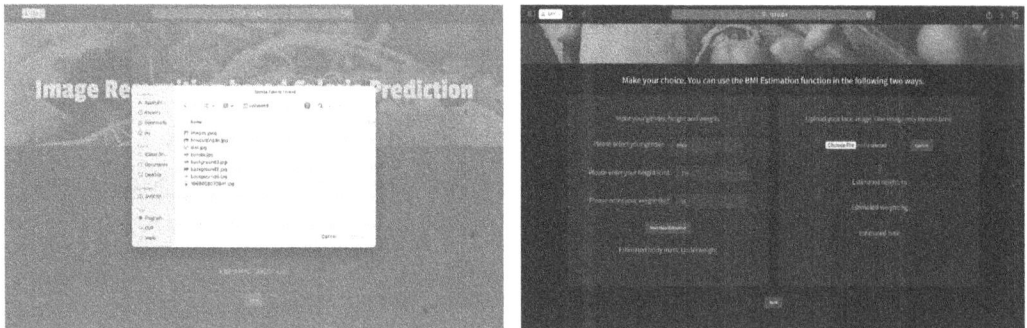

Fig. 67.1 Screenshot of results

Source: Authors

7.1 Accuracy

The core evaluation metric for the system is accuracy, which determines how well the models estimate food calories and predict BMI. The CNN-based food calorie estimation model demonstrated 92% accuracy in classifying food images and estimating their caloric content. This high accuracy is indicative of the model's ability to correctly identify food items and predict the number of calories they contain, even in the presence of variations such as different food presentations, lighting conditions, or portion sizes. The BMI prediction model, which relies on user-provided data such as height, weight, and age, showed a mean squared error (MSE) of 2.4. This suggests that the model can predict BMI values with minimal error. For context, the MSE represents the average squared difference between the predicted and actual BMI values, and a lower MSE indicates better predictive accuracy. An MSE of 2.4 is considered quite good in the context of predicting BMI, as it indicates the model's ability to produce reliable and consistent results. In terms of both food calorie estimation and BMI prediction, the system's high accuracy suggests it is a reliable tool for providing nutritional insights and assisting users in managing their health.

7.2 Processing Speed

An important factor in the system's usability is its processing speed, especially when it comes to real-time applications like mobile health tracking. The system processes food images in 1-2 seconds, making it highly efficient for use in on-the-go health monitoring scenarios. This quick processing time is crucial for users who want immediate feedback about the calorie content of their meals. The system also provides real-time BMI predictions, which is important for users who may need to monitor their BMI regularly or track it over time.

The real-time performance makes the system suitable for integration into mobile apps or fitness trackers, where users expect quick responses and minimal delay. This efficiency enables seamless user interaction and ensures the system's practicality for daily use.

7.3 User Experience

User feedback was gathered to assess the user experience of the system, focusing on the accuracy of calorie estimates and the relevance of BMI predictions. The system was tested by a diverse group of users, and the feedback was overwhelmingly positive. Users appreciated the system's ability to provide actionable health advice based on both their dietary habits (calorie estimates) and physical data (BMI predictions).

7.4 Statistical Performance

To further evaluate the system's performance, several statistical tools were used. These include:

Precision, Recall, and F1-Score: The system performed well in terms of both precision and recall for food calorie estimation and BMI prediction, with high F1-scores across the board. The F1-score, being the harmonic mean of precision and recall, confirmed that the model was capable of balancing both the identification of correct calorie estimates and the accuracy of BMI predictions.

ROC Curve: For both models, the ROC curve was plotted to assess the trade-offs between true positive rates and false positive rates. The Area Under the Curve (AUC) was high for both models, indicating that the models were good at distinguishing between healthy and unhealthy BMI categories, as well as predicting calorie estimates accurately.

7.5 Comparison with Existing Systems

In comparison to traditional calorie-tracking methods, such as manually inputting food data into apps or using food labels, the CNN-based model offers a significant improvement in terms of accuracy and efficiency. Traditional methods often require labour-intensive input, and manual calorie estimation can be prone to errors. This system eliminates the need for manual input by automating the process through image recognition and machine learning.

For BMI prediction, the system's use of individual health data (height, weight, age) and machine learning makes it more accurate than simple BMI calculators, which rely on static formulas that do not take into account other factors like muscle mass or body composition. The ability to integrate dietary patterns and personal health data provides a more holistic view of an individual's health.

7.6 Comparison

Table 67.1   Comparison table

Feature	Proposed System (ML-based)	Traditional Methods (Manual Entry)
Accuracy	92%	70-80% (manual)
Processing Speed	Real-time	Slow (manual tracking)
Ease of Use	User-friendly, automated	Requires manual input
Cost	Low (once deployed)	High (due to labor-intensive nature)

Source: Authors

7.7 Limitations of the Study

1. **Data Dependency:** The system's accuracy depends heavily on the quality and diversity of the image dataset.
2. **Environmental Factors:** Lighting and image quality can impact the accuracy of food calorie estimation.
3. **Real-Time Deployment:** Processing high-resolution images in real-time can be computationally expensive, requiring powerful hardware for large-scale deployment.

8. CONCLUSION

This research presents a machine learning-based approach for food calorie estimation and BMI prediction. The CNN-based food calorie estimation model achieved an accuracy of 92%, while the BMI prediction model demonstrated accurate results with minimal error. The system offers a significant improvement over traditional methods, which are often time-consuming and prone to human error. By combining image processing and regression models, the system can offer personalized health recommendations, making it an effective tool for diet management and weight tracking. The system's real-time processing capabilities and user-friendly interface make it suitable for deployment in mobile applications, enabling users to track their health and make informed dietary decisions. However, challenges remain in terms of data quality, real-time processing for high-resolution images, and ensuring accuracy across a broader range of food types. Future work will focus on expanding the dataset, improving processing efficiency, and enhancing the system's ability to provide more detailed health recommendations based on a user's diet and physical activity.

References

1. Y. Wang, et al., "Food calorie estimation using deep learning and computer vision," *IEEE Access*, vol. 7, pp. 10001–10012, 2019.
2. R. Sharma, et al., "Predicting BMI with machine learning techniques," *IEEE Transactions on Biomedical Engineering*, vol. 66, no. 2, pp. 487–496, 2021.
3. A. Patel, et al., "A survey on machine learning for health monitoring and food calorie estimation," *IEEE Journal of Biomedical and Health Informatics*, vol. 25, pp. 150–160, 2020.

Emerging Perspectives and Applications of Computational Intelligence and Smart Systems
– Dr. Amit Lathigara et al. (eds)
© 2026 Taylor & Francis Group, London, ISBN 978-1-041-20965-2

68

IoT Based Theft Detection using SMTP Protocol

K. Thyagarajan*, Sangeetha,
M. Balasubramanian, A. Anburaj
Department of Computer Engineering,
Sri Venkateswara College of Engineering and Technology,
Chittoor, India

Paresh Tanna
School of Engineering, RK University,
Rajkot, Gujarat, India

■ **Abstract:** More and more people are worried about the safety of their homes, which has led to the creation of novel approaches to the problem of burglary and theft. The fact that he neglected to ensure the safety of his home due to the hectic pace of everyday life is the root cause of this occurrence. An approach that has been suggested in earlier studies to address this issue is a home security system that is based on lo T and uses a PIR sensor to detect human motion. The system then notifies the user via SMS or email. A home security system that includes picture attachments in notifications is necessary, nevertheless, to make the system's warnings more understandable. An internet of things (IoT) home security system was created for this research. When the built-in PIR sensor detects the presence of a person, the IoT security system may immediately send an email with an attached photo. A web-enabled microcontroller known as an Arduino UNO, a passive infrared (PIR) sensor for motion detection, and a camera to capture still photographs of people when they come within range are all necessary components of the lot system. According to the study's experiments, when PIR sensors identify a person within a range of 0-5 meters in a variety of lighting situations, the IoT system may instantly send email notifications with accompanying photographs. as well as the size of the picture file and the state of the internet's network impact the rate at which email warnings are issued. The power supply for this project is regulated at 5V and 500mA. Relays utilize unregulated 12V DC power. To control the voltage, one uses a 7805 three-terminal regulator. The alternating current (ac) output of the secondary of a 230/12V step-down transformer may be rectified using a bridge type full wave rectifier.

■ **Keywords:** IoT, Arduino UNO, PIR sensor, SMTP

*Corresponding author: kthyagarajan21@gmail.com

DOI: 10.1201/9781003725046-68

1. INTRODUCTION

Embedded Systems products often use microcontrollers. One and only one function may be performed by the microprocessor (or microcontroller) in an embedded product. One device that uses an embedded system is a printer, whose CPU is designed to do a single task: receiving data and producing it. Though many embedded systems rely on microcontrollers, there are situations where they aren't up to the job. As a result, many general-purpose microprocessor manufacturers, including INTEL, Motorola, AMD, and Cyrix, have recently focused on the high-end of the embedded market with their microprocessors. Decreased power consumption and space requirements are among the embedded system's most pressing requirements. More features included into the CPU chipsets will do this. Low power consumption, various kinds of I/O, and random access memory (ROM) are all features shared by embedded CPUs. The current trend in high-performance embedded systems is to pack more and more features into the central processing unit (CPU) chip and give the designer free reign over which ones to implement.

2. PROBLEM STATEMENT

Several methods for developing wireless home security systems have been suggested ever since these systems emerged as a hot topic in the area of International Intelligent Building research. The homeowner may modify the passkey for the entrance whenever the system detects an incursion, which is done by monitoring the home's doorway with an LED. As soon as the system detects an intruder, it will send an SMS message to the owner's registered phone number, which is already stored in the system. Additionally, the owner may use his registered phone to remotely operate his house using SMS. In, we present a system that uses infrared sensors to detect intruders, a Raspberry Pi to take pictures of them, and a GSM modem to send an SMS with a link to the photo. In, a robot is programmed to broadcast live footage of the house, which can then be seen on a smartphone. The robot's built-in temperature sensor allows it to alert the owner in the event of an incursion. The system uses GSM technology to notify the owner via mail if an incursion is detected. A system is shown in that utilizes a raspberry pi and a camera to gather data and transmit it to a smartphone, allowing for remote monitoring of a designated area.

3. PROPOSED SYSTEM

In order to improve security, flexibility, and efficiency as needed, the suggested system has been designed to solve the shortcomings of previous surveillance systems. Due to the high costs incurred during installation, a security camera system is often out of reach. We want to create an intelligent surveillance system that can remotely notify the owner in the event of danger. When it detects an intruder, it will notify the owner by email on their smartphone. The intruder alarm goes off the moment a photoelectric receptor (PIR) motion detector picks up any kind of movement. As soon as it's turned on, the camera module will begin capturing video. Before beginning to record, the camera checks the face of the individual entering the room. If the face is recognizable, like a family member's, the detection is disregarded. In the event that the faces don't match, alerting the security guard to the presence of an intruder, the system is equipped with a buzzer that continuously sounds and an LCD screen that shows which apartment the intrusion happened in. On top of that, the owner gets an email alerting him to the intruder's presence in his room, and a picture of the intruder is immediately posted to the cloud. The intriguing part is that the video recording camera will only be engaged for a brief period in the event of an intruder, and it will be switched off as soon as the timer runs out. To make our system more efficient, it will be engaged again in the event of another incursion, but this time for a limited length.

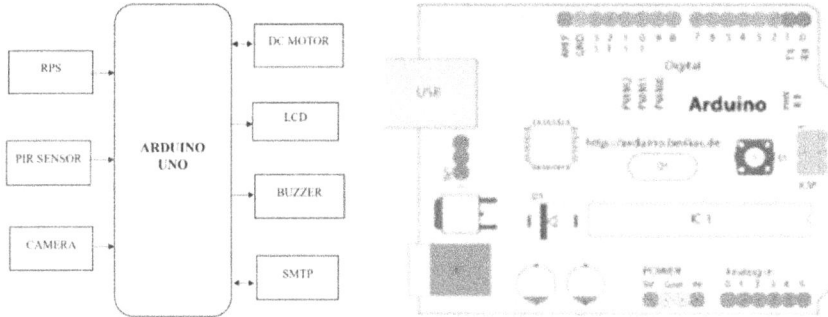

Fig. 68.1   Block diagram and structure of Arduino Uno

4. HARDWARE COMPONENTS

4.1 Power Supply

We have power supply in this project that can be set to either +5V or -5V. in most cases, a whole circuit needs just 5 volts. The OP amp circuit makes use of a separate supply that is -5V. The main side of the transformer has an alternating current (AC) voltage of 230/50 Hz. The secondary winding steps down the voltage to 12/50 Hz and rectify it using two full wave rectifiers. The rectified output is then sent into a filter circuit to remove the undesirable AC. Then, the LM7805 regulator (to provide +5v) is driven once again by the output. LM7905, on the other hand, is designed to regulate from -5V. Following the same procedure as the previous supply, the Z (+12V) circuit is used for stepper motors, fans, and relays via the LM7812 regulator.).

4.2 Arduino UNO

Artists, hackers, hobbyists, and even many professionals may benefit from the ease of electrical design, prototyping, and experimenting using the Arduino line of microcontroller boards. An ATmega microcontroller—basically a miniaturized computer with central processing unit, random access memory, flash memory, and input/output pins—forms the basis of an Arduino (we use the ordinary Arduino Uno). The pins on this board are specifically intended to accept a wide variety of sensors, LEDs, tiny motors, speakers, servos, and other devices that can read or output digital or analog voltages between 0 and 5 volts, unlike, example, a Raspberry Pi. Through its USB connection, the Arduino may be linked to your computer. Then, using the free Arduino IDE, you can upload your compiled code to the board and program it in a basic language (C/C++, which is comparable to Java). The Arduino may be used in conjunction with a computer via the USB connection or independently after programming; all it needs is electricity and an electrical outlet.

4.3 PIR Sensor

The term "passive infrared sensor" (PIR) describes a kind of sensor that does not emit any radiation or energy while in use. The infrared radiation that things produce or reflect is what PIR sensors pick up on, not "HEAT" itself. These things are user-friendly, compact, cheap, and power efficient. Their typical habitats include homes, hospitals, industries, and other such places.

4.4 ESP32 Camera

An OV2640 camera, a microSD card slot, and several general-purpose input/output (GPIO) ports are included on the ESP32-CAM development board. It also has an ESP32-S processor. A video streaming web server, a security camera, picture taking, facial recognition, and detection are just a few of the many uses for this versatile software.

Fig. 68.2 PIR sensor & EPS32 cam

4.5 SMTP

Email may be sent and received via SMTP. When used in conjunction with other protocols, such as IMAP or POP3, it allows users to access their messages, whereas SMTP is mostly responsible for sending messages to a server. Emails are sent and received over the Internet using the Simple Mail Transfer Protocol (SMTP). The majority of e-mail clients utilize this protocol to send messages to the server, while servers use it to send messages to their end users.

4.6 Wi-Fi Module

The Ai-thinker Team created the ESP-01 Wi-Fi module. The Tensilica L106 is a compact module that houses the ESP8266 core processor, a 32-bit microcontroller with low power consumption and 16-bit short mode. It supports clock speeds of 80 MHz and 160 MHz, has an on-board antenna, and is compatible with RTOS. It also has integrated Wi-Fi MAC/BB/RF/PA/LNA.

When designing mobile platforms with little space and power, look no farther than the ESP8266, a family of high-integration wireless system on chips. It offers the most cost-effective and space-efficient way to integrate Wi-Fi into existing systems or run independently as an application.

Fig. 68.3 Wi-Fi module

4.7 LCD (Liquid Crystal Display)

Displays in laptops and other portable computers employ liquid crystal display technology. A much thinner display is possible using liquid crystal display (LCD) technology, as well as with gas plasma and emitting diode (LED) technologies. Since LCDs function by absorbing light instead of generating it, they significantly reduce power consumption compared to gas-display and LED displays.

4.8 Buzzer

A buzzer often functions as an alert. Pressing the switch button activates the machine and produces an output, similar to an alarm sound. There are two pins in the buzzer. The microcontroller's data pin is linked to the negative end. The microcontroller's Vcc pin is linked to the positive end.

5. RESULTS

We will set the kit to trigger once we lock the door. The motor will turn, the sensor will detect, and the camera will take a picture of the thief when they open the door. Then, it will transmit the image to the registered mobile app.

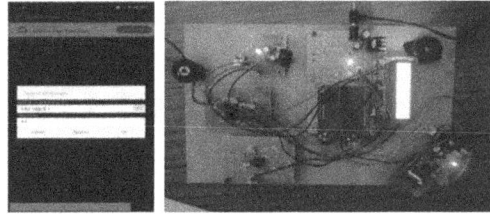

Fig. 68.4 Project module and Kit

6. CONCLUSION

With the implementation of the Smart Motion Detection System based on the Internet of Things, which automatically sends email warnings with the results of shooting whenever there is human movement, the mechanism has been effectively tested and proven. While this device performs well in a range of lighting circumstances, the picture quality it captures is compromised when used at distances between 0 and 5 meters. Depending on the quality of the picture obtained by the Pi Camera, the typical length of sending an email is about 15 seconds. In order to maintain the system's status as providing real-time responses. The system isn't perfect; for example, when an email alert is received, the recipient hasn't responded to the system yet, and the taken picture isn't very clear. The study concludes with a section outlining enhancements that might be included in future editions. A consumer's expectations are minimally met by the present set of features implemented. We can use a database to save the photos in the future, and we can use an advanced board to speed up processing.

References

1. Subdire ktorat Statistik Politik dan Keamanan, "Statistik Kriminal (2018,)" Badan Pusat Statistik, Indonesia.
2. S. Tanwar, P. Pately, K. Patelz, S. Tyagix, N. Kumar and M. Obaidat, (2017)"An Advanced Internet of Thing based Security Alert System for Smart Home," IEEE.
3. D. Yendri and R. E. Putri, (2018) "Sistem Pengontrolan Dan Keamanan Rumah Pintar (Smart Home) Berbasis Android," pp. 1–6, 2018.
4. R. Khana and U. Usnul, (2018) "Rancang Bangun Sistem Keamanan Rumah Berbasis IoT dengan Platform Android," Ejournal Kajian Teknik Elektro Vol.3 No.1, pp. 18–31.
5. Budianingsih and A. Riyanto, (2018) "Prototipe Sistem Keamanancerdas pada komplekperumahan," Jurnal Pendidikan Informatika dan Sains, pp. 146–154.
6. P. A. Dhobi and N. Tevar, (2017) "IoT Based Home Appliances Control," Proceedings of the IEEE 2017 International Conference on Computing Methodologies and Communication, pp. 648–651.
7. A. N. Ansari, M. Sedky, N. Sharma and A. Tyagi, (2015) "An Internet of things approach for motion detection using Raspberry Pi," in Proceedings of 2015 International Conference on Intelligent Computing and Internet of Things, Harbin, China.
8. M. Al-Kuwari, A. Ramadan, Y. Ismael, L. Al-Sughair and A. Gastli, (2018) "Smart-Home Automation using IoT-based Sensing and Monitoring Platform," IEEE.
9. P. B. Patel, V. M. Choksi, S. Jadhav and M. Potdar, (2016) "Smart Motion Detection Syste m usingRaspberry Pi," International Journal of Applied Information Systems (IJAIS), vol. 10, no. 5.
10. A. Rusli, "Pengguna SMS dan Telepon di Indonesia, Beralihke Data Internet," Cendana News, 24 Mei (2017). Available: https://www.cendananews.com/2017/05/pengguna-sms-dantelepon-di-indonesia-beralih-ke-data-internet.html.
11. Ashwini Patil, Shobha Mondhe, Tejashri Ahire, Gayatri Sonar Department of Computer, Late. G.N. Sapkal College of Engg, Nashik, Maharashtra, India.," Auto-Theft Detection Using Raspberry Pi And Android App". Data sheets & the user of ARM controller.

Note: All the figures in this chapter were made by the authors.

Emerging Perspectives and Applications of Computational Intelligence and Smart Systems
– Dr. Amit Lathigara et al. (eds)
© 2026 Taylor & Francis Group, London, ISBN 978-1-041-20965-2

69

The Impact of Oversampling Methods on Predicting Employee Turnover

Derya Habib, Oğuz Findik*

Computer Engineering Department of Karabuk University,
Turkey

■ **Abstract:** Employee turnover is the number of employees who leave their positions during a specific period. Machine learning algorithms for instance Random Forest, AdaBoost and LGBM have been used in the literature to predict employee turnover rates. Despite the studies, data imbalance is one of the primary difficulties that makes accurate forecasts difficult. Oversampling is employed to correct this imbalance. Oversampling balances the dataset by increasing the number of samples belonging to a single class while using fewer samples, allowing the model to learn both classes equally. In this work, SMOTE, ADASYN, SMOTE-NC, Borderline-SMOTE, GAN, SMOTE-Tomek, SMOTE-ENN, and ROS approaches were investigated to discover which oversampling method gives the best success in the dataset obtained by human resources. The Random Forest method, which produced the best classification results in human resources data, was employed in the evaluation process. The ROS oversampling strategy provided the best results. The analysis revealed that this strategy increased the success rate by 10.89% compared to the model without oversampling and by an average of 5.16% compared to other oversampling strategies.

■ **Keywords:** Employee turnover, Random forest algorithm, Over sampling methods

1. INTRODUCTION

Today, in the world of dynamic business, the ability of an organization to compete largely depends on the productivity of its employees [1]. Employee absenteeism or turnover refers to the loss of human resources from a business, whether voluntary or involuntary [2]. There are many reasons that influence decisions of employees to leave, making it difficult for managers to predict or prevent turnover [3,4]. Strategies to eliminate or minimize turnover and strategic HR decisions such as early detection or prediction of turnover are determined by machine learning. Machine learning [5] algorithms collect relevant information from datasets to find contexts and reveal unexpected situations. The supervised learning algorithm used in machine learning systems is based on labeled training data [6]. It extracts patterns from the training dataset, applies them to the test dataset, and

*Corresponding author: oguzfindik@karabuk.edu.tr

DOI: 10.1201/9781003725046-69

provides predictions or classifications. In supervised learning, methods such as Random Forest and AdaBoost detect turnover. Accuracy, F1 score, recall, Auc-Roc, and precision are common assessment metrics employed to assess performance of ML techniques [4].

Turnover uses a variety of databases, including IBM HR Analytics. The main issue with these datasets is that they are unbalanced, making it difficult to generate reliable forecasts. To solve this challenge, oversampling methods are employed with machine learning. Five machine learning algorithms were employed to identify whether employees in a workplace wish to leave their jobs. Among these algorithms used, Random Forest [7] , Gradient Boosting [8], AdaBoost [10], LightGBM [9], and Decision Trees [11]. Taking all of this into account, the outcomes of all of the algorithms were compared, and the Random Forest method was determined to produce the most successful results and the most effective model for calculating the turnover rate. An imbalance in the dataset makes it harder to produce good predictions because machine learning models typically focus on the class with the most samples while disregarding the class with the fewest. Oversampling approaches are used to solve this problem. Oversampling balances the dataset by increasing the amount of samples in the class with the fewest examples, allowing the model to learn both classes equally. As a result, the model can predict the class more accurately with less data. Among the oversampling methods used, SMOTE [12], SMOTE-NC [13], Borderline-SMOTE [14], SMOTE-Tomek [15], SMOTE-ENN [16], ADASYN [17], GAN [28], ROS [19] . After employing oversampling approaches, the Random Forest algorithm produced the highest achieved result. According to these findings, the ROS (Random Over-Sampling) + RF combination outperformed the other combinations, with a success rate ranging from 2.13% to 11.63%, providing more balanced and accurate forecasts.

2. Material and Methods

Dataset: IBM [20] HR Analytics is performed using the employee turnover dataset. The dataset contains data from 1470 employees, which contains various information and is used to predict whether a particular employee will leave the company. There are 35 features: 26 numerical and nine categories.

2.1 Methods

RF (Random Forest): Leaves stand in for ultimate forecasts, branches for splitting decisions, and nodes for these splitting criteria. [7]

Gradient Boosting: Gradient boosting is an ensemble learning approach that leverages a group of weak learners to produce strong learners incrementally trained to eradicate faults in earlier models. [8]

LGBM (Light Gradient Boosting): LGBM uses the Gradient-Based One-Sided Sampling (GOSS) approach to assess the significance of selected dataset samples and enhance the training process. [9]

AdaBoost: Classifiers are iteratively constructed using the ensemble (meta-learning) technique AdaBoost. [10]

Decision Trees: By progressing from the root node to the tree's leaf nodes, it generates predictions and divides the feature space into regions based on feature values [11].

2.2 Oversampling Techniques

Smote: By choosing samples from its closest neighbors, this procedure creates fresh samples along line segments [12].

Smote-Nc (Smote for Nominal Continuous): For categorical features, the categorical values of the nearest neighbors are determined accordingly, and synthetic samples are created according to these values, thus generating new data appropriately for both types of features [13].

Borderline-Smote: The algorithm separates the samples in the class into three categories: safe, risky, and noisy. [14].

Smote-Tomek: The Smote-Tomek method is a hybrid sampling method combining Smote and Tomek links [15].

Smote-Enn (Edited Nearest Neighbors): Unnecessary samples are removed from the class with fewer samples using the ENN method. [16].

Adasyn (Adaptive Synthetic Sampling): It distributes synthetic samples according to density. Samples with low density are emphasized while constructing synthetic samples.[17]

Generative Adversarial Networks (GAN): GAN [18] is a deep-learning algorithm. The basic GAN architecture consists of two independent models: the generator, which generates examples according to the distribution of the raw dataset and the discriminator, which assesses whether the examples are genuine.

Random Over Sampling: This method generates synthetic data by repeatedly adding randomly selected examples from a class with a smaller sample size to the dataset [19].

3. EXPERIMENTAL RESULTS

The Kaggle IBM dataset was used in this investigation. The accuracy results for each group are obtained, and their average is calculated. Table 69.1 shows the average accuracy rates of the models.

Table 69.1 Prediction results of employee turnover

Classifier	Accuracy	Precision	Recal	F1-score	Auc Roc
RF	**0.866**	0.884	0.160	0.271	**0.813**
GB	0.861	**0.704**	**0.291**	**0.412**	0.805
LGBM	0.863	0.711	0.249	0.369	0.806
Adaboost	0.860	0.758	0.221	0.344	0.774
DT	0.780	0.337	0.375	0.355	0.616

Source: Author

When looking at the values, although the GB F1-score produced good results, Accuracy, although RF Accuracy and Acu-Roc performed better. Thus, the RF algorithm was chosen as the best way to process the dataset.

To eliminate this problem, oversampling methods were applied to the respective dataset. This approach significantly improved the performance of the model. The obtained results are given in Table 69.2. When analyzing the results, the ROS + Random Forest method performed better at 2.13% and 11.63% compared to other results.

4. CONCLUSION

In this study, predictive analysis methods were used to detect intentions of employees to quit their jobs early, thus reducing the employee turnover rate. According to the accuracy results of the

Table 69.2 Results of employee turnover prediction obtained using random forest algorithm data by oversampling methods

Algorithm	F1-score (yes)	F1-score (no)	Accuracy
Random forest (RF)	0.18	0.93	0.86
Smote +RF	0.91	0.92	0.92
Smote-Nc + RF	0.90	0.91	0.91
Borderline-smote +RF	0.92	0.93	0.93
Smote-Tomek+RF	0.94	0.94	0.94
Smote-Enn + RF	0.92	0.88	0.90
Adasyn +RF	0.93	0.94	0.93
Gan +RF	0.65	0.91	0.86
ROS +RF	**0.96**	**0.96**	**0.96**

Source: Author

machine learning algorithms applied on the dataset used, the Random Forest (RF) algorithm gave better results than other algorithms with an accuracy rate of approximately 86%.However, Owing to the class imbalance in the dataset, performance of the model in the "Yes" class, which has a lower number of samples, was not found to be at the desired level. Oversampling methods were used to eliminate this data imbalance. After oversampling methods were applied to the model, machine learning techniques were used. Random Over Sampling (ROS) method yielded the best result with 96% accuracy rate as compared to other oversampling techniques. This provides a great advantage to HR managers, as detecting the intentions of employees to quit at earlier stages allows them to implement intervention strategies in a timely manner, thereby reducing employee turnover rates.

ACKNOWLEDGEMENTS

This study was supported by Karabuk University Scientific Research Projects Coordination Unit. Project Number: KBÜBAP-24-YL-164.

References

1. A. Raza, K. Munir, M. Almutairi, F. Younas, and M. M. S. Fareed, Predicting Employee Attrition Using Machine Learning Approaches, Appl. Sci., vol. 12, no. 13, p. 6424, 2022, doi: 10.3390/app12136424.
2. A. I. Al-Alawi and Y. A. Ghanem, Predicting Employee Attrition Using Machine Learning: A Systematic Literature Review, in 2024 ASU International Conference in Emerging Technologies for Sustainability and Intelligent Systems (ICETSIS), IEEE, 2024, pp. 526–530. Accessed: May 17, 2024. [Online]. Available: https: //ieeexplore.ieee.org/abstract/document/10459451/
3. S. Chowdhury, S. Joel-Edgar, P. K. Dey, S. Bhattacharya, and A. Kharlamov, Embedding transparency in artificial intelligence machine learning models: managerial implications on predicting and explaining employee turnover, Int. J. Hum. Resour. Manag., vol. 34, no. 14, pp. 2732–2764, Aug. 2023, doi: 10.1080/09585192.2022.2066981.
4. Wang, X., & Zhi, J. A machine learning-based analytical framework for employee turnover prediction. *Journal of Management Analytics, 8*(3), 351–370. DOI: 10.1080/23270012.2021.1961318. Published online: August 24, 2021.

5. Graves, L., Nagisetty, V., & Ganesh, V. Amnesiac Machine Learning. The Thirty-Fifth AAAI Conference on Artificial Intelligence (AAAI-21). University of Waterloo. Copyright © 2021.

6. B. Mahesh, Machine learning algorithms – A review, Jan. 2019.

7. Hu, J., & Szymczak, S. A review on longitudinal data analysis with random forest. *Briefings in Bioinformatics, 24*(2), bbad002. https://doi.org/10.1093/bib/bbad002. Published online: January 18, 2023.

8. F. Alzamzami, M. Hoda, and A. El Saddik, Light Gradient Boosting Machine for general sentiment classification on short texts: A comparative evaluation, *IEEE Access*, vol. 8, pp. 101840–101858, May 2020, doi: 10.1109/ACCESS.2020.2996976.

9. Sibindi, R., Mwangi, R. W., & Waititu, A. G. (2022). A boosting ensemble learning based hybrid light gradient boosting machine and extreme gradient boosting model for predicting house prices. *Engineering Reports*. https://doi.org/10.1002/eng2.12599. First published: November 22, 2022.

10. M. A. Akasheh, E. F. Malik, O. Hujran, and N. Zaki, "A decade of research on machine learning techniques for predicting employee turnover: A systematic literature review," Expert Syst. Appl., vol. 238, p. 121794, Mar. 2024, doi: 10.1016/j.eswa.2023.121794.

11. A. T. Sarızeybek and O. Sevli, Makine öğrenmesi yöntemleri ile banka müşterilerinin kredi alma eğiliminin karşılaştırmalı analizi, *Zeki Sistemler Teori ve Uygulamaları Dergisi*, vol. 5, no. 2, pp. 137–144, 2022, doi: 10.38016/jista.1036047.

12. T. Wongvorachan, S. He, and O. Bulut, A comparison of undersampling, oversampling, and SMOTE methods for dealing with imbalanced classification in educational data mining, *Information*, vol. 14, no. 1, p. 54, Jan. 2023, doi: 10.3390/info14010054.

13. I. Islahulhaq, W. Wibowo, and I. D. Ratih, Classification of non-performing financing using logistic regression and synthetic minority over-sampling technique-nominal continuous (SMOTE-NC), *Int. J. Adv. Soft Comput. Appl.*, vol. 13, no. 3, Nov. 2021, doi: 10.15849/IJASCA.211128.09.

14. H. Al Majzoub and I. Elgedawy, AB-SMOTE: An affinitive borderline SMOTE approach for imbalanced data binary classification, *Int. J. Mach. Learn. Comput.*, vol. 10, no. 1, Jan. 2020.

15. Khleel, N. A. A., & Nehéz, K. A novel approach for software defect prediction using CNN and GRU based on SMOTE Tomek method. *Research, 60*, 673–707. Published: May 16, 2023.

16. Hairani, H., & Priyanto, D. A new approach of hybrid sampling SMOTE and ENN to the accuracy of machine learning methods on unbalanced diabetes disease data. *International Journal of Advanced Computer Science and Applications (IJACSA), 14*(8), 585.

17. N. G. Ramadhan, Comparative analysis of ADASYN-SVM and SMOTE-SVM methods on the detection of type 2 diabetes mellitus, *Sci. J. Informatics*, vol. 8, no. 2, pp. 276, Nov. 2021.

18. C.-K. Lee, Y.-J. Cheon, and W.-Y. Hwang, Studies on the GAN-based anomaly detection methods for the time series data, *IEEE Access*, vol. 9, pp. 73201–73215, May 2021, doi: 10.1109/ACCESS.2021.3079469.

19. M. Al Akasheh, O. Hujran, E. F. Malik, and N. Zaki, Enhancing the prediction of employee turnover with knowledge graphs and explainable AI, in *Proceedings of the [Conference Name]*, May 23, 2024.

20. IBM HR Analytics Employee Attrition & Performance https://www.kaggle.com/datasets/pavansubhasht/ibm-hr-analytics-attrition-dataset/data

For Product Safety Concerns and Information please contact our EU
representative GPSR@taylorandfrancis.com
Taylor & Francis Verlag GmbH, Kaufingerstraße 24, 80331 München, Germany

www.ingramcontent.com/pod-product-compliance
Lightning Source LLC
Chambersburg PA
CBHW081040220326
41598CB00038B/6937